T0310627

A Gentle Course in Local Class Field Theory

Local Number Fields, Brauer Groups, Galois Cohomology

This book offers a self-contained exposition of local class field theory, serving as a second course on Galois theory. It opens with a discussion of several fundamental topics in algebra, such as profinite groups, p-adic fields, semisimple algebras and their modules, and homological algebra with the example of group cohomology. The book culminates with the description of the abelian extensions of local number fields, as well as the celebrated Kronecker-Weber theorem, in both the local and global cases. The material will find use across disciplines, including number theory, representation theory, algebraic geometry, and algebraic topology. Written for beginning graduate students and advanced undergraduates, this book can be used in the classroom or for independent study.

PIERRE GUILLOT is a lecturer at the University of Strasbourg and a researcher at the Institut de Recherche Mathématique Avancée (IRMA). He has authored numerous research papers in the areas of algebraic geometry, algebraic topology, quantum algebra, knot theory, combinatorics, the theory of Grothendieck's dessins d'enfants, and Galois cohomology.

A Gentle Course in Local Class Field Theory

Local Number Fields, Brauer Groups, Galois Cohomology

Pierre Guillot
University of Strasbourg

CAMBRIDGE
UNIVERSITY PRESS

CAMBRIDGE
UNIVERSITY PRESS

University Printing House, Cambridge CB2 8BS, United Kingdom

One Liberty Plaza, 20th Floor, New York, NY 10006, USA

477 Williamstown Road, Port Melbourne, VIC 3207, Australia

314-321, 3rd Floor, Plot 3, Splendor Forum, Jasola District Centre, New Delhi - 110025, India

79 Anson Road, #06-04/06, Singapore 079906

Cambridge University Press is part of the University of Cambridge.

It furthers the University's mission by disseminating knowledge in the pursuit of education, learning and research at the highest international levels of excellence.

www.cambridge.org
Information on this title: www.cambridge.org/9781108421775
DOI: 10.1017/9781108377751

First published 2018

A catalogue record for this publication is available from the British Library

Library of Congress Cataloging in Publication data
Names: Guillot, Pierre, 1978– author.
Title: A gentle course in local class field theory : local number fields,
Brauer groups, Galois cohomology / Pierre Guillot (University of Strasbourg).
Description: Cambridge ; New York, NY : Cambridge University Press, 2019.
Identifiers: LCCN 2018026580| ISBN 9781108421775 (hardback : alk. paper) |
ISBN 9781108432245 (pbk. : alk. paper)
Subjects: LCSH: Class field theory–Textbooks. | Brauer groups–Textbooks. |
Galois theory–Textbooks. | Galois cohomology–Textbooks.
Classification: LCC QA247 .G8287 2019 | DDC 512.7/4–dc23
LC record available at https://lccn.loc.gov/2018026580

ISBN 978-1-108-42177-5 Hardback
ISBN 978-1-108-43224-5 Paperback

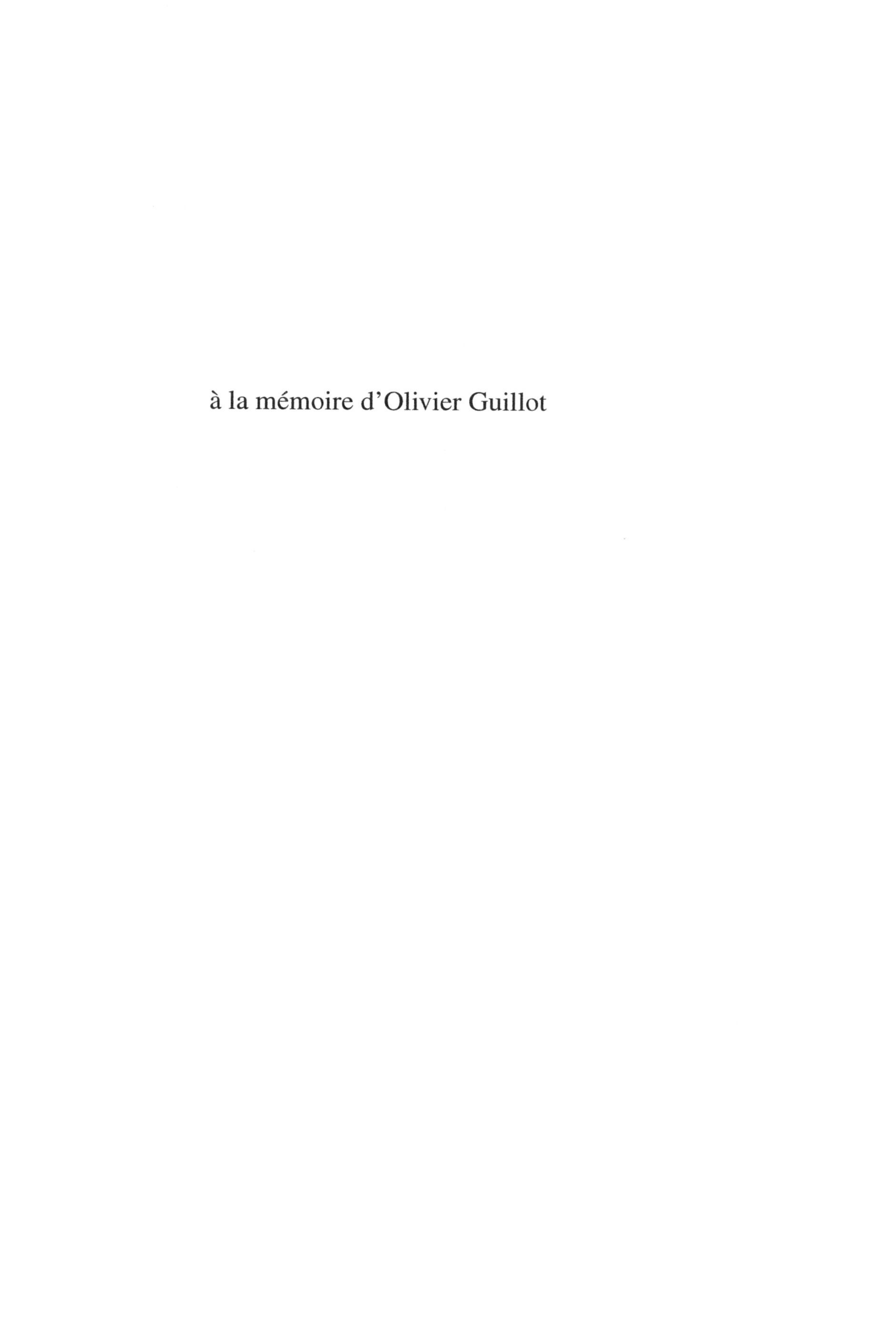

à la mémoire d'Olivier Guillot

Je reconnaissais ce genre de plaisir qui requiert, il est vrai, un certain travail de la pensée sur elle-même, mais à côté duquel les agréments de la nonchalance qui vous font renoncer à lui, semblent bien médiocres. Ce plaisir, dont l'objet n'était que pressenti, que j'avais à créer moi-même, je ne l'éprouvais que de rares fois, mais à chacune d'elles il me semblait que les choses qui s'étaient passées dans l'intervalle n'avaient guère d'importance et qu'en m'attachant à sa seule réalité je pourrais commencer enfin une vraie vie.

Marcel Proust,
À l'ombre des jeunes filles en fleurs

I recognized the kind of pleasure which, admittedly, requires some positive work of the mind upon itself, but compared to which the charms of idleness, that invite you to abandon the effort, seem mediocre. I have felt this pleasure, whose object I could only suspect, and which I had to create myself, only a few times, but it seemed to me that everything which had taken place between these occasions mattered very little, and that I could at last start a true life by clinging to its reality alone.

Contents

Preface

I have taught Galois theory at the undergraduate level for a number of years, with great pleasure. Usually I would follow, with more or less liberty, the first two chapters of Patrick Morandi's book, *Field and Galois theory* [Mor96], taking my favorite detours here and there. At the end of the semester, obviously, I am precisely aware of what the students know and do not know yet, and this is why I have often been embarrassed when asked for advice on choosing a book dealing with Galois theory beyond an introduction. The students I have in mind are not, by a long shot, ready for Jean-Pierre Serre's *Galois cohomology* [Ser02], nor can they start with Serre's *Local fields* [Ser79], to name two classic, beautiful textbooks in the area.

Thus I decided to write an exposition of some topics in Galois cohomology. After much hesitation, I resolved to pick the *Kronecker–Weber theorem* as a final destination, and to include only those facts which are useful for its proof. This celebrated result says that any finite abelian extension of $\mathbb{Q}$ is contained in a cyclotomic extension, a statement which my readers should be able to understand now (see the introduction to Part I for a list of prerequisites). To give another statement that can be appreciated immediately, let me state an easy consequence. Let $P \in \mathbb{Z}[X]$ be a monic polynomial, and assume that the splitting field of P is an *abelian* extension of $\mathbb{Q}$. Then there exists an integer m with the following property: for a prime number p, the question of deciding whether the reduction of P mod p splits into a product of linear factors has an answer that depends only on p mod m (with finitely many exceptions). For $P = X^2 - q$, where q is another prime, one can recover from this the *quadratic reciprocity law*, which says that, in order to decide whether q is a square mod p, you only have to know whether p is a square mod q (and whether p and q are ± 1 mod 4, assuming they are both odd); the generalization is a deep one.

There are several ways to prove the Kronecker–Weber theorem, even if *class field theory*, the theory of abelian extensions of global and local fields (here $\mathbb{Q}$ and its completions are in view), seems inevitable. For example, Childress in [Chi09] gives an account which stays elementary, and is oriented toward students with a strong preference for number theory. We shall follow, by contrast, what is known as the "cohomological approach", here developed from scratch. Let me try to argue in favor of this approach; that it gives me a perfect excuse to include some of my favorite topics should not be seen as its only virtue. My main point is that the various techniques to be discussed will be of interest to many more students and mathematicians than just number theorists. The material in Part I, Part II, and Part III will be useful in many other contexts, and I hope that my readers will find its study rewarding. Let me go through this in more detail, as I give a road map, of sorts, for the book.

- In Part I, we mostly deal with p-adic fields: the various completions $\mathbb{Q}_p$ of $\mathbb{Q}$, and their finite extensions. The field $\mathbb{Q}_p$, and its subring $\mathbb{Z}_p$, should be known to all students who wish to study algebra; and many people do analysis over p-adic fields, too. The p-adic fields form a heaven for Galois theory, their extensions being under very good control (and yet nontrivial): for example, we prove that there are only finitely many extensions $K/\mathbb{Q}_p$ of a given degree, and that $\mathrm{Gal}(K/\mathbb{Q}_p)$ is always a solvable group. Among other preliminaries, we give the basics of topological groups, study briefly vector spaces over complete fields (we discover that $\mathbb{Q}_p$ can replace $\mathbb{R}$ or $\mathbb{C}$ in the classical theorems of analysis), and provide a basic inspection of inverse limits, which appear everywhere in algebra.
- Part II is devoted to *skewfields*, or "noncommutative fields", a topic that is avoided in undergraduate classes, given the maturity that it requires, although students usually ask about the existence of these very early on. As it turns out, the study of skewfields leads us to semisimple algebras, and we end up proving the fundamental results of representation theory over a general field. Of course, complex representations of a finite group can be understood well using characters, but over more complicated fields, semisimple algebras cannot be avoided. We also cover the concept of an extension of a group by another, and explain how these are controlled by a *cohomology group*; this is basic group theory. The first three chapters of Part II can be read independently from Part I, but in the final chapter we fix a p-adic field F and consider the set $\mathrm{Br}(F)$ of all skewfields whose center is precisely F; this set is an abelian group, the *Brauer group* of F, and we prove that it is isomorphic to $\mathbb{Q}/\mathbb{Z}$. This is the first genuinely difficult result in the book.
- Part III deals with *group cohomology*: these are abelian groups written $\mathrm{H}^n(G,M)$, associated with a group G and a G-module M, generalizing the group $\mathrm{H}^2(G,M)$ that appeared in Part II. Collectively, these have astonishing properties. This part is an introduction to the more general phenomena of *homological algebra*, including a discussion of Ext and Tor. Students continuing in algebraic topology or algebraic geometry will face homological algebra all over the place, and group cohomology is a nice first example, on the comparatively concrete side. For students of representation theory, it is quite a revelation to understand that, when one considers a p-group acting on vector spaces of characteristic p, absolutely *nothing* of the usual theory remains useful, and in its place we have the mod p cohomology groups of G. (We shall not explain this connection to representation theory in this book, but we do provide the basic tools which will be needed for it.) Highlights for Part III include Hilbert's Theorem 90, which says that $\mathrm{H}^1(\mathrm{Gal}(K/F),K^\times)=0$, a fact with many consequences.

- Part IV finally studies class field theory, and proves the Kronecker–Weber theorem. Things are brought together in the following way. We prove Tate's Theorem, which gives a criterion for the existence of isomorphisms of the form $\mathrm{H}^n(G,M) \cong \mathrm{H}^{n-2}(G,\mathbb{Z})$ (in a precise, technical sense). Two ingredients are needed: one related to H^1, and provided by Hilbert's Theorem 90 when $G = \mathrm{Gal}(K/F)$ is a Galois group, and one related to H^2. When looked at the right way, this second, required ingredient turns out to be exactly what Part II was all about. The conclusions of Tate's Theorem can be translated into statements of Galois theory, and are strong enough for us to be able, with some work, to classify all the abelian extensions of p-adic fields – this is called "local class field theory". A "local" version of the Kronecker–Weber theorem follows, about abelian extensions of $\mathbb{Q}_p$. In the final chapter, we go back and forth between number fields (the finite extensions of $\mathbb{Q}$) and their completions, which are p-adic fields, and deduce the "global" Kronecker–Weber theorem. The facts explored in this chapter are the basics of algebraic number theory.

More about the organization of the book can be gathered from the table of contents, and the individual parts have their own introductions, providing some guidance. At the end of each part, references for further reading are given. These include, of course, *Local fields* and *Galois cohomology* by Serre – if I have offered a useful preparation for these masterful expositions, then my work was not in vain. For considerably different reasons, I have also tremendous respect for Blanchard's book, *Les corps non-commutatifs* [Bla72], and for *Cohomology of number fields* [NSW08] by Neukirch, Schmidt, and Wingberg.

The interdependence of chapters is given by Figure 1, in which an arrow from n to m indicates that chapter n must be read in order to understand chapter m. Still, I suggest that you read the chapters from 1 to 14, turning the pages in the usual fashion.

Several people have offered words of encouragement or advice, while this book was being written, and I thank them all warmly. The exposition was, in particular, improved by suggestions of Pierre Baumann, Filippo Nuccio, Chloé Perin, and Olivier Wittenberg. Special thanks are also due to Diana Gillooly at Cambridge University Press for being always so tactful when I had to be convinced to rely on the expertise of others. Finally, anonymous reviewers should know that their work is much appreciated.

Pierre Guillot
Strasbourg, April 2018

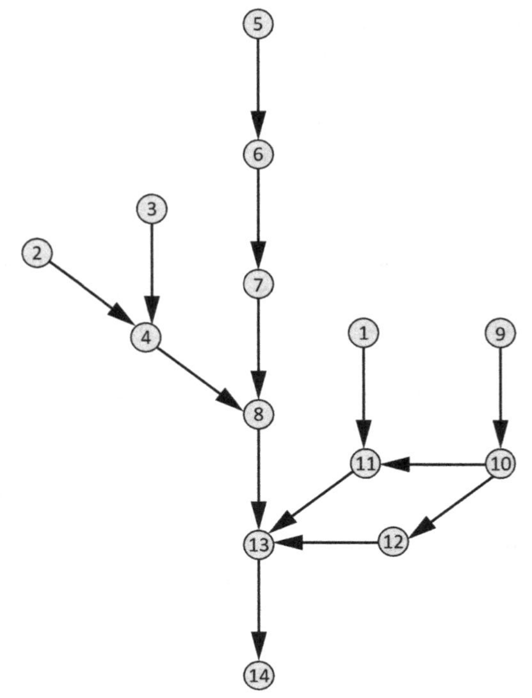

Figure 1. *Leitfaden*

Part I

Preliminaries

This part of the book sets the stage for later developments. We start with *Kummer theory*, which is the study of abelian extensions K/F of fields under the crucial assumption that F contain sufficiently many roots of unity. We include "equivariant Kummer theory", which has been part of the folklore around Galois theory for a long time, but does not seem to be treated in any textbook.

While the theory is very satisfying, it is natural to wish to remove the hypothesis about roots of unity. One of the final achievements of the book is the description, in a fashion directly analogous to Kummer theory, of the abelian extensions of *local number fields* (see Part IV). These fields, useful in many areas of mathematics, are introduced in Chapter 2, and their study is continued in Chapter 4. Chapter 3 is an interlude on topological groups and fields.

This may be the place to list the prerequisites for this book. We expect the reader to know a few basic facts from topology/analysis: Cauchy sequences, complete metric spaces, general topological spaces, compactness. We require the basics of linear algebra, and just a few things from commutative algebra: prime ideals, the notion of quotient ring, the Chinese Remainder theorem. We expect the reader to know that a finitely generated abelian group is a direct sum of cyclic groups, and that a subgroup of a finitely generated abelian group is itself finitely generated. In general, the theory of modules over euclidean domains will be required a couple of times (more will be said in due course).

Quite importantly, it is assumed that the reader knows about the Galois theory of finite extensions: separable, normal, and Galois extensions, the primitive element theorem (a separable, finite extension K/F is of the form $K = F(x)$ for some $x \in K$), Dedekind's lemma on the independence of characters, the fundamental theorem of Galois theory giving a bijection between subgroups of the Galois group and intermediate fields in a finite Galois extension. (In Chapter 3, we extend this to infinite Galois extensions.) You should know about cyclotomic extensions and cyclotomic polynomials. Basic facts about finite fields are assumed (you should be able to give a "list" of all the finite fields in the world).

We shall also rely on norms and traces. Most of the time, one can get away by simply knowing this: when K/F is a finite Galois extension, then we define for $a \in K$:

$$\mathrm{N}_{K/F}(a) = \prod_{\sigma \in \mathrm{Gal}(K/F)} \sigma(a), \qquad \mathrm{Tr}_{K/F}(a) = \sum_{\sigma \in \mathrm{Gal}(K/F)} \sigma(a).$$

These are called the norm and the trace of a, respectively. There are a few occasions (very rare in this book) when we need to talk about the norm or trace in an extension K/F which is not assumed Galois; and in the final chapter of this book, we shall have use for norms and traces when K is not even assumed to be a field. This is not always part of the standard, undergraduate treatment of Galois theory, and so for convenience we include a discussion in the Appendix.

Generally speaking, we point out that the first two chapters of Morandi's book [Mor96] are a great reference for the material which we assume is known.

1 Kummer theory

We begin gently with a warm-up chapter on Kummer theory, which is the study of abelian extensions of fields containing "enough" roots of unity. This will serve as a motivation for the rest of the book, where abelian extensions of *local number fields* are eventually described, in full generality. Besides, Kummer theory should be known to all students of Galois theory (and no doubt many readers will have already seen this).

We take the opportunity to setup some notation that will accompany us throughout the book.

Some basics

We start by recalling some basic material which, in principle, every reader will already know. We do this to fix the notation, and give an idea of where we take off from (in case the list of prerequisites, in the previous pages, left too much to the imagination).

Whenever R is a ring, we write $R^\times$ for the (multiplicative) group of invertible elements of R, also called units. When F is a field, of course $F^\times = F \setminus \{0\}$. Note that the letters of the alphabet used for fields in this book will usually be F, E, K, L, M, sometimes (but rarely) $\mathbb{F}$ or $\mathbb{K}$ (never k).

A very important fact about the group $F^\times$ is:

Lemma 1.1 *Let F be a field. Then every finite subgroup of $F^\times$ is cyclic.*

Proof. Let G be such a finite subgroup. By the classification of finite abelian groups, there is an isomorphism

$$G \cong \mathbb{Z}/a_1\mathbb{Z} \times \cdots \times \mathbb{Z}/a_k\mathbb{Z}$$

with $a_i > 1$ an integer dividing a_{i+1}. So the order of G is $a_1 a_2, \ldots a_k$, while every $g \in G$ satisfies $g^{a_k} = 1$. However, the equation $X^{a_k} - 1$ has at most a_k solutions in the field F, and the order of G must be $\leq a_k$. From $a_1 \ldots a_k \leq a_k$ we draw $k = 1$. □

A crucial subgroup of $F^\times$, in this chapter and elsewhere, is

$$\mu_n(F) = \{x \in F : x^n = 1\},$$

for any $n \geq 1$, the group of nth *roots of unity*. Thus $\mu_n(F)$ is a cyclic group of order $\leq n$; this order may well be strictly less than n, even if we try to enlarge F, as follows from:

Lemma 1.2 *Suppose the characteristic of F is the prime number p, and that $n = p^r m$. Then $\mu_n(F) = \mu_m(F)$, and the order of this group is thus $\leq m$. If n is a power of p, then $\mu_n(F)$ is trivial.*

Proof. Indeed, we simply write

$$X^n - 1 = (X^m - 1)^{p^r},$$

from which the result is clear. □

If, on the other hand, the integer n is prime to the characteristic of F, then the roots of $X^n - 1$ (in an algebraic closure of F) are distinct. If K is the splitting field of $X^n - 1$ over F, then $\mu_n(K)$ has order n, as does $\mu_n(L)$ for any field containing K. (In other words, unless the characteristic gets in the way, we can enlarge any field to have the "right" number of roots of unity.) The extension K/F is called a cyclotomic extension.

Remark 1.3 Many people write informally μ_n instead of $\mu_n(\overline{F})$, where $\overline{F}$ is an algebraic closure of F. With this notation, the field K just mentioned is $F(\mu_n)$. We shall refrain from employing this shorthand in this chapter (but we will have some use for it later). ■

A *primitive nth root of unity* is an element $\omega \in F^\times$ of order n (in the sense of group theory, that is, n is the smallest positive integer such that $\omega^n = 1$). It is important to keep in mind that the existence of such an element implies in particular that $\mu_n(F)$ has order n, and so the characteristic of F does not divide n. Conversely, if $\mu_n(F)$ has order n, then primitive nth roots of unity exist by Lemma 1.1.

In this chapter we will frequently refer to the subgroup

$$F^{\times n} = \{f^n : f \in F\} \subset F^\times .$$

More precisely, we shall encounter quite often the quotient $F^\times/F^{\times n}$. The image of $x \in F^\times$ in $F^\times/F^{\times n}$ will be denoted $[x]$ or $[x]_F$ if there is any ambiguity. (Note that n does not appear in the notation.)

Example 1.4 Take $F = \mathbb{Q}$. It takes some time to get used to the following example. We have

$$\mathbb{Q}^\times = \{\pm 1\} \times A,$$

where A is a free abelian group, with a basis consisting of the set of prime numbers. Indeed, any element of $\mathbb{Q}^\times$ can be written uniquely $\pm p_1^{n_1} p_2^{n_2} \cdots$ where $n_i \in \mathbb{Z}$ and $p_1, p_2, \ldots,$ is an enumeration of the primes. Now

$$\mathbb{Q}^\times / \mathbb{Q}^{\times 2} = \{\pm 1\} \times (A/2A),$$

a vector space over $\mathbb{Z}/2\mathbb{Z}$, with basis given by the elements $[\ell]$ where ℓ is a prime number, together with $[-1]$. On the other hand, when p is an odd prime we have

$$\mathbb{Q}^\times / \mathbb{Q}^{\times p} = A/pA,$$

since $-1 = (-1)^p$.

Cyclic extensions

cyclic extension

Definition 1.5 An extension of fields K/F will be called **cyclic extension** when it is finite and Galois, with $\mathrm{Gal}(K/F)$ cyclic.

When F has enough roots of unity, we will be able to characterize cyclic extensions completely. The main ingredient for this is the next lemma, which is a first version of Hilbert's Theorem 90. Later in the book, we shall discover more sophisticated versions of the same result.

Lemma 1.6 (Hilbert 90) *Let F be a field containing a primitive nth root of unity ω, for some $n \geq 1$, and let K/F be a cyclic extension of degree n. If σ is a generator for $\mathrm{Gal}(K/F)$, then there exists $x \in K^\times$ such that*

$$\omega = \frac{\sigma(x)}{x}.$$

Proof. We want to find $x \neq 0$ with $\sigma(x) = \omega x$, so what we want is to show that ω is an eigenvalue of the F-linear endomorphism σ (the given form is here for "historical" reasons; it will come up naturally when we generalize this result).

Since σ^n is the identity, the minimal polynomial P of σ (in the sense of linear algebra) divides $X^n - 1$. However, this minimal polynomial must have degree n, for the distinct automorphisms $1, \sigma, \sigma^2, \ldots, \sigma^{n-1}$ are linearly independent, by Dedekind's lemma. So $P = X^n - 1$. On the other hand, the characteristic polynomial χ of σ is a multiple of P, which also has degree n, and we conclude that $\chi = P = X^n - 1$. Thus, ω is indeed a root of χ. □

Proposition 1.7 *Let F be a field containing a primitive nth root of unity ω.*

1. *For $a \in F^\times$, consider $K = F(\sqrt[n]{a})$, where we write $\sqrt[n]{a}$ for some root of $X^n - a$ in an algebraic closure of F. Then K/F is a cyclic extension. Its degree $m = [K : F]$ is the order of $[a]_F$ in $F^\times / F^{\times n}$, which divides n.*
2. *Let K/F be a cyclic extension of degree n. Then $K = F(\sqrt[n]{a})$ for some $a \in F^\times$. The order of $[a]_F$ in $F^\times / F^{\times n}$ is n.*

Remark 1.8 (on notation) When K/F is a field extension and $\alpha \in K$, the (completely standard) notation $F[\alpha]$ is for the smallest subring containing F and α, while $F(\alpha)$ is the smallest subfield containing F and α; when α is algebraic over F, we have $F[\alpha] = F(\alpha)$, so that one has to choose. The author admits his agnosticism in the matter, implying that $F[\alpha]$ and $F(\alpha)$ alternate in this book. In absolutely all examples, the element α will be algebraic, so that no confusion can possibly arise (a field is always meant); the only exception is when we form a polynomial ring, always clearly identified by the use of capital letters, such as $F[X]$ or $F[Y]$. ■

Proof of the proposition. (1) The roots of $X^n - a$ are $\omega^k \sqrt[n]{a}$ for $0 \leq k < n$, so if $\sigma \in \mathrm{Gal}(K/F)$, we must have $\sigma(\sqrt[n]{a}) = \omega^{k(\sigma)} \sqrt[n]{a}$ for some integer $k(\sigma)$ whose class in $\mathbb{Z}/n\mathbb{Z}$ is well defined. The map $\mathrm{Gal}(K/F) \to \mathbb{Z}/n\mathbb{Z}$ taking σ to $k(\sigma)$ is readily seen to be a group homomorphism, which is injective since K is generated by $\sqrt[n]{a}$. Thus, $\mathrm{Gal}(K/F)$ is isomorphic to a subgroup of $\mathbb{Z}/n\mathbb{Z}$, and we see that it is cyclic, of order m dividing n. We turn to the alternative description of m.

We allow ourselves to write $a^{r/n}$ instead of $(\sqrt[n]{a})^r$, when r is an integer. We first claim that

$$a^{r/n} \in F^\times \iff a^r \in F^{\times n}. \qquad (*)$$

The implication $\Longrightarrow$ is trivial. If $a^r = (a^{r/n})^n = f^n$ with $f \in F^\times$, then $a^{r/n} f^{-1}$ is an nth root of unity, and so belongs to F, showing the converse. Next, let σ be a generator for $\mathrm{Gal}(K/F)$, and let $k = k(\sigma)$. We have

$$\sigma(a^{1/n}) = \omega^k a^{1/n},$$

so that

$$\sigma(a^{r/n}) = \omega^{kr} a^{r/n}.$$

An element of K lies in F if and only if it is fixed by σ, so

$$a^{r/n} \in F^\times \iff n \text{ divides } kr. \qquad (**)$$

Comparing (*) and (**) shows that m, defined above to be the smallest integer such that n divides km, is also the smallest integer such that $a^m \in F^{\times n}$, as we wanted to show.

(2) This is where we use Lemma 1.6 ("Hilbert 90"), giving us the existence of $x \in K^\times$ with $\sigma(x) = \omega x$, where σ is a generator for K/F. It follows that $\sigma(x^n) = x^n$, so the element $a := x^n$ belongs to F, and we may write $x = \sqrt[n]{a} = a^{1/n}$. We have $F \subset F(a^{1/n}) \subset K$ and it is enough to prove that $[F(a^{1/n}) : F] = n$. By part (1), we must show that the order of $[a]$ in $F^\times/F^{\times n}$ is n, and by (*), we must show that the smallest integer r so that $x^r \in F$ is $r = n$. Examining the relation $\sigma(x^r) = \omega^r x^r$, this appears clearly true. □

The two parts of the proposition are almost converses for one another, but not quite. In the important case when n is a prime number at least, we obtain a clean-cut result:

Corollary 1.9 *Let p be a prime, and let F be a field containing a root of unity of order p^2. Then the Galois extensions of F of degree p (which are automatically cyclic) are precisely those of the form $F(\sqrt[p]{a})/F$ where $a \in F^\times \smallsetminus F^{\times p}$.* □

Some classical applications

The field $\mathbb{R}$ has a *finite* extension which is algebraically closed, namely $\mathbb{C} = \mathbb{R}[\sqrt{-1}]$. We could alternatively emphasize that the algebraically closed field $\mathbb{C}$ has an automorphism of finite order. Do algebraically closed fields possess automorphisms of arbitrary finite order, and do they admit complicated finite groups of automorphisms? Using the material on cyclic extensions just obtained, we shall see that the answer is, in a precise sense, "no". The case of $\mathbb{C}/\mathbb{R}$ is as complicated as can be.

Lemma 1.10 *Let p be a prime number, and let F be a field containing a primitive p^2th root of unity. Then for any Galois extension K/F with $[K : F] = p$, there exists a field L containing K such that L/F is cyclic of degree p^2.*

Proof. The extension K/F is cyclic of degree p, so by Proposition 1.7 we must have $K = F[a^{1/p}]$ for some $a \in F^\times$. Let $L = F[a^{1/p^2}]$, so L/F is cyclic, $K \subset L$, and $[L : F]$ is the order of $[a]$ in F^*/F^{*p^2}. Suppose this order were to divide p. Then we would have $a^p = f^{p^2}$ for some $f \in F^\times$, so $(a/f^p)^p = 1$ and $a = f^p\omega^k$, where ω is a primitive pth root of unity and k is some integer. Using that ω has a pth root in F, we see that a has a pth root: this is a contradiction, however, as the order of $[a]$ in $F^\times/F^{\times p}$ is $[K : F] = p$. So the only possibility is that $[L : F] = p^2$. □

Lemma 1.11 *Let p be a prime number, and let F be a field of characteristic $\neq p$. Suppose K/F is a Galois extension of degree p with K algebraically closed. Then $p = 2$ and $K = F[\sqrt{-1}]$.*

Proof. The field K, being algebraically closed and of characteristic $\neq p$, contains a primitive pth root of unity ω. However, ω is a root of

$$1 + X + \cdots + X^{p-1} \in F[X],$$

so $F(\omega)/F$ has degree prime to p, and from $F(\omega) \subset K$ we deduce $F(\omega) = F$, that is, $\omega \in F$.

On the other hand, we claim that F does not have a primitive p^2th root of unity: if it did, then Lemma 1.10 would give us the existence of an extension of K of degree p, which is absurd. Let us write $\omega^{1/p}$ for such a root, which lives in $K \smallsetminus F$. Clearly, we must have $K = F(\omega^{1/p})$.

Let σ be a generator of the group $\mathrm{Gal}(K/F)$. There is an integer k such that $\sigma(\omega^{1/p}) = \omega^k\omega^{1/p}$, and for clarity let us put $\zeta = \omega^k$, another (primitive) pth root of unity.

Now suppose p is odd. The following computation is classical:

$$\mathrm{N}_{K/F}(\omega^{1/p}) = (\omega^{1/p}) \cdot (\zeta\omega^{1/p}) \cdot (\zeta^2\omega^{1/p}) \cdots (\zeta^{p-1}\omega^{1/p}) = \zeta^{p(p-1)/2}\omega = \omega,$$

using that p divides $p(p-1)/2$ when p is odd. This leads to a contradiction. Indeed, as K is algebraically closed, every element of K has a pth root in K, including $\omega^{1/p}$; taking norms down to F, we see that ω has a pth root in F, contrary to what we have shown.

Thus $p = 2$, and $\omega = -1$. We have seen that $K = F[\omega^{1/p}] = F[\sqrt{-1}]$. □

Theorem 1.12 *Let F be a field of characteristic 0, and suppose that K/F is a finite Galois extension with K algebraically closed. Then either $K = F$, or $K = F[\sqrt{-1}]$.*

Equivalently, if G is a finite group of automorphisms of a field K, which is algebraically closed and of characteristic 0, then either $G = \{1\}$ or G has order 2; in the latter case, the fixed field F of G is such that $K = F[\sqrt{-1}]$.

Proof. That the two statements are equivalent follows from Artin's theorem (which asserts that, when F is the fixed field of G, then $G = \mathrm{Gal}(K/F)$). We work with the first formulation.

Let σ be an element of $\mathrm{Gal}(K/F)$ of prime order p. Consider the fixed field F_σ of σ; the extension K/F_σ has degree p, so by Lemma 1.11 we have $p = 2$. It follows that the order of $\mathrm{Gal}(K/F)$ is a power of 2, as this group does not have elements of odd prime order.

Next, consider the element $\sqrt{-1} \in K$, and the field $F[\sqrt{-1}]$. Let σ be an element of $\mathrm{Gal}(K/F[\sqrt{-1}])$ of order 2, if there is one, and let F_σ be its fixed field. Then $[K : F_\sigma] = 2$ while $\sqrt{-1} \in F_\sigma$, and this contradicts Lemma 1.11. So no such element σ exists. Since the group $\mathrm{Gal}(K/F[\sqrt{-1}])$ has an order which is a power of 2, but does not have elements of order 2, we are compelled to conclude that it is trivial, and that $K = F[\sqrt{-1}]$. □

Remark 1.13 (1) The hypothesis on the characteristic of F is here for simplicity. In fact, one can prove that, if K/F is finite and Galois, with K algebraically closed, and $K \neq F$, then F (and K) are automatically of characteristic 0. See [Lan02, VI, corollary 9.3]. The argument raises subtle points about separability which we do not want to review, as they are not particularly relevant for the rest of the book.

(2) The group $\mathrm{Aut}(K)$ may have several elements of order 2, of course. However, the theorem asserts that if σ and τ are two such elements with $\sigma \neq \tau$, then the group they generate in $\mathrm{Aut}(K)$ is infinite. ■

Kummer extensions

n-Kummer extension

Definition 1.14 A finite Galois extension K/F is said to be an ***n*-Kummer extension** when F contains an nth primitive root of unity, and $\mathrm{Gal}(K/F)$ is abelian of exponent dividing n. (In other words, $g^n = 1$ for all $g \in \mathrm{Gal}(K/F)$.) ■

It is an important fact that there is an alternative definition of n-Kummer extensions: they are the extensions which can be obtained by adjoining to F finitely many elements of the form $\sqrt[n]{a}$ for $a \in F^\times$. Before we turn to this, let us give a useful proposition about the Galois group of a compositum of two fields. It is more general than is immediately needed.

Proposition 1.15 *Let F be a field, and let K and L be two finite extensions of F contained in a common algebraic closure, as the following diagram indicates.*

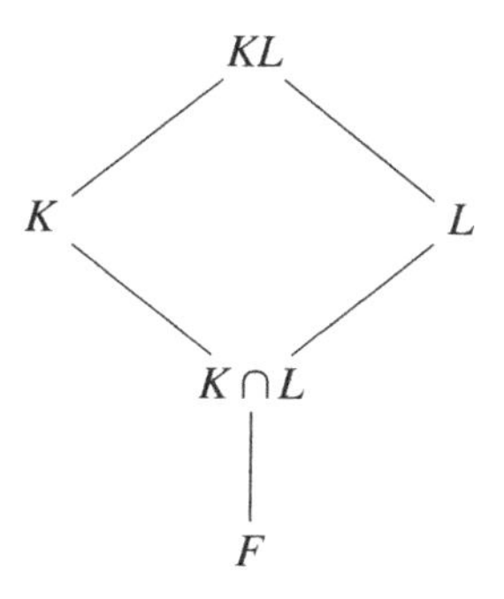

1. *If K/F is Galois, then so is KL/L, and moreover*

$$\mathrm{Gal}(KL/L) \cong \mathrm{Gal}(K/K \cap L).$$

2. *If K/F and L/F are both Galois, then so is KL/F, and*

$$\mathrm{Gal}(KL/F) \cong \{(\sigma, \tau) \in \mathrm{Gal}(K/F) \times \mathrm{Gal}(L/F) : \sigma|_{K\cap L} = \tau|_{K\cap L}\}.$$

The group appearing in (2) is sometimes called the *fiber product* of $\mathrm{Gal}(K/F)$ and $\mathrm{Gal}(L/F)$ "above" $\mathrm{Gal}(K \cap L/F)$.

Proof. (1) Since K is obtained by adjoining to F the roots of a collection of separable polynomials, the field KL can be obtained from L by adding the same roots, so KL/L is Galois, by a standard criterion. An element $\sigma \in \mathrm{Gal}(KL/L) \subset \mathrm{Gal}(KL/F)$ must map K to itself, since K/F is normal by assumption, so the restriction $\sigma \mapsto \sigma|_K$ defines a homomorphism r: $\mathrm{Gal}(KL/L) \to \mathrm{Gal}(K/K \cap L)$. The latter is visibly injective.

Let E be the intermediate field such that $\mathrm{Im}(r) = \mathrm{Gal}(K/E)$. Certainly $K \cap L \subset E \subset K$, so it remains to prove $E \subset L$ to obtain $E = K \cap L$. However, E is a subfield of KL which is fixed by $\mathrm{Gal}(KL/L)$, so $E \subset L$ as desired.

(2) That KL/F is Galois is obvious. An element $\gamma \in \mathrm{Gal}(KL/F)$ must preserve K and L, and is determined by $\sigma = \gamma|_K$ and $\tau = \gamma|_L$. Hence, we have an injective homomorphism

$$\mathrm{Gal}(KL/L) \longrightarrow \mathrm{Gal}(K/F) \times \mathrm{Gal}(L/F)$$

whose image is contained in the group Γ described in the proposition. It suffices to establish that the order of Γ is $[KL : F]$. However, the second projection $\Gamma \longrightarrow \mathrm{Gal}(L/F)$ is surjective, and its kernel is

$$\{(\sigma, 1) \in \mathrm{Gal}(K/F) \times \mathrm{Gal}(L/F) : \sigma|_{K\cap L} = 1\} \cong \mathrm{Gal}(K/K \cap L),$$

so

$$|\Gamma| = [K : K \cap L][L : F] = [KL : L][L : F] = [KL : F],$$

using (1). □

Proposition 1.16 *Let F be a field containing a primitive nth root of unity, and let K/F be an extension. The following properties are equivalent.*

1. *K/F is n-Kummer.*
2. *There are elements $a_1, \ldots, a_k \in F^\times$ such that $K = F[\sqrt[n]{a_1}, \ldots, \sqrt[n]{a_k}]$.*

Proof. (2) $\Longrightarrow$ (1). Each extension $F[\sqrt[n]{a_i}]/F$ is cyclic of order dividing n by Proposition 1.7, so it is n-Kummer. The field K is the compositum of all the $F[\sqrt[n]{a_i}]/F$ as i varies; so by (2) of the last proposition, applied repeatedly, we see that K/F is Galois and that $\mathrm{Gal}(K/F)$ is a subgroup of a product of cyclic groups, each of order dividing n. Clearly, K/F is then itself n-Kummer.

(1) $\Longrightarrow$ (2). Write

$$\mathrm{Gal}(K/F) = C_1 \times C_2 \times \cdots \times C_k,$$

where each C_i is cyclic, of order dividing n. Also, let

$$H_i = \prod_{j \neq i} C_j \subset \mathrm{Gal}(K/F),$$

and let E_i be the fixed field of H_i. Then $\mathrm{Gal}(E_i/F) \cong C_i$, so E_i/F is cyclic, and it follows from Proposition 1.7 that $E_i = F[\sqrt[n]{a_i}]$ for some $a_i \in F^\times$. There remains only to prove that K is the compositum of the fields E_i. This compositum L, being the smallest field containing all the E_i, is associated in the Galois correspondence with the largest subgroup contained in all the H_i, that is, the intersection $\cap_i H_i$. This intersection is trivial, so $L = K$. □

The Kummer pairing

When studying cyclic extensions above, we have come across a certain computational remark several times. Namely, if K/F is a Galois extension, $\sigma \in \mathrm{Gal}(K/F)$, and $\sqrt[n]{a} \in K$ is an n-th root for $a \in F$, then $\sigma(\sqrt[n]{a})/\sqrt[n]{a}$ is an n-th root of unity. In this section we shall study the association

$$\sigma \mapsto \frac{\sigma(\sqrt[n]{a})}{\sqrt[n]{a}},$$

seen as a homomorphism $\mathrm{Gal}(K/F) \to \mu_n(F)$. This requires some preparation.

Pontryagin dual

Definition 1.17 Let G be a finite abelian group. Its **Pontryagin dual** is

$$G' = \mathrm{Hom}(G, \mathbb{C}^\times),$$

the (finite abelian) group of homomorphisms $G \to \mathbb{C}^\times$, often called (linear) characters. (The notation $\widehat{G}$ is also common.) ■

Lemma 1.18 *Let G be a finite abelian group.*

1. *If C is a cyclic group, of finite order divisible by the exponent of G, then $G' \cong \mathrm{Hom}(G,C)$.*
2. *There is an isomorphism $G \cong G'$ (but there is no canonical one!).*
3. *There is a canonical isomorphism $G \cong G''$.*

Proof. We leave (1) as a very easy exercise. Property (2) is easy when G is cyclic of order n, generated by g, as G' is then cyclic of order n, generated by $\chi : G \to \mathbb{C}^\times$ with $\chi(g) = \exp(2i\pi/n)$. The general case follows, as $(G_1 \times G_2)' \cong G_1' \times G_2'$.

Turning to (3), consider the homomorphism $G \to G''$ taking $g \in G$ to $\mathrm{ev}_g : G' \to \mathbb{C}^\times$, the "evaluation at g", with $\mathrm{ev}_g(\chi) = \chi(g)$. The groups G and G'' have the same order by (2), and $g \mapsto \mathrm{ev}_g$ is easily seen to be injective, so it is an isomorphism. □

Suppose G_1 and G_2 are two finite abelian groups, with a specified isomorphism $G_1 \cong G_2'$. Then we also have canonically $G_1' \cong G_2'' \cong G_2$, so that the roles of G_1 and G_2 can be naturally interchanged. It is also true that G_1 and G_2 are isomorphic, but not canonically. The next definition attempts to place G_1 and G_2 on equal footing.

bilinear pairing

Definition 1.19 Let G_1 and G_2 be abelian groups. A **bilinear pairing** is a map

$$B\colon G_1 \times G_2 \longrightarrow C,$$

where C is another abelian group, such that

$$B(g_1g_1', g_2) = B(g_1, g_2)\, B(g_1', g_2)$$

and

$$B(g_1, g_2g_2') = B(g_1, g_2)\, B(g_1, g_2'),$$

for $g_i, g_i' \in G_i$. The pairing is said to be **nondegenerate**, or perfect, when the following property is satisfied. For any $g_1 \in G_1$ with $g_1 \neq 1$, there exists some $g_2 \in G_2$ such that $B(g_1, g_2) \neq 1$; and for any $g_2 \in G_2$ with $g_2 \neq 1$, there exists some $g_1 \in G_1$ such that $B(g_1, g_2) \neq 1$. ■

Lemma 1.20 *Suppose G_1 and G_2 are abelian groups, with a perfect pairing $B\colon G_1 \times G_2 \to C$, where C is a cyclic group of order divisible by the exponents (assumed finite) of both G_1 and G_2. Finally, assume that one of G_1 or G_2 is finite.*

Then G_1 and G_2 are both finite, we have canonical isomorphisms $G_1 \cong \mathrm{Hom}(G_2, C)$ and $G_2 \cong \mathrm{Hom}(G_1, C)$, and a non-canonical isomorphism $G_1 \cong G_2$.

Proof. Using the pairing, we define $G_1 \to \mathrm{Hom}(G_2, C)$ by taking $g_1 \in G_1$ to the character $g_2 \mapsto B(g_1, g_2)$. The nondegeneracy of the pairing implies that this is injective. Likewise, G_2 is isomorphic to a subgroup of $\mathrm{Hom}(G_1, C)$.

If, say, G_1 is finite, then $|G_2| \leq |\mathrm{Hom}(G_1, C)| = |G_1'| = G_1$, using the previous lemma, so G_2 is also finite. Reversing the argument, we see that G_1 and G_2 have the same order, and that the homomorphisms $G_1 \to \operatorname{Hom}(G_2, C)$ and $G_2 \to \operatorname{Hom}(G_1, C)$ are isomorphisms.

The last statement follows, since $G_1 \cong G_1' \cong \operatorname{Hom}(G_1, C) \cong G_2$. □

Let us finally apply this to Galois theory. For the rest of this section, let K/F be an n-Kummer extension. The group $G_1 = \operatorname{Gal}(K/F)$ will appear in a perfect pairing.

The second group in the pairing will be

$$G_2 = \frac{F^\times \cap K^{\times n}}{F^{\times n}} \subset \frac{F^\times}{F^{\times n}}.$$

Note that $F^\times$ and $K^{\times n}$ are both subgroups of $K^\times$, so it makes sense to speak of their intersection. The group G_2 consists thus of the elements of $F^\times$ which have an n-th root in K, modulo those which already have an n-th root in F. Also, G_2 is the kernel of

$$F^\times / F^{\times n} \longrightarrow K^\times / K^{\times n}.$$

Kummer pairing

Definition 1.21 The **Kummer pairing** is the map

$$B\colon \operatorname{Gal}(K/F) \times \frac{F^\times \cap K^{\times n}}{F^{\times n}} \longrightarrow \mu_n(F)$$

defined by

$$B(\sigma, [a]) = \frac{\sigma(\sqrt[n]{a})}{\sqrt[n]{a}}.$$

■

Here the notation $\sqrt[n]{a}$ means that an arbitrary n-th root of a is to be taken in K. The next theorem will resolve any ambiguity.

Theorem 1.22 *When K/F is an n-Kummer extension, the Kummer pairing is well-defined, bilinear, and nondegenerate. Moreover, suppose we have elements $a_1, \ldots, a_k \in F^\times$ such that $K = F[\sqrt[n]{a_1}, \ldots, \sqrt[n]{a_k}]$. Then the group $\frac{F^\times \cap K^{\times n}}{F^{\times n}}$ is generated by $[a_1], \ldots, [a_k] \in F^\times / F^{\times n}$.*

Proof. First note that the proposed expression for $B(\sigma, [a])$ does belong to $\mu_n(F)$. We show that this element is independent of all choices. First, if $[a] = [a']$, then $a' = af^n$ for some $f \in F^\times$. An n-th root for af^n is of the form $\zeta \sqrt[n]{a} f$, where $\zeta \in \mu_n(F)$, and $\sqrt[n]{a}$ is our previous choice. So

$$\frac{\sigma(\zeta \sqrt[n]{a} f)}{\zeta \sqrt[n]{a} f} = \frac{\zeta \sigma(\sqrt[n]{a}) f}{\zeta \sqrt[n]{a} f} = \frac{\sigma(\sqrt[n]{a})}{\sqrt[n]{a}}.$$

Thus the map B is well-defined.

We check linearity on the left. Write

$$B(\sigma\tau, [a]) = \frac{\sigma\tau(\sqrt[n]{a})}{\tau(\sqrt[n]{a})} \cdot \frac{\tau(\sqrt[n]{a})}{\sqrt[n]{a}}.$$

We have $\sigma\tau = \tau\sigma$ as $\mathrm{Gal}(K/F)$ is assumed abelian. Thus

$$\frac{\sigma\tau(\sqrt[n]{a})}{\tau(\sqrt[n]{a})} = \tau\left(\frac{\sigma(\sqrt[n]{a})}{\sqrt[n]{a}}\right) = \frac{\sigma(\sqrt[n]{a})}{\sqrt[n]{a}},$$

using that $\sigma(\sqrt[n]{a})/\sqrt[n]{a} \in F$. This shows $B(\sigma\tau,[a]) = B(\sigma,[a])B(\tau,[a])$.

We turn to linearity on the right. This time we write

$$B(\sigma,[a][b]) = \frac{\sigma(\sqrt[n]{a}\sqrt[n]{b})}{\sqrt[n]{a}\sqrt[n]{b}} = \frac{\sigma(\sqrt[n]{a})}{\sqrt[n]{a}} \cdot \frac{\sigma(\sqrt[n]{b})}{\sqrt[n]{b}},$$

picking $\sqrt[n]{a}\sqrt[n]{b}$ as an n-th root for ab. As desired, this reads $B(\sigma,[a][b]) = B(\sigma,[a])$ $B(\sigma,[b])$.

Now we prove that the pairing is perfect. We do a little more; let $G_1 = \mathrm{Gal}(K/F)$ and let G_2 be *either* $F^\times \cap K^{\times n}/F^{\times n}$ or the subgroup generated by $[a_1],\dots,[a_k]$, where the a_i are as in the theorem. Regardless of which group G_2 is used, we prove that

$$B\colon G_1 \times G_2 \longrightarrow \mu_n(F)$$

is nondegenerate. By Lemma 1.20, the orders of G_1 and G_2 must then agree. As a result, the two definitions of G_2 produce the same group.

Indeed, suppose $\sigma \in \mathrm{Gal}(K/F)$ is such that $B(\sigma,[a_i]) = 1$ for all i. This means that σ fixes all the elements $\sqrt[n]{a_i}$, and so σ is the identity of K. Dually, suppose that $[a] \in G_2$ is such that $B(\sigma,[a]) = 1$ for all $\sigma \in \mathrm{Gal}(K/F)$. Then the element $\sqrt[n]{a}$ is fixed by $\mathrm{Gal}(K/F)$, so it belongs to F. In other words, the element $a \in F^\times$ lies in $F^{\times n}$, so $[a] = [1]$. □

The next corollary, which is merely a summary, is easy to commit to memory.

Corollary 1.23 *Suppose F is a field containing a primitive n-th root of unity, and let*

$$K = F[\sqrt[n]{a_1},\dots,\sqrt[n]{a_k}],$$

where $a_i \in F$. Then $\mathrm{Gal}(K/F)$ *is isomorphic to the subgroup of $F^\times/F^{\times n}$ generated by $[a_1],\dots,[a_k]$.*

Proof. The extension K/F is n-Kummer by Proposition 1.16. The theorem applies, and with the help of Lemma 1.20, we derive the result. □

Example 1.24 Take $F = \mathbb{Q}$, and $n = 2$ (certainly $-1 \in \mathbb{Q}$, and $-1 \neq 1$). Let $p_1,\dots,p_k$ be distinct prime numbers in $\mathbb{Z}$, and let $K = \mathbb{Q}[\sqrt{p_1},\dots,\sqrt{p_k}]$. If we return to Example 1.4 which describes $\mathbb{Q}^\times/\mathbb{Q}^{\times 2}$, we deduce that $\mathrm{Gal}(K/\mathbb{Q})$ is isomorphic to $(\mathbb{Z}/2\mathbb{Z})^k$.

But Theorem 1.22 is more precise than that. It gives an explicit isomorphism between $\mathrm{Gal}(K/\mathbb{Q})$ and $\mathrm{Hom}(V,\{\pm 1\})$, where V is the subgroup (subspace, really!)

spanned by $[p_1], \ldots, [p_k]$ in $\mathbb{Q}^\times/\mathbb{Q}^{\times 2}$. Thus, there are elements $\sigma_1, \ldots, \sigma_k \in \mathrm{Gal}(K/\mathbb{Q})$ with

$$\sigma_i(\sqrt{p_i}) = -\sqrt{p_i}, \qquad \sigma_i(\sqrt{p_j}) = \sqrt{p_j} \quad (\text{for } j \neq i)$$

The elements σ_i form a basis for the $\mathbb{Z}/2\mathbb{Z}$-vector space $\mathrm{Gal}(K/\mathbb{Q})$; it is sometimes called the "Kummer dual" of the basis $[p_1], \ldots, [p_k]$. ■

The fundamental theorem

In this book, we shall meet several "fundamental theorems": apart from the fundamental theorem of Kummer theory in this chapter, there will be the fundamental theorem of Galois theory (3.23), the fundamental theorem of local class field theory (13.1), and the fundamental theorem of global class field theory (14.40). We deem "fundamental" those results which summarize a great deal (though not all) of the surrounding theory in very few words. All four results express the existence of a bijection between seemingly unrelated sets of objects.

Theorem 1.25 (Fundamental theorem of Kummer theory) *Let $n \geq 1$, and let F be a field containing a primitive nth root of unity. Fix an algebraic closure $\overline{F}$ of F. There exists a one-to-one, order-preserving correspondence between the n-Kummer extensions K/F contained in $\overline{F}$, and the finite subgroups of $F^\times/F^{\times n}$. The correspondence maps K/F to $F^\times \cap K^{\times n}/F^{\times n}$, and maps $A \subset F^\times/F^{\times n}$ to $F[\sqrt[n]{a} : [a] \in A]$. Moreover, when K/F and A correspond to each other, then* $\mathrm{Gal}(K/F)$ *and A are related by the Kummer pairing (and in particular, are isomorphic).*

Proof. Let $f(A) = F[\sqrt[n]{a} : [a] \in A]$ and $g(K/F) = F^\times \cap K^{\times n}/F^{\times n}$ (the letters f and g are for "field" and "group" respectively). We have $g(f(A)) = A$ by Theorem 1.22. Moreover, any K/F is of the form $f(A)$ for some A, by Proposition 1.16. So:

$$f(g(K/F)) = f[g(f(A))] = f(A) = K/F.$$

Thus, f and g are inverses of each other. □

What happens when F is not the smallest field in sight? That is, if F/F_0 is a Galois extension, can we decide which of the fields K produced by the theorem are actually Galois over F_0, and not just F? The answer is given by "equivariant Kummer theory".

Theorem 1.26 (Equivariant Kummer theory) *Let F be as in the fundamental theorem, and suppose F/F_0 is a finite Galois extension, with $G = \mathrm{Gal}(F/F_0)$. The fundamental correspondence restricts to a bijection between the n-Kummer extensions K/F such that K/F_0 is Galois, and the subgroups of $F^\times/F^{\times n}$ which are preserved by the action of G.*

Moreover, let K/F and A correspond to each other. The action of $\mathrm{Gal}(K/F_0)$ *on the abelian, normal subgroup* $\mathrm{Gal}(K/F)$ *by conjugation factors through an action* *of G. Using this, and the natural G-action on* $A = F^\times \cap K^{\times n}/F^{\times n}$ *and on* $\mu_n(F)$*, the Kummer pairing*

$$B\colon \operatorname{Gal}(K/F) \times \frac{F^\times \cap K^{\times n}}{F^{\times n}} \longrightarrow \mu_n(F)$$

is G-equivariant. Explicitly, one has

$$B(\tau\sigma\tau^{-1}, [\tau(a)]) = \tau(B(\sigma, [a])),$$

where $\sigma \in \operatorname{Gal}(K/F)$*,* $a \in F^\times \cap K^{\times n}$*, and* $\tau \in \operatorname{Gal}(K/F_0)$*.*

Proof. If A is stable under the action of G, and generated by $[a_1], \ldots, [a_k]$, then the field $f(A)$ is generated by F and by the roots of

$$P = \prod_{\substack{\sigma \in \operatorname{Gal}(F/F_0) \\ 1 \leq i \leq k}} (X^n - \sigma(a_i)) \in F_0[X].$$

It follows that K/F_0 is separable and normal, hence Galois. (We leave it as an exercise to check that, when a_i is fixed, the product of all $X^n - \sigma(a_i)$ is a power of a polynomial in $F_0[X]$ with distinct roots; hence P is separable.)

Conversely, suppose K/F_0 is Galois and let $A = g(K/F) = F^\times \cap K^{\times n}/F^{\times n}$. If $[a] \in A$, then a has an nth root in K, say $a = x^n$ with $x \in K$. For any $\sigma \in G$, pick an extension $\widetilde{\sigma} \in \operatorname{Gal}(K/F_0)$, and simply write that $\sigma(a) = \widetilde{\sigma}(x)^n \in K^{\times n}$, showing that $[\sigma(a)] \in A$.

It remains to prove the equivariance of the pairing. First notice that $\tau(\sqrt[n]{a})$ is an nth root for $\tau(a)$, when $a \in F$, so:

$$B(\tau\sigma\tau^{-1}, [\tau(a)]) = \frac{\tau\sigma\tau^{-1}(\tau(\sqrt[n]{a}))}{\tau(\sqrt[n]{a})} = \tau\left(\frac{\sigma(\sqrt[n]{a})}{\sqrt[n]{a}}\right) = \tau(B(\sigma, [a])),$$

which was the requested formula. □

Remark 1.27 When $n = p$ is a prime number, the results of the last two sections can be profitably given an air of linear algebra. The group $F^\times/F^{\times p}$ is a vector space over $\mathbb{Z}/p\mathbb{Z}$, as is $\mu_p(F)$. If we are willing to fix a choice of primitive pth root of unity ω, then $k \mapsto \omega^k$ gives a linear isomorphism between $\mu_p(F)$ and $\mathbb{Z}/p\mathbb{Z}$ itself.

Kummer theory identifies explicitly the vector space $\operatorname{Gal}(K/F)$, when K/F is a p-Kummer extension, with $\operatorname{Hom}(A, \mu_p(F))$ where $A = F^\times \cap K^{\times p}/F^{\times p}$, and this is none other than the dual A^*, when $\mu_p(F)$ is identified with the base field, as above.

In the equivariant setting, with $G = \operatorname{Gal}(F/F_0)$, the last theorem states that the action of $\tau \in G$ on $f\colon A \longrightarrow \mathbb{Z}/p\mathbb{Z}$ is given, for $x \in A$, by

$$(\tau \cdot f)(x) = \tau \cdot f(\tau^{-1}x).$$

On the right-hand side, a reference is made to the action of G on $\mathbb{Z}/p\mathbb{Z} \cong \mu_p(F)$. In the important case when the pth roots of unity are actually in F_0, rather than merely F, this action is trivial. We get the formula

$$(\tau \cdot f)(x) = f(\tau^{-1}x).$$

This is the natural G-action to define on the dual A^* of a vector space A which is itself endowed with a G-action. ∎

Examples

We present a few samples of Kummer theory in action. By the end, the details of the arguments are a little involved, but these should not frighten the reader: in the next chapter we return to a gentle pace. Besides, in the rest of the book we shall only make a few references to the easier parts of the discussion below, so this section is optional.

Consider the following simple statement.

Proposition 1.28 *Let F be a field of characteristic $\neq 2$, and let $a, b \in F^\times$ be such that $[a]$ and $[b]$ are linearly independent in $F^\times/F^{\times 2}$. Suppose the equation*

$$x^2 - ay^2 = b$$

has a solution with $x, y \in F$. Then there exists a Galois extension L/F with $\mathrm{Gal}(L/F) \cong D_8$, the dihedral group of order 8, such that $F[\sqrt{a}, \sqrt{b}] \subset L$.

Incidentally, this result will come back several times in this book, but this is a point to be made later. The first proof we give uses only a small amount of Kummer theory, just to show how one can rely on it to cut corners in familiar situations. One could stick to completely elementary arguments, if desired. In the rest of this section, however, we will provide a more general proof, which yields a stronger result (valid for all primes, rather than $p = 2$), and for this, equivariant Kummer theory will be fully required.

Proof. Let $K = F[\sqrt{a}, \sqrt{b}]$. By Kummer theory, the group $G := \mathrm{Gal}(K/F)$ is isomorphic to $(\mathbb{Z}/2\mathbb{Z})^2$; we let σ, τ be the basis which is Kummer dual to the basis $[a], [b]$ for $A := F^\times \cap K^{\times 2}/F^{\times 2}$. (That is $\sigma(\sqrt{a}) = -\sqrt{a}$ and $\sigma(\sqrt{b}) = \sqrt{b}$; for τ the roles of a and b are exchanged.) Also, put $F_a = F[\sqrt{a}]$. A preliminary remark is that b is not a square in F_a. The argument for this is one of constant use: suppose that we had, on the contrary,

$$F[\sqrt{a}, \sqrt{b}] = F_a[\sqrt{b}] = F_a = F[\sqrt{a}];$$

then Kummer theory shows that $[a]$ and $[b]$ span a group of order ≤ 2, contrary to the hypothesis. (*Aside:* it is in the spirit of the fundamental theorem to write an equality of fields, here $F[\sqrt{a}] = F[\sqrt{a}, \sqrt{b}]$, and deduce an equality of subgroups of $F^\times/F^{\times 2}$. Of course, a possible, slightly different argument is that, if $[b]_{F_a} = 1$, then $[b]_F$ is in the kernel of $F^\times/F^{\times 2} \to F_a^\times/F_a^{\times 2}$, and this kernel is generated by $[a]_F$, as we proved in Theorem 1.22.)

Now let $\alpha = x + y\sqrt{a} \in F_a$, where x and y are as in the proposition. We have

$$\mathrm{N}_{F[\sqrt{a}]/F}(\alpha) = (x + y\sqrt{a})(x - y\sqrt{a}) = b.$$

It follows that α is not a square in F_a (otherwise, after taking norms down to F, we would wrongly conclude that b is a square in F). Let us show that, in fact, α is not a square in K. Indeed, suppose $\alpha \in K^{\times 2}$. Then

$$F_a[\sqrt{b}, \sqrt{\alpha}] = K[\sqrt{\alpha}] = K = F_a[\sqrt{b}],$$

and so by Kummer theory applied to F_a, the classes $[b]_{F_a}$ and $[\alpha]_{F_a}$ are proportional. These two classes are nonzero in the $\mathbb{Z}/2\mathbb{Z}$-vector space $F_a^{\times}/F_a^{\times 2}$, so in the end $[b]_{F_a} = [\alpha]_{F_a}$. We rewrite this as $\alpha = be^2$, with $e \in F_a$. Take norms down to F and obtain $b = b^2 f^2$ (with $f = \mathrm{N}_{F_a/F}(e)$), so $[b]_F = [1]_F$ contrary to the hypothesis.

We have established that $[\alpha]_K \neq [1]_K$. Let us study the action of G on this element. Certainly $\tau(\alpha) = \alpha$, so $\tau[\alpha]_K = [\alpha]_K$. As for σ, we write

$$\sigma(x + y\sqrt{a}) = x - y\sqrt{a} = \frac{b}{x + y\sqrt{a}} = \alpha \cdot \left(\frac{\sqrt{b}}{x + y\sqrt{a}}\right)^2.$$

That is, $\sigma(\alpha) = \alpha k^2$ for some $k \in K$, and $\sigma[\alpha]_K = [\alpha]_K$. In the end, the group G fixes $[\alpha]_K$. If we put $L = K[\sqrt{\alpha}]$, then equivariant Kummer theory (with F, K, L playing the roles of F_0, F, K in the notation of Theorem 1.26) shows that L/F is Galois. Its degree is $[L : K][K : F] = 2 \times 4 = 8$.

It remains to identify $\mathrm{Gal}(L/F)$. We still write σ and τ for (arbitrary) extensions of these two elements to $\mathrm{Gal}(L/F)$. Since $\tau(\alpha) = \alpha$, we have $\tau(\sqrt{\alpha}) = \pm\sqrt{\alpha}$ (both sides square to α), so

$$\tau^2(\sqrt{\alpha}) = \pm\tau(\sqrt{\alpha}) = (\pm 1)^2\sqrt{\alpha} = \sqrt{\alpha}.$$

Since $\tau^2|_K$ is the identity, in the end we have $\tau^2 = 1$.

We treat σ^2 similarly. Indeed, $\sigma(\alpha) = b\alpha^{-1}$ as observed above, so $\sigma(\sqrt{\alpha}) = \pm\sqrt{b}/\sqrt{\alpha}$. It follows that

$$\sigma^2(\sqrt{\alpha}) = \frac{\pm\sqrt{b}}{\pm\sqrt{b}/\sqrt{\alpha}} = \sqrt{\alpha}.$$

We conclude again that $\sigma^2 = 1$.

Let us finally study the commutator $\sigma\tau\sigma\tau$. Since $\sigma(\sqrt{\alpha}) = \varepsilon_1\sqrt{b}/\sqrt{\alpha}$ and $\tau(\sqrt{\alpha}) = \varepsilon_2\sqrt{\alpha}$, with $\varepsilon_i = \pm 1$, a direct computation shows that $\sigma\tau\sigma\tau(\sqrt{\alpha}) = -\varepsilon_1^2\varepsilon_2^2\sqrt{\alpha} = -\sqrt{\alpha}$. So $\sigma\tau\sigma\tau$ is not the identity, or in other words, σ and τ do not commute. (Note that this elementary computation shows that $L \neq K$, so we recover that α is not a square in K.)

We can conclude. The element $R = \sigma\tau$ is not of order 2 by the last computation, and it cannot be of order 8, lest we should conclude that $\mathrm{Gal}(L/F)$ is abelian. So the order of R is 4, and $\sigma R\sigma^{-1} = \tau\sigma = R^{-1}$, from which it is clear that $\mathrm{Gal}(L/F)$ is isomorphic to D_8. □

We proceed to generalize this statement to an arbitrary prime p. First, we need a substitute for the group D_8. Observe that the group $(\mathbb{Z}/p\mathbb{Z})^p$ has an automorphism φ of order p defined by

$$\varphi(x_1,x_2,\ldots,x_p)=(x_2,x_3,\ldots,x_p,x_1)\,.$$

We can then form the semi-direct product $(\mathbb{Z}/p\mathbb{Z})^p \rtimes \langle\varphi\rangle$. A more common notation for this group is $(C_p)^p \rtimes_\varphi C_p$, where C_p denotes a cyclic group of order p (not having a specified generator). In fact, one usually writes $C_p \wr C_p$ and calls it the "wreath product of C_p with itself". You will see easily that for $p=2$, we have $C_2 \wr C_2 \cong D_8$. In general, the order of $C_p \wr C_p$ is p^{p+1}.

Our goal is then to prove the next proposition.

Proposition 1.29 *Let p be a prime number, let F be a field containing a primitive pth root of unity, and let $a,b \in F^\times$ be such that $[a]$ and $[b]$ are linearly independent in $F^\times/F^{\times p}$. Suppose there is an element $\alpha \in F[\sqrt[p]{a}]$ such that*

$$\mathrm{N}_{F[\sqrt[p]{a}]/F}(\alpha)=b\,.$$

Then there exists a Galois extension L/F with $\mathrm{Gal}(L/F) \cong C_p \wr C_p$, and such that $F[\sqrt[p]{a},\sqrt[p]{b}] \subset L$.

We will prove this using equivariant Kummer theory, and we will make every effort to present the argument as a piece of linear algebra. This begins with the decision, once and for all, to write $\mathbb{F}_p$ instead of $\mathbb{Z}/p\mathbb{Z}$. Also, the group $F^\times/F^{\times p}$ will be written additively, so $[ab]=[a]+[b]$, and $[a]=0$ when a is a pth power in F.

When G is a finite group, an "$\mathbb{F}_p[G]$-module" is a vector space V over $\mathbb{F}_p$ on which G acts linearly; that is, there is an action $G\times V \to V$, and each map $v \mapsto \sigma\cdot v$, for $\sigma \in G$, is $\mathbb{F}_p$-linear. (Later in the book we define $\mathbb{F}_p[G]$ as an algebra, see page 87.) For example, in the setting of equivariant Kummer theory, the vector space $F^\times/F^{\times p}$ is an $\mathbb{F}_p[\mathrm{Gal}(F/F_0)]$-module.

The following remark comes back all the time in the study of $\mathbb{F}_p[G]$-modules.

Lemma 1.30 *Suppose the order of G is a power of p. Then any $\mathbb{F}_p[G]$-module V with $V \neq 0$ contains a nonzero vector which is fixed by the action of G.*

Proof. The vector space V certainly contains a nonzero subspace which is finite-dimensional over $\mathbb{F}_p$, and which is an $\mathbb{F}_p[G]$-module: for example, the span of all the elements $\sigma\cdot v$, for all $\sigma\in G$ and some $v\neq 0 \in V$. Thus, it is enough to prove the lemma when V is finite-dimensional, say $n=\dim_{\mathbb{F}_p} V$.

The action of G translates into a homomorphism $G\to \mathrm{GL}(V)$, and if we pick a basis for V, we obtain $h\colon G\to \mathrm{GL}_n(\mathbb{F}_p)$; different choices of basis lead to conjugate homomorphisms. The Sylow theorems imply, then, that we can choose a basis such that $h(G)$ is contained in any given p-Sylow subgroup of $\mathrm{GL}_n(\mathbb{F}_p)$. However, the order of $\mathrm{GL}_n(\mathbb{F}_p)$ being $(p^n-1)(p^n-p)\cdots(p^n-p^{n-1})$, a possible p-Sylow subgroup is given by the matrices, which are upper-triangular with 1s on the diagonal; they form a group B of order $p^{n(n-1)/2}$. The inclusion $h(G)\subset B$ means in particular that the first vector in the basis is fixed by the action of G. □

An $\mathbb{F}_p[G]$-module V will be called "free of rank 1" when there is an element $v \in V$ such that the various $\sigma \cdot v$, for $\sigma \in G$, constitute a basis for V over $\mathbb{F}_p$. (In this case, we say that V is "spanned by v as a free module".) It is easy to see, but helpful to notice, that free modules of rank 1 exist. To construct one, consider the vector space of functions $G \to \mathbb{F}_p$. Define the function f_σ by $f_\sigma(\sigma) = 1$ and $f_\sigma(\tau) = 0$ for $\tau \neq \sigma$; then any function can be written $\sum_\sigma \lambda_\sigma f_\sigma$, with $\lambda_\sigma \in \mathbb{F}_p$. The action of $\sigma \in G$ on f is given by $(\sigma \cdot f)(\tau) = f(\sigma^{-1}\tau)$. The module is free, spanned by f_1. Later in the book, we will write $\mathbb{F}_p[G]^1$ for this "canonical" free module of rank 1.

Here is a helpful criterion to recognize free modules.

Lemma 1.31 *Suppose the order of G is a power of p, and let V be an $\mathbb{F}_p[G]$-module. Let $v \in V$ be such that*

$$\sum_{\sigma \in G} \sigma \cdot v \neq 0.$$

Then v spans a free $\mathbb{F}_p[G]$-module of rank 1.

Proof. Let W be a free module of rank 1, spanned by w. Define a linear map $\pi : W \to V$ by $\pi(\sigma \cdot w) = \sigma \cdot v$, for $\sigma \in G$. Then π is certainly surjective, and it suffices to prove that $\ker(\pi) = \{0\}$.

However, we see also that π is compatible with the action of G, so that $\ker(\pi)$ is itself an $\mathbb{F}_p[G]$-module. We suppose it is nonzero, and work toward a contradiction. We apply Lemma 1.30, and deduce that there is some element $x \neq 0$ in $\ker(\pi)$ which is fixed by all elements of G. As x lives inside a free module, spanned by w, we have $x = \sum_\sigma \lambda_\sigma \sigma \cdot w$, for some scalars $\lambda_\sigma \in \mathbb{F}_p$; the invariance of x under the action of G implies that the coefficients are all equal, that is, $x = \lambda \sum_\sigma \sigma \cdot w$ for some nonzero $\lambda \in \mathbb{F}_p$. This is absurd, as $\pi(x) = \lambda \sum_\sigma \sigma \cdot v \neq 0$ by assumption. This contradiction shows that $\ker(\pi) = \{0\}$. □

One last, easy fact about $\mathbb{F}_p[G]$-modules:

Lemma 1.32 *Let V be an $\mathbb{F}_p[G]$-module, and let $V^* = \mathrm{Hom}(V, \mathbb{F}_p)$ be its dual.*

1. *V^* can be turned into an $\mathbb{F}_p[G]$-module, with action given by $(\sigma \cdot f)(v) = f(\sigma^{-1}v)$.*
2. *If V is free of rank* 1, *so is V^*.*

Proof. (1) is straightforward to check (note that we use σ^{-1} rather than σ in order to have an action on the left). As for (2), suppose V is spanned by v as a free module. Let $v^* \in V^*$ be defined by $v^*(v) = 1$ and $v^*(\sigma \cdot v) = 0$ when $\sigma \neq 1$. Then one checks readily that v^* spans V^* as a free module of rank 1. □

With this general material on $\mathbb{F}_p[G]$-modules at hand, we can turn to the proof of the proposition.

Proof of Proposition 1.29. Let $K = F[\sqrt[p]{a}]$, and $G = \mathrm{Gal}(K/F)$, which is cyclic of order p. Let V be the $\mathbb{F}_p[G]$-module spanned by $[\alpha]_K$ in $K^\times / K^{\times p}$. We wish to use the criterion of Lemma 1.31, so we compute

$$\sum_{\sigma \in G} \sigma[\alpha]_K = [\mathrm{N}_{K/F}(\alpha)]_K = [b]_K\,.$$

If we had $[b]_K = 0$, then we would draw

$$F[\sqrt[p]{a}] = K = K[\sqrt[p]{b}] = F[\sqrt[p]{a}, \sqrt[p]{b}]\,,$$

implying by Kummer theory that $[a]_F$ and $[b]_F$ span a subspace of dimension ≤ 1 in $F^\times/F^{\times p}$, which is absurd, given our hypothesis. So $[b]_K \neq 0$, Lemma 1.31 applies, and V is free of rank 1.

Now we let $L = K[\sqrt[p]{\sigma(\alpha)} : \sigma \in G] = K[\sqrt[p]{v} : [v] \in V]$, noting that $\sqrt[p]{b} \in L$, as b is the product of all the elements $\sigma(\alpha)$ for $\sigma \in G$.

The conclusions of equivariant Kummer theory are very strong. First, the extension L/F is Galois, since V is G-stable by construction. Next, the normal subgroup $\mathrm{Gal}(L/K)$ is isomorphic to $\mathbb{F}_p^p$ (and the order of $\mathrm{Gal}(L/F)$ is p^{p+1}). More precisely, the conjugation action of $\mathrm{Gal}(L/F)$ on $\mathrm{Gal}(L/K)$ factors through an action of the cyclic group G, and as an $\mathbb{F}_p[G]$-module, we have $\mathrm{Gal}(L/K) \cong V^*$, the dual of V. By Lemma 1.32, we know that $\mathrm{Gal}(L/K)$ is a free $\mathbb{F}_p[G]$-module. Thus the action of $G \cong C_p$ on $\mathrm{Gal}(L/K) \cong \mathbb{F}_p^p$ is, in a certain basis, given by the "cyclic" formula presented during the definition of $C_p \wr C_p$.

It remains to prove that $\mathrm{Gal}(L/F)$ is a semi-direct product $\mathrm{Gal}(L/K) \rtimes \mathrm{Gal}(K/F)$. This is a purely group-theoretic fact, and later in the book we will see it as an application of "Shapiro's lemma". For now, we use a trick to finish. We wish to show the existence of an element $\widetilde{\tau} \in \mathrm{Gal}(L/F)$ which has order p, and restricts to a generator $\tau \in G$. Fixing τ, the number of its "lifts" $\widetilde{\tau} \in \mathrm{Gal}(L/F)$ is the order of $\mathrm{Gal}(L/K)$, which is p^p, and we must show that one of these lifts has order p. Let ω be a primitive pth root of unity in F. For any lift $\widetilde{\tau}$, we have whenever $0 \leq i < p$:

$$\widetilde{\tau}\left(\sqrt[p]{\tau^i(\alpha)}\right) = \omega^{n_i} \sqrt[p]{\tau^{i+1}(\alpha)}\,,$$

for some integer n_i with $0 \leq n_i < p$, as the pth power of either side of this equation is $\tau^{i+1}(\alpha)$. Now, $\widetilde{\tau}$ is determined by these integers n_i, and there are p^p choices for them at the most. If τ is to have p^p different lifts, all the potential choices must actually occur. So there is a lift $\widetilde{\tau}$ with

$$\widetilde{\tau}\left(\sqrt[p]{\tau^i(\alpha)}\right) = \sqrt[p]{\tau^{i+1}(\alpha)}\,,$$

for all i. As a result,

$$\widetilde{\tau}^p\left(\sqrt[p]{\tau^i(\alpha)}\right) = \sqrt[p]{\tau^{i+p}(\alpha)} = \sqrt[p]{\tau^i(\alpha)}\,.$$

The element $\widetilde{\tau}^p$ is the identity on K (because τ^p is), and it fixes the elements $\sqrt[p]{\tau^i(\alpha)}$, so it is the identity on all of L. This concludes the proof. □

Problems

1.1. Let p be a prime, let F be a field of characteristic $\neq p$, and let $a \in F^\times \smallsetminus F^{\times p}$. We make no assumption on the presence of roots of unity in F, and we wish to prove that $[F(\alpha) : F] = p$, where α is a root of $X^p - a$. For this, we assume $[F(\alpha) : F] \neq p$, and work toward a contradiction:

1. Consider the composite field $F(\mu_p,\alpha)$, and show that $F(\alpha) \subset F(\mu_p)$.
2. Deduce that $F(\alpha)/F$ is Galois, and then that $F(\alpha) = F(\mu_p)$.
3. From the fact that $[F(\alpha) : F]$ does not depend on the choice of root α, conclude.

1.2. Let F be a field of characteristic $\neq 2$, let $a \in F^\times$ and $K = F[\sqrt{a}]$, and assume that $a \notin F^{\times 2}$. We do not assume that $\sqrt{-1} \in F$.

1. Suppose there exists an extension L/F with $K \subset L$, such that L/F is Galois and $\mathrm{Gal}(L/F)$ is cyclic of order 4. Show that there exists $\alpha \in K$ such that $\mathrm{N}_{K/F}(\alpha) = -1$.
   Hint: try $\alpha = (\theta - \sigma^2(\theta))/(\sigma(\theta) - \sigma^3(\theta))$, in notation which is left to the reader to figure out.
2. Conversely, suppose there exists $\alpha \in K$ with $\mathrm{N}_{K/F}(\alpha) = -1$. Show that K/F is contained in a Galois extension L/F which is cyclic, of degree 4.
   Hint: the case when $\alpha \in F$ must be treated separately. When $\alpha \notin F$, try $L = K\left[\sqrt{1+\alpha^2}\right]$.
3. Show that the two properties below are equivalent:
   (a) There exists $\alpha \in K$ with $\mathrm{N}_{K/F}(\alpha) = -1$.
   (b) a is a sum of two squares in F.

1.3. Generalize Proposition 1.28 as follows. Let F be a field of characteristic $\neq 2$, and let $a,b \in F^\times$. Suppose the equation $x^2 - ay^2 = b$ has a solution with $x,y \in F$. Then there exists a Galois extension L/F with $\mathrm{Gal}(L/F)$ isomorphic to a subgroup of D_8, such that $F[\sqrt{a},\sqrt{b}] \subset L$.

Hint: there are many cases to consider. Some of them can be dealt with using the previous problem; for this, notice that when $a = b$ and we have a solution to $x^2 - ay^2 = b$, then a is a sum of two squares in F.

1.4. Suppose F and K are fields such that $F \subset K \subset \mathbb{R}$. We assume that K/F is finite and Galois, and that $f \in F[X]$ is a polynomial whose splitting field is K. Show that it is possible to choose $c \in F$ such that, if L is the splitting field of $f(X^2 + c)$, then $K \subset L$ and $\mathrm{Gal}(L/F)$ is a semidirect product $C_2^n \rtimes \mathrm{Gal}(K/F)$. Here, n is the degree of f, and the action of $\mathrm{Gal}(K/F)$ on C_2^n is by permuting the factors, following the natural action on the roots of f.

Hint: complete the following sketch. By general arguments, the group $\mathrm{Gal}(L/K)$ *is of the form C_2^m for some $m \leq n$. The trick is to pick c such that $f(X+c)$ has exactly one negative root. Then consider complex conjugation, and all its conjugates within the group* $\mathrm{Gal}(L/F)$*, in order to show that $n = m$. Finish with a counting argument, similar to that in the proof of Proposition 1.29. This problem is adapted from a paper by Gow [Gow86].*

1.5. Let F be a field of characteristic $p > 0$, and let K/F be a cyclic extension of degree p. We let σ denote a generator of $\mathrm{Gal}(K/F)$, and we write $T = \sigma - I$, an F-linear operation on the F-vector space K.

1. Show that $\ker(T) = F$, that $T^p = 0$, and that $\mathrm{Im}(T^{p-1}) = F$.
2. Let $c \in K$ be such that $T^{p-1}(c) = 1$, and let $\alpha = T^{p-2}(c)$. Compute $\sigma(\alpha)$ and deduce that $\alpha^p - \alpha \in F$.
3. Conclude that K is obtained from F by adjoining a root of a polynomial of the form $X^p - X - a \in F[X]$, which is irreducible. *This is called an Artin–Schreier polynomial.*

1.6. (A converse for the previous problem.) Let F be a field of characteristic $p > 0$, let $a \in F$ be an element which is **not** of the form $\beta^p - \beta$ for $\beta \in F$, let $f = X^p - X - a$, and let K be a splitting field for f.

1. If $\alpha \in K$ is a root of f, what are the other roots? Deduce that $K = F[\alpha]$.
2. Show that there exists $\sigma \in \mathrm{Gal}(K/F)$ which has order p. Deduce that f is irreducible, and conclude.

1.7. Develop a version of Kummer theory for p-Kummer extensions of a field of characteristic p, using the two preceding problems. Replacing $\sqrt[p]{-}$ by $\mathscr{P}^{-1}$, where $\mathscr{P} = X^p - X$, this should be formally similar to the case in which F has a primitive pth root of unity, as in this chapter. You should have a Kummer pairing, and a fundamental theorem.

A solution is given in Lang [Lan02], see Theorem 8.3 in that book, for example.

2 Local number fields

The field $\mathbb{C}$ of complex numbers is, in many ways, easier to work with than the abstract algebraic closure $\overline{\mathbb{Q}}$ of the rational field. For one thing, the complex numbers have a pleasant geometric interpretation, since we can think of them as arranged in a plane. Also, and this will be the emphasis of this chapter, we have analytical tools at our disposal, like the exponential, for example. Is it not reassuring, when dealing with roots of unity, to think of $e^{\frac{2\pi i}{n}}$?

When constructing $\mathbb{C}$, you must remember that the hard step was to construct $\mathbb{R}$. (The rest is a simple matter of adding a square root for -1.) It takes a moment of pondering to convince oneself that all the qualities of $\mathbb{C}$ are essentially consequences of one fact: the field $\mathbb{R}$ is *complete* for its natural, metric topology. For example, in order to prove that $\mathbb{C}$ is algebraically closed, one of the easiest arguments uses that a continuous function on a closed, bounded subset of the complex plane must reach its minimum and its maximum, and this ultimately relies on completeness. Likewise, the exponential is afforded by a power series, the convergence of which is guaranteed, at its heart, by a Cauchy sequence argument.

In fact, since $\mathbb{Q}$ is *dense* in $\mathbb{R}$, it is possible to *define* the field of real numbers as the completion of $\mathbb{Q}$ for the metric induced by the usual absolute value. This may or may not have been the very definition given to you as an undergraduate. It is also quite possible to define $\mathbb{R}$ in some other way, using Dedekind cuts, for example, or to keep things simple, using decimal expansions. One proves the completeness afterwards.

In this chapter, we shall construct other completions of $\mathbb{Q}$, for other absolute values. Indeed, let p be a prime number. Any nonzero $x \in \mathbb{Q}$ can be written $x = p^n \frac{a}{b}$ where $n \in \mathbb{Z}$ and a, b are integers prime to p. This expression is even unique if we take a and b coprime and $b > 0$, and in any case the number n is uniquely defined by x. We write it $v_p(x)$ and call it the *p-adic valuation of* x. Now put $|x|_p = p^{-v_p(x)}$. As we will explain below, this formula defines a new absolute value $|\cdot|_p$ on $\mathbb{Q}$. It is decidedly different from the usual one: An integer is close to 0, now, when it is divisible by a large power of p.

The field $\mathbb{Q}_p$ of *p-adic numbers* can be defined as the completion of $\mathbb{Q}$ for this new absolute value. However, we shall avoid relying on existence theorems for completions, and give a direct construction instead.

There are very few algebraic extensions of $\mathbb{R}$ (in fact, just two). The field $\mathbb{Q}_p$, while not as nearly algebraically closed, is also under good control in terms of Galois theory: In this book we shall prove that it has only *finitely many* algebraic extensions of a given degree (Theorem 4.13), and that whenever $K/\mathbb{Q}_p$ is finite and Galois, the group $\mathrm{Gal}(K/\mathbb{Q}_p)$ is solvable (Theorem 2.58). The study of p-adic fields – the name for a field such as K – provides an ideal playground for us, providing field extensions that are not trivial, yet fairly well understood.

We also point out that many questions about $\mathbb{Q}$ can be translated into questions about its completions. We will not dwell much on these topics in this book (unfortunately), but we can quote (without proof!) the famous *Hasse–Minkowski* theorem. This result pertains to quadratic forms over $\mathbb{Q}$, that is polynomials of the form

$$q(X_1,\ldots,X_n) = \sum_{i,j} a_{ij} X_i X_j$$

with $a_{ij} \in \mathbb{Q}$. Many of the interesting questions about quadratic forms can be reduced to the following: Do there exist $x_1,\ldots,x_n \in \mathbb{Q}$, not all 0, such that $q(x_1,\ldots,x_n) = 0$? The Hasse–Minkowski theorem asserts that the x_i can be found in $\mathbb{Q}$ if and only if (1) for each prime number p, we can find $y_1,\ldots,y_n \in \mathbb{Q}_p$, not all 0, with $q(y_i) = 0$, and (2) the same over $\mathbb{R}$, that is, we can find $r_1,\ldots,r_n \in \mathbb{R}$, not all 0, such that $q(r_i) = 0$.

Construction of $\mathbb{Q}_p$

Let p be a prime number. As the reader must know, any positive integer x can be written uniquely "in base p", that is, in the form

$$x = a_0 + a_1 p + a_2 p^2 + \cdots + a_n p^n \,, \tag{2.1}$$

where each a_k is an integer with $0 \le a_k < p$. Here, uniqueness means in particular that the numbers $a_0, a_1, \ldots, a_n$ are data extracted from x; one could represent all integers by finite sequences of such "digits". The definition of the set $\mathbb{Z}_p$ below is beautifully simple, merely allowing us (formally) to perform infinite sums.

Definition 2.1 We write $\mathbb{Z}_p$ for the set of formal sums

$$\sum_{k=0}^{\infty} a_k p^k \,,$$

where a_k is an integer satisfying $0 \le a_k < p$.

A "formal sum" is thus nothing more than a collection of integers a_0, a_1, ..., as specified, written as a sum for purposes of intuition. We will soon see that we can, if we so wish, interpret this as an actual sum, but for now the intuitive notation serves us to recognize instantly that $\mathbb{N} \subset \mathbb{Z}_p$, identifying $x \in \mathbb{N}$ with the sum as in (2.1), extended by $a_k = 0$ for $k > n$. (The formal sum for 0 has only zero coefficients.)

Our first goal is to show that $\mathbb{Z}_p$ is a ring. In the next chapter we shall present an alternative approach for this, which is more abstract but perhaps also more clean-cut;

here we want to work this out "by hand". The idea should be to mimic the addition and multiplication of $\mathbb{N}$ when performed with pencil and paper, with carries and so on. However, to make it easier to prove that the operations behave as expected (for example, to check associativity), we argue as follows. There is a map

$$\mathbb{Z}_p \xrightarrow{\pi_n} \mathbb{Z}/p^n\mathbb{Z}$$

for all $n \geq 1$, which takes a formal sum

$$x = \sum_{k=0}^{\infty} a_k p^k$$

to the number $\pi_n(x) = x_n \bmod p^n$ where

$$x_n = \sum_{k=0}^{n-1} a_k p^k$$

with the sum performed in $\mathbb{Z}$. Note that $0 \leq x_n < p^n$, so that one recovers x_n from $\pi_n(x) \in \mathbb{Z}/p^n\mathbb{Z}$ if needed; the first few terms of x are then recovered from x_n by writing it in base p. So x is determined by all the $\pi_n(x)$, for $n \geq 1$. (Again, in the next chapter the relationship between $\mathbb{Z}_p$ and the various $\mathbb{Z}/p^n\mathbb{Z}$ will be described more conceptually.)

Now, given $x, y \in \mathbb{Z}_p$, we can consider the sum $\pi_n(x) + \pi_n(y)$ as well as the product $\pi_n(x)\pi_n(y)$, all computed in $\mathbb{Z}/p^n\mathbb{Z}$. Moreover, there is a unique integer, say z_n, such that $0 \leq z_n < p^n$, and representing $\pi_n(x) + \pi_n(y) \bmod p^n$. Finally, writing z_n in base p, we view it as a formal sum in $\mathbb{Z}_p$, denoted by $(x+y)_n$. Define $(xy)_n$ similarly.

The reader should prove the following easy, but crucial, lemma. It follows simply from the fact that the natural map

$$\mathbb{Z}/p^{n+1}\mathbb{Z} \longrightarrow \mathbb{Z}/p^n\mathbb{Z}$$

is a ring homomorphism.

Lemma 2.2 *The sum $(x+y)_{n+1}$ is obtained from $(x+y)_n$ by adding a single term of degree p^n, and likewise for $(xy)_{n+1}$ and $(xy)_n$.* □

So we may define

$$x + y = \sum_{k=0}^{\infty} c_k p^k$$

where c_k is the kth coefficient of $(x+y)_n$ for any n large enough. The product xy is defined similarly. Crucially, we have $\pi_n(x+y) = \pi_n(x) + \pi_n(y)$, and $\pi_n(xy) = \pi_n(x)\pi_n(y)$, for all $n \geq 1$. Since each $\mathbb{Z}/p^n\mathbb{Z}$ is a commutative ring, we obtain immediately:

Lemma 2.3 *With the operations defined above, the set $\mathbb{Z}_p$ is turned into a commutative ring. For any $n \geq 1$, the map π_n is a ring homomorphism. The kernel of π_n consists of all formal sums of the form*

$$\sum_{k=p^n}^{\infty} a_k p^k .$$

Moreover, it is the ideal $p^n\mathbb{Z}_p$.

Proof. The first statements are true by construction: for example, the associativity $(x+y)+z = x+(y+z)$ follows if for each $n \geq 1$ we have $\pi_n((x+y)+z) = \pi_n(x+(y+z))$; however,

$$\pi_n((x+y)+z) = \pi_n(x+y) + \pi_n(z) = \pi_n(x) + \pi_n(y) + \pi_n(y),$$

and $\pi_n(x+(y+z))$ can be expanded to the same expression. All the other verifications are as trivial as this one.

We turn to the claim about the kernel of π_n. For this, with the above notation we see that $\pi_n(x) = 0$ implies $x_n = 0$ (as already pointed out), and the sum for x does start at $a_n p^n$ in this case, as stated. All such sums are in the kernel, conversely. For the very last statement, it remains to observe directly that

$$p\Big(\sum_{k=0}^{\infty} a_k p^k\Big) = \sum_{k=1}^{\infty} a_{k-1} p^k .$$

This follows right from the definitions, but must be verified nonetheless. □

In symbols, the lemma says in particular $\mathbb{Z}_p/p^r\mathbb{Z}_p \cong \mathbb{Z}/p^r\mathbb{Z}$, an isomorphism which we will mostly treat as the identity. One can replace the notation $\pi_n(x)$ by $x \bmod p^n\mathbb{Z}_p$ or even $x \bmod p^r$. If $x \in \mathbb{Z}$, the notation $x \bmod p^r$ means the same regardless of whether x is seen in $\mathbb{Z}_p$ or not.

Example 2.4 Here are a few calculations. With $\mathbb{Z}_p$ being a ring, it must possess a ring map $\iota\colon \mathbb{Z} \to \mathbb{Z}_p$. Our first observation is that this map extends our identification of $\mathbb{N}$ with a subset of $\mathbb{Z}_p$. In particular, $\mathbb{Z}_p$ must have characteristic 0, so that ι is injective, and $\mathbb{Z}_p$ contains a copy of $\mathbb{Z}$. But where are all the negative numbers? Consider the identity

$$-1 = (p-1) + (p-1)p + (p-1)p^2 + \cdots + (p-1)p^{n-1} - p^n .$$

This is an identity in $\mathbb{Z}$, so we know "abstractly" that we may see it as an identity in $\mathbb{Z}_p$, and apply π_n on both sides. The last lemma gives $\pi_n(p^n) = 0$, of course, and we deduce the first few terms of -1; since n is arbitrary, our conclusion is

$$-1 = \sum_{k=0}^{\infty} (p-1)p^k \in \mathbb{Z}_p .$$

The next proposition shows the importance of the ideal $p\mathbb{Z}_p$.

Proposition 2.5 *An element $x \in \mathbb{Z}_p$ is a unit if and only if it does not belong to the ideal $p\mathbb{Z}_p$. In symbols*

$$\mathbb{Z}_p^\times = \mathbb{Z}_p \smallsetminus p\mathbb{Z}_p.$$

It follows that $p\mathbb{Z}_p$ is the unique maximal ideal of $\mathbb{Z}_p$.

Proof. A proper ideal never contains units. For the converse, recall first that the elements of $\mathbb{Z}/p^n\mathbb{Z}$ which are invertible are precisely those represented by integers prime to p. However, if $x \in \mathbb{Z}_p$ does not belong to $p\mathbb{Z}_p$, we see that each $x_n \in \mathbb{Z}$ is prime to p, so that $\pi_n(x)$ has a unique inverse, represented by a unique integer y_n with $0 \leq y_n < p^n$. Again we observe that y_{n+1}, when written in base p, is obtained from the expression for y_n by adding a single term of degree p^n. Thus we may combine the y_n to define $y \in \mathbb{Z}_p$. We have $\pi_n(xy) = 1$ for all n, so that $xy = 1$, and x is a unit.

The next statement is a general fact. Suppose that R is a commutative ring, and that I is an ideal such that $R^\times = R \smallsetminus I$. For any ideal $J \neq R$, the fact that J cannot contain units shows that $J \subset I$, and thus I contains *all* the proper ideals of R. It follows clearly that it is the unique maximal ideal if R. □

Example 2.6 The element $1 - p \in \mathbb{Z} \subset \mathbb{Z}_p$ is not in $p\mathbb{Z}_p$, so it is invertible. One guesses that its inverse should be the sum of the p^k, and indeed the usual argument applies: Write

$$1 = (1 + p + \cdots + p^{n-1})(1 - p) + p^n$$

so that, multiplying by $(1-p)^{-1} \in \mathbb{Z}_p$, we get

$$(1-p)^{-1} = 1 + p + \cdots + p^{n-1} \bmod p^n\mathbb{Z}_p,$$

so

$$\frac{1}{1-p} = \sum_{k=0}^{\infty} p^k.$$

■

Corollary 2.7 *Any nonzero $x \in \mathbb{Z}_p$ can be written uniquely $p^\nu u$ where $u \in \mathbb{Z}_p^\times$ and $\nu \in \mathbb{N}$. Moreover, the ring $\mathbb{Z}_p$ is an integral domain.*

Proof. We use again the identity

$$p\Big(\sum_{k=0}^{\infty} a_k p^k\Big) = \sum_{k=1}^{\infty} a_{k-1} p^k$$

to see inductively that $x = p^\nu u$ where ν is the smallest index such that $a_\nu \neq 0$, and u is an element which is not in $p\mathbb{Z}_p$. By the proposition, u is a unit, and we have the existence. Uniqueness is now clear (ν being as just described).

If $x = p^{v_1}u_1$ and $y = p^{v_2}u_2$ are both nonzero, then $xy = p^{v_1+v_2}u_1u_2$. The element u_1u_2 is a unit, so if we had $xy = 0$ then we would draw $p^{v_1+v_2} = 0$, which is not the case. □

Now that we know that $\mathbb{Z}_p$ is an integral domain, it makes sense to make the following definition.

Definition 2.8 We define $\mathbb{Q}_p$ to be the field of fractions of $\mathbb{Z}_p$.

Notice that, as $\mathbb{Z}_p$ contains $\mathbb{Z}$, the field $\mathbb{Q}_p$ contains $\mathbb{Q}$.

Lemma 2.9 *Any nonzero $x \in \mathbb{Q}_p$ can be written uniquely as $p^v u$ where $u \in \mathbb{Z}_p^\times$ and $v \in \mathbb{Z}$. Moreover,*

$$\mathbb{Q}_p = \mathbb{Z}_p\left[\frac{1}{p}\right],$$

where the right-hand side denotes the smallest subring of $\mathbb{Q}_p$ containing $\mathbb{Z}_p$ and $\frac{1}{p}$.

Proof. The first point follows from the last corollary, since any element of $\mathbb{Q}_p$ is a quotient of two elements of $\mathbb{Z}_p$ (uniqueness is an easy exercise). The second point is obvious. □

The integer v in this lemma is called the *p-adic valuation* of x, and is denoted by $v_p(x)$. For good measure, we throw in $v_p(0) = +\infty$. We can write any $x \in \mathbb{Q}_p^\times$ as

$$x = p^{v_p(x)} \sum_{k=0}^{\infty} a_k p^k,$$

with $a_0 \neq 0$.

Now of course $v_p(x)$ makes sense for $x \in \mathbb{Q} \subset \mathbb{Q}_p$. And indeed, if we write $x = p^v \frac{a}{b}$ where the integers a and b are prime to p, then we can see a and b as units in $\mathbb{Z}_p$, and we have put x in the required form, so $v_p(x) = v$. In other words v_p is the function used in the introduction to this chapter.

For future reference, let us introduce some terminology.

valuation

Definition 2.10 A **valuation** on a field F is a map

$$v\colon F \longrightarrow \mathbb{R} \cup \{+\infty\}$$

such that

1. $v(x) = +\infty \Leftrightarrow x = 0$,
2. $v(xy) = v(x) + v(y)$ for all $x, y \in F$,
3. $v(x+y) \geq \min(v(x), v(y))$ for all $x, y \in F$.

The additive group $v(F^\times) \subset \mathbb{R}$ is called the valuation group of v, and v is called **discrete** if the valuation group is of the form $\alpha\mathbb{Z}$ for some number $\alpha \geq 0$.

For example, the field $\mathbb{Q}$ possesses all the valuations v_p, for all primes p. On $\mathbb{Q}_p$ we have the single valuation v_p (are there others?). These are all discrete.

absolute value

Definition 2.11 An **absolute value** on a field F is a map

$$F \longrightarrow \mathbb{R}, \qquad x \mapsto |x|,$$

such that

1. $|x| \geq 0$ for all $x \in F$, and $|x| = 0 \Leftrightarrow x = 0$,
2. $|xy| = |x||y|$ for all $x, y \in F$,
3. $|x + y| \leq |x| + |y|$ for all $x, y \in F$, the **triangle inequality**.

non-Archimedean

If, moreover, we have $|x + y| \leq \max(|x|, |y|)$ for $x, y \in F$, which is stronger than the triangle inequality, then the absolute value is said to be **non-Archimedean** (and Archimedean otherwise).

For example, on $\mathbb{R}$, and so also on $\mathbb{Q}$, we have the usual absolute value $|x| = \max(x, -x)$, which is Archimedean. To produce more examples, one proves the next (obvious) lemma.

Lemma 2.12 *If v is a valuation on F, and if $0 < c < 1$, then one defines a non-Archimedean absolute value on F by $|x|_v = c^{v(x)}$.*

Moreover, given any absolute value at all $|\cdot|$ on F, one defines a metric by $d(x, y) = |x - y|$.

The topology induced by the metric corresponding to $|\cdot|_v$ is independent of the choice of c. □

When dealing with v_p, one traditionally picks $c = {}^1/_p$, and thus defines $|x|_p = p^{-v_p(x)}$ for $x \in \mathbb{Q}_p$.

In the rest of this section, we leave the generalities and collect a few initial facts about this absolute value $|\cdot|_p$ on $\mathbb{Q}_p$. Still, the formal similarity with the case of $\mathbb{R}$, equipped with the good old absolute value that we know so well, is of great help. For example, we shall use below that multiplication by a fixed $\lambda \in \mathbb{Q}_p$ gives a *continuous* map $\mathbb{Q}_p \to \mathbb{Q}_p$; not only can this be easily proved directly, but the statement itself should be obvious to us, because we are used to the analogous result in $\mathbb{R}$, and it should be intuitive that it only depends on the axioms for an absolute value.

The definitions are so arranged that $\mathbb{Z}_p$ is the subset of $\mathbb{Q}_p$ of those x such that $v_p(x) \geq 0$, or equivalently $|x|_p \leq 1$ (a fact that does not depend on our choice of constant c, just like the rest, and we will stop commenting on that). So *$\mathbb{Z}_p$ is the closed unit ball in $\mathbb{Q}_p$*. Likewise, the ideal $p\mathbb{Z}_p$ is comprised of those x with $v_p(x) > 0$, or $|x|_p < 1$, and *the unique maximal ideal in $\mathbb{Z}_p$ is the open unit ball*, while *the units of $\mathbb{Z}_p$ comprise precisely the unit sphere*.

Proposition 2.13 *The metric induced by $|\cdot|_p$ has the following properties.*

1. *The field $\mathbb{Q}_p$ is complete.*
2. *The subfield $\mathbb{Q}$ is dense in $\mathbb{Q}_p$, and the closure of $\mathbb{Z}$ is $\mathbb{Z}_p$.*

In other words, we can see $\mathbb{Q}_p$ as the completion of $\mathbb{Q}$ for $|\cdot|_p$.

Proof. Let $(s_n)_{n\geq 0}$ be a Cauchy sequence in $\mathbb{Z}_p$. For any given integer N, we have for large values of n that $v_p(s_{n+m} - s_n) \geq N$, so that $s_{n+m} - s_n \in p^N\mathbb{Z}_p$, and the first N terms of s_n and s_{n+m}, written as formal sums, are the same. Since N is arbitrary here, we see in particular that for each $k \geq 0$, the term $a_k p^k$ found in s_n is the same for all n large enough; this defines a_k for all k. Let $\ell = \sum_k a_k p^k \in \mathbb{Z}_p$. Then $v_p(\ell - s_n) > N$ when n is large enough, showing that s_n converges to ℓ.

In general, we consider a Cauchy sequence $(s_n)_{n\geq 0}$ in $\mathbb{Q}_p$. Then it is a bounded sequence, and so $v_p(s_n)$ is bounded below. Multiplying by an appropriate power of p, we can arrange for all terms to have a nonnegative valuation, that is, the sequence $(u_n) = (p^\alpha s_n)_{n\geq 0}$ is a sequence in $\mathbb{Z}_p$, for some α large enough. The new sequence (u_n) is still a Cauchy sequence, clearly, so it converges to some ℓ, and (s_n) converges to $p^{-\alpha}\ell$ (because multiplication by $p^{-\alpha}$ is continuous). This proves (1).

For (2), we use the notation introduced above: for $x \in \mathbb{Z}_p$, we have the elements x_n for $n \geq 1$, which are in $\mathbb{Z}$ (in fact in $\mathbb{N}$) and converge to x. So $\mathbb{Z}$ (or even $\mathbb{N}$) is dense in $\mathbb{Z}_p$ (which is closed). The other statement is similarly established. $\square$

In the spirit of this last proof, we realize, perhaps with some awe, that

$$\lim_{n\to\infty} \sum_{k=0}^{n} a_k p^k = \sum_{k=0}^{\infty} a_k p^k .$$

Here, on the left-hand side, we have a limit of integers, taken with respect to the metric induced by $|\cdot|_p$, while on the right-hand side we have *a priori* a formal sum defining an element in $\mathbb{Z}_p$. And lo, it appears that the formal sums were actually convergent series all along.

The definition of $\mathbb{Z}_p$ in terms of formal sums was a piece of rope thrown over the river we had to cross. Now that a bridge has been fully built, it is probably best to let ourselves be driven smugly by the definition of $\mathbb{Z}_p$ and $\mathbb{Q}_p$ as the completions of $\mathbb{Z}$ and $\mathbb{Q}$ respectively. For example, we have used above the formula

$$p\left(\sum_{k=0}^{\infty} a_k p^k\right) = \sum_{k=1}^{\infty} a_{k-1} p^k .$$

Initially, to verify this we had to think of the very definition of the multiplication in $\mathbb{Z}_p$, going via all the $\mathbb{Z}/p^n\mathbb{Z}$ and so on. Now, you should mentally picture the identity

$$p\left(\sum_{k=0}^{n} a_k p^k\right) = \sum_{k=1}^{n+1} a_{k-1} p^k ,$$

which takes place in $\mathbb{Z}$ and deserves no comment; then, take limits on both sides as $n \to +\infty$, and argue to yourself that multiplication by p is continuous.

There is never much work to do in order to prove that limits exist, for in the non-Archimedean case we have the following liberating lemma:

Lemma 2.14 *Suppose F is a field that is complete with respect to a non-Archimedean absolute value. If $a_n \to 0$, then the series $\sum_{n=0}^{\infty} a_k$ converges.*

Proof. Indeed, the non-Archimedean condition implies that

$$\left|\sum_{k=n}^{n+m} a_k\right| \leq \max_{n\leq k\leq n+m} |a_k|,$$

from which we draw that the partial sums form a Cauchy sequence. □

Around Hensel's lemma

In this section we study Hensel's lemma and some of its consequences. This is a result pertaining to fields that are complete with respect to a discrete valuation, of which our sole example is that of $\mathbb{Q}_p$ so far. (Here, of course, completeness is meant with respect to the *metric* associated to the valuation, as the reader will have guessed; we will use such shorthands frequently.) However, Hensel's lemma is also instrumental in building more examples, so we must turn to it now.

Our discussion begins with a few general properties of fields that are equipped with a discrete valuation. Mostly, we seek to establish in the general case some facts that we have observed when working with $\mathbb{Q}_p$, this time chiefly on the assumption that there is a valuation around. Should the reader find this a little frivolous, we have taken the opportunity to go a little further and describe new properties not mentioned above.

So let F be endowed with the discrete valuation v (no completeness assumption is made yet). First, a general definition.

Definition 2.15 The set

$$\mathcal{O}_F = \{x \in F : v(x) \geq 0\}$$

ring of integers

is called **ring of integers**. We usually write

$$\mathfrak{p} = \{x \in F : v(x) > 0\}.$$

■

Example 2.16 When $F = \mathbb{Q}$, equipped with v_p, we find that $\mathcal{O}_F = \mathbb{Z}_{(p)}$, which is the ring of all elements of the form $\frac{a}{b}$ with b prime to p, usually called the localization of $\mathbb{Z}$ at p. When $F = \mathbb{Q}_p$, also with v_p, we have $\mathcal{O}_F = \mathbb{Z}_p$ (which contains $\mathbb{Z}_{(p)}$). In either case, $\mathfrak{p}$ is the principal ideal generated by p. ■

Another useful piece of terminology is this: When $v(F^\times) = \mathbb{Z}$, one says that v is *normalized*. When we have a discrete valuation which is *not* normalized, that is $v(F^\times) = \alpha\mathbb{Z}$ with $\alpha > 0$, then we may replace v by $v' = \frac{1}{\alpha}v$. The topologies, using v and v', are the same (by Lemma 2.12); also $\mathcal{O}_F$ and $\mathfrak{p}$ are unchanged if we define them using v' instead of v. Some statements are more conveniently formulated with normalized valuations, as in the next proposition.

Proposition 2.17 *Let F be a field with a discrete valuation v.*

1. *$\mathcal{O}_F$ is a ring. Its group of units consists of those elements having zero valuation.*
2. *$\mathfrak{p}$ is its unique maximal ideal.*
3. *There exists $\pi \in \mathcal{O}_F$ such that $\mathfrak{p} = (\pi)$, and then all the ideals of $\mathcal{O}_F$ are of the form (π^n) for some $n \in \mathbb{N}$. In particular, $\mathcal{O}_F$ is a principal ideal domain. When v is normalized, one may take any element π such that $v(\pi) = 1$.*
4. *A nonzero element $x \in \mathcal{O}_F$ can be written uniquely $x = \pi^n u$ where $n \in \mathbb{N}$ and $u \in \mathcal{O}_F^\times$ is a unit. A nonzero element $x \in F$ can be written uniquely $x = \pi^n u$ with $n \in \mathbb{Z}$ and u a unit.*

An element such as π in this proposition is called a *uniformizer*, or a *prime element* (why?), or a *local parameter*. Some authors only use these terms when the valuation is normalized. In this book, we will often just announce that we pick π such that $(\pi) = \mathfrak{p}$.

Proof. That $\mathcal{O}_F$ is a ring follows from the properties of a valuation, and we see similarly that $\mathfrak{p}$ is an ideal. Next we show that $x \in \mathcal{O}_F$ is a unit if and only if $v(x) = 0$. Indeed, $x \neq 0$ always has an inverse $x^{-1} \in F$, and $v(xx^{-1}) = v(x) + v(x^{-1}) = v(1)$, and $v(1) = 0$ can be derived directly from the axioms, so $v(x^{-1}) = -v(x)$; we conclude that x and x^{-1} are both in $\mathcal{O}_F$ if and only if $v(x) = v(x^{-1}) = 0$.

As a consequence, the elements of $\mathcal{O}_F \smallsetminus \mathfrak{p}$ are all invertible, and we deduce that $\mathfrak{p}$ is the unique maximal ideal, as we did during the proof of Proposition 2.5. We have (1) and (2).

Replacing v by v' if necessary (as in the comment before the proposition), we may assume that v is normalized. Let us pick $\pi \in \mathcal{O}_F$ such that $v(\pi) = 1$. Let I be any ideal, and let $n = \min(v(x) : x \in I)$. Then $\pi^{-n} I$ is contained entirely in $\mathcal{O}_F$ since the valuation of any of its elements is ≥ 0, and it is obviously still an ideal. However, it contains units, so $\pi^{-n} I = \mathcal{O}_F$ and $I = \pi^n \mathcal{O}_F = (\pi^n)$. This shows (3).

Finally, for $x \in F^\times$, let $n = v(x)$, so that $v(\pi^{-n} x) = 0$. This shows that $u = \pi^{-n}(x)$ is a unit, and the existence of the product $x = \pi^n u$. Uniqueness is easy. □

Remark 2.18 This trick of going from v to v' is so simple that one is tempted to restrict the study of valuations to the normalized ones. However, a typical situation for us will be that of an extension K/F, where K has a valuation w and we call v the restriction of w to F. Then the trick is not enough: we may arrange for one of v or w to be normalized, but not always both. For this reason, one must get used to dealing with general (discrete) valuations.

Now that we know that $\mathfrak{p}$ is maximal, we make a definition.

Definition 2.19 The field

$$\mathbb{F} = \mathcal{O}_F / \mathfrak{p}$$

residue field

is called the **residue field** of F. (Or, more properly, of (F, v).)

Example 2.20 When $F=$ either $\mathbb{Q}$ or $\mathbb{Q}_p$, with the valuation v_p, the residue field is $\mathbb{F}=\mathbb{F}_p$, the field with p elements.

Given an element $x \in \mathcal{O}_F$, we will sometimes write $\overline{x} \in \mathbb{F}$ for its image in the residue field. Likewise, when $f \in \mathcal{O}_F[X]$ is a polynomial, we write $\overline{f} \in \mathbb{F}[X]$ for the polynomial obtained by reducing the coefficients mod $\mathfrak{p}$. This operation is a homomorphism, so in particular, a factorization $f=gh$ gives $\overline{f}=\overline{g}\overline{h}$. Hensel's lemma is about the converse.

Theorem 2.21 (Hensel's lemma) *Let F be a field that is complete with respect to a discrete valuation. Let $f \in \mathcal{O}_F[X]$ be a polynomial such that there is a factorization $\overline{f}=f_1f_2$ with $f_1,f_2 \in \mathbb{F}[X]$ relatively prime (and nonzero). Then there is a factorization $f=gh$ with $g,h \in \mathcal{O}_F[X]$ satisfying $\overline{g}=f_1$, $\overline{h}=f_2$, and* $\deg(g)=\deg(\overline{g})$.

Note that, of course, we may have $\deg(\overline{f})<\deg(f)$ when f is an arbitrary polynomial with coefficients in $\mathcal{O}_F$.

Proof. ([Neu99, chap. II, 4.6]) We start with some choice of $g_0, h_0 \in \mathcal{O}_F[X]$ with $\overline{g}_0=f_1$ and $\overline{h}_0=f_2$; in fact, let us be economical and arrange to have $\deg(g_0)=\deg(\overline{g}_0)$, $\deg(h_0)=\deg(\overline{h}_0)$.

Assuming for notational convenience that the valuation in F is normalized, we pick a uniformizer π. We look for g and h in the form

$$g=g_0+p_1\pi+p_2\pi^2+\cdots,$$

$$h=h_0+q_1\pi+q_2\pi^2+\cdots,$$

recalling that $\pi^n \to 0$ as n increases. More precisely, let us write $d=\deg(f)$ and $m=\deg(f_1)$, so that $\deg(f_2)\leq d-m$. We shall prove by induction on n that we may find $p_n, q_n \in \mathcal{O}_F[X]$ with $\deg(p_n)<m$ and $\deg(q_n)\leq d-m$, so that if we put

$$g_n=g_0+p_1\pi+\cdots+p_n\pi^n,$$

$$h_n=h_0+q_1\pi+\cdots+q_n\pi^n,$$

then

$$f\equiv g_nh_n \bmod \pi^{n+1}. \tag{2.2}$$

Suppose we had proved this. We wish then to take $g=\lim g_n$. It is helpful to keep in mind that the degree of all the polynomials g_n stays bounded (by m), so that it is easy to make sense of the limit: It is taken coefficient by coefficient, and there are finitely many of those. Each individual limit exists, say by Lemma 2.14, and this is where we use the assumption that F is complete. Likewise, it makes sense to define $h=\lim h_n$. The claimed congruence implies then that g_nh_n converges to f, but it also clearly converges to gh, proving the theorem.

The polynomial $f-g_0h_0$ reduces to zero in $\mathbb{F}[X]$, by definition, and this proves that (2.2) holds for $n=0$. We suppose that it holds for $n-1$, and now select p_n and q_n

to arrange the congruence for n. A simple rearrangement shows that the condition to be met is

$$f - g_{n-1}h_{n-1} \equiv (g_{n-1}q_n + h_{n-1}p_n)\pi^n \bmod \pi^{n+1} .$$

Put $f_n = \pi^{-n}(f - g_{n-1}h_{n-1})$, a polynomial in $\mathcal{O}_F[X]$ (by induction). The last congruence is equivalent to

$$f_n \equiv g_{n-1}q_n + h_{n-1}p_n \equiv g_0q_n + h_0p_n \bmod \pi .$$

Note that, since $\overline{g}_0$ and $\overline{h}_0$ are relatively prime, by Bézout we may select $a, b \in \mathcal{O}_F$ such that $\overline{g}_0\overline{a} + \overline{h}_0\overline{b} = 1$. We have $g_0a + h_0b \equiv 1 \bmod \pi$, so

$$g_0(af_n) + h_0(bf_n) \equiv f_n \bmod \pi .$$

It is tempting to put $q_n = af_n$ and $p_n = bf_n$, but we need to worry about the degrees. The fact that $\deg(g_0) = \deg(\overline{g}_0)$ implies that the leading coefficient of g_0 is a unit in $\mathcal{O}_F$, so we may perform a long division in $\mathcal{O}_F[X]$ and obtain:

$$bf_n = g_0s + p_n ,$$

defining p_n of degree less than m. We put temporarily $q_n = af_n + h_0s$, so that

$$g_0q_n + h_0p_n = g_0af_n + h_0(g_0s + p_n) = g_0af_n + h_0bf_n \equiv f_n \bmod \pi ,$$

exactly as needed. The equality $\overline{g}_0\overline{q}_n + \overline{h}_0\overline{p}_n = \overline{f}_n$ implies easily that $\deg(\overline{q}_n) \leq d - m$, so to complete the induction step we only need to replace q_n by a polynomial of degree $\leq d - m$ having the same reduction mod π. □

Example 2.22 A classical example is that of $f = X^{p-1} - 1 \in \mathbb{Z}_p[X]$. Its reduction in $\mathbb{F}_p[X]$ factors as a product of distinct linear factors, so that Hensel's lemma applied several times shows that f has $p - 1$ distinct roots in $\mathbb{Z}_p$. Thus *the $(p - 1)$-st roots of unity are in $\mathbb{Q}_p$* (a fact we shall revisit several times in this book). One has to marvel about the fact that, $\mathbb{Z}$ being dense in $\mathbb{Z}_p$, there is an integer $x \neq 1$ such that x^{p-1} is very close to 1 – that is, up to a high power of p. While the reader may try to think of a direct argument, the way we have obtained this using a general principle must be appreciated. ■

Example 2.23 (squares in $\mathbb{Q}_p$ for p odd) Let $a = p^n u$ with $n = v_p(a)$ and $u \in \mathbb{Z}_p^\times$ be a nonzero element of $\mathbb{Q}_p$. Then a necessary condition for a to possess a square root in $\mathbb{Q}_p$ is that n must be even; and if $n = 2m$, then a has a square root if and only if u has one. Also, if $x^2 = u$ then the valuation of x is half that if u, so x is itself a unit. In order to study squares and square roots in $\mathbb{Q}_p$, we may thus restrict attention to units. Let $a \in \mathbb{Z}_p^\times$.

When a is a square, the polynomial $f = X^2 - a \in \mathbb{Z}_p[X]$ factors as a product of two linear factors, so the same can be said of its reduction in $\mathbb{F}_p[X]$, and we see that $\overline{a} \in \mathbb{F}_p$ has a square root in $\mathbb{F}_p$. When p is odd, Hensel's lemma gives us the exact converse. In the end *$a \in \mathbb{Z}_p^\times$ has a square root if and only if $\overline{a} \in \mathbb{F}_p$ has a square root*, when p

is odd. For example, the number $7 \in \mathbb{Z}$ has a square root in $\mathbb{Z}_3$ since $\overline{7} = \overline{1}$ has a square root in $\mathbb{F}_3$. We often write $\sqrt{7} \in \mathbb{Z}_3$. The squares in $\mathbb{Q}_2$ will be studied in Example 4.9.

The next lemma is another classic.

Lemma 2.24 *Suppose F is complete with respect to a discrete valuation, and let*

$$f = a_n X^n + a_{n-1} X^{n-1} + \cdots + a_0 \in F[X].$$

If we suppose that f is irreducible in $F[X]$, and that a_0, a_n are both in $\mathcal{O}_F$, then $f \in \mathcal{O}_F[X]$.

Proof. Let π be a uniformizer, and let k be the smallest nonnegative integer such that $\pi^k f \in \mathcal{O}_F[X]$. We suppose that $k > 0$ and work toward a contradiction. The first and last coefficients of the polynomial $P = \pi^k f$ are then in the ideal $\mathfrak{p}$, so $\deg(\overline{P}) < \deg(P)$ and X divides $\overline{P}$. Let us write

$$\overline{P} = X^m Q,$$

where $0 < m < n$, and X does not divide Q. Note that $\overline{P} \neq 0$ by minimality of k, so Hensel's lemma applies. It follows that there is a polynomial $g \in \mathcal{O}_F[X]$ of degree m which divides f in $F[X]$. This contradicts the fact that f is irreducible. □

Of the utmost importance for the rest of this chapter (and this book) is the following theorem.

Theorem 2.25 *Let F be a field which is complete with respect to a discrete valuation v, and let K/F be an extension with $[K : F] = n$. Then there exists a unique discrete valuation v_K on K extending v. It is given by the rule*

$$v_K(\alpha) = \frac{1}{n} v(\det(m_\alpha))$$

where m_α is multiplication by α, that is the F-linear map $K \to K$ mapping x to αx. Moreover, if the minimal polynomial of α is

$$\min(F, \alpha) = X^d + a_{d-1} X^{d-1} + \cdots + a_0,$$

then $v_K(\alpha) = \frac{1}{d} v(a_0)$.

The reader will probably recognize that $v_K(\alpha) = \frac{1}{n} v(\mathrm{N}_{K/F}(\alpha))$, by definition of the norm map, and will deduce that the two formulae given for $v_K(\alpha)$ agree (see page 281). However, much later in this book it will be necessary to apply this theorem when K is only a *skewfield*, that is, without assuming that multiplication in K is commutative. The following proof is designed to work even in this case (and thus avoids references to well-known facts from field theory).

Proof. ([Bla72, Théorème V-1]) It is clear that the proposed v_K extends v, and that $v_K(\alpha) = +\infty$ only happens when $\alpha = 0$. The axiom $v_K(\alpha\beta) = v_K(\alpha) + v_K(\beta)$ is also clear, from the basic properties of determinants. We must prove that $v_K(\alpha+\beta) \geq \min(v_K(\alpha), v_K(\beta))$ for all α, β, and that is clearly equivalent to proving that $v_K(1+\alpha) \geq 0$ if $v_K(\alpha) \geq 0$ for all α.

For this, assume first that we have proved the second formula, that is $v_K(\alpha) = \frac{1}{d}v(a_0)$ as above. A minimal polynomial being irreducible, Lemma 2.24 applies, showing that $P \in \mathcal{O}_F[X]$, where $P = \min(F, \alpha)$. Now $\min(F, 1+\alpha) = P(X-1) \in \mathcal{O}_F[X]$, and looking at the constant coefficient we see that $v_K(1+\alpha) \geq 0$, as needed.

Still postponing the proof that the two formulae at our disposal agree, we turn to the uniqueness. Letting v'_K denote a putative other valuation extending v, we first show that $v_K(\alpha)$ and $v'_K(\alpha)$ are of the same sign, for any $\alpha \in K$; replacing α by α^{-1} if needed reduces to proving that $v_K(\alpha) \geq 0$ implies $v'_K(\alpha) \geq 0$. To see this, we use that $\min(F, \alpha) \in \mathcal{O}_F[X]$ as above, so $v(a_i) \geq 0$, while

$$\alpha = -a_{d-1} - a_{d-2}\alpha^{-1} - \cdots - a_0\alpha^{1-d}.$$

If we had $v'_K(\alpha) < 0$, we would have $v'_K(a_i\alpha^{-k}) > 0$ whenever $k > 0$, and $v'_K(\alpha) \geq 0$ by the last equality – a contradiction. Thus v_K and v'_K have the same sign.

To conclude with the uniqueness, pick any $f \in F$ such that $v(f) > 0$ (there is nothing further to prove if v is constantly 0). Pick $\alpha \in K$. For any integer $q > 0$, we may find $p \in \mathbb{Z}$ such that

$$pv(f) \leq qv_K(\alpha) \leq (p+1)v(f).$$

The first inequality here can be rewritten $v_K(\alpha^q f^{-p}) \geq 0$, so that $v'_K(\alpha^q f^{-p}) \geq 0$, and likewise for the second. In the end we have

$$pv(f) \leq qv'_K(\alpha) \leq (p+1)v(f).$$

It follows that

$$|v_K(\alpha) - v'_K(\alpha)| \leq \frac{1}{q}v(f),$$

so that $v_K(\alpha) = v'_K(\alpha)$, since q is arbitrary.

It remains to prove that

$$\det(m_\alpha) = \pm a_0^{\frac{n}{d}},$$

which upon taking valuations gives $\frac{1}{n}v(\det(m_\alpha)) = \frac{1}{d}v(a_0)$. For this, consider the basis $1, \alpha, \ldots, \alpha^{d-1}$ of $F[\alpha]$ as an F-vector space, and let $\varepsilon_1, \ldots, \varepsilon_{n/d}$ be a basis for K as an $F[\alpha]$-vector space. The collection of all $\varepsilon_i\alpha^j$ forms a basis for K as an F-vector space. Putting it in lexicographical order, the matrix of m_α in this basis becomes block diagonal, each of the $\frac{n}{d}$ blocks being

$$\begin{pmatrix} 0 & 0 & \cdots & 0 & -a_0 \\ 1 & 0 & \cdots & 0 & -a_1 \\ 0 & 1 & \ddots & 0 & -a_2 \\ \vdots & \ddots & \ddots & \ddots & \vdots \\ 0 & 0 & \cdots & 1 & -a_{d-1} \end{pmatrix}.$$

The determinant of this block is $\pm a_0$, concluding the proof. □

Corollary 2.26 *Suppose that K/F is moreover Galois, let $\sigma \in \mathrm{Gal}(K/F)$, and let $x \in K$. Then*

$$v_K(\sigma(x)) = v_K(x),$$

or in other words the Galois action preserves the valuation of elements.

Proof. The map defined by $x \mapsto v_K(\sigma(x))$ is a valuation extending v. □

In the next chapter we shall establish a nice complement:

Proposition 2.27 *Let the notation be as in Theorem 2.25, and assume that $F = \mathbb{Q}_p$. Then K is complete with respect to the valuation v_K.*

Our reason for postponing the proof is that it relies on topological ideas, and we focus on the algebra for now. (In the rest of this chapter, the implication is just that we will be able to use Hensel's lemma for K.) There is nothing exciting about the argument: we shall reinforce the belief that $\mathbb{Q}_p$ and $\mathbb{R}$ are close cousins (mostly by showing that $\mathbb{Q}_p$ is locally compact), so that familiar facts about $\mathbb{R}$ remain true with $\mathbb{Q}_p$, usually with the same proof. For example, finite-dimensional, normed vector spaces over $\mathbb{R}$ are always complete, and the same is true over $\mathbb{Q}_p$ by the same argument. But K is just that, if we consider the absolute value $|\cdot|_K$ corresponding to the valuation v_K, playing the role of the norm. For the details, see Theorem 3.28.

Local number fields

local number field

p-adic field

Definition 2.28 A **local number field** is a finite-dimensional extension K of $\mathbb{Q}_p$, for some prime number p. One also says that K is a ***p*-adic field**.

Note that, at this stage, we offer no justification for calling these fields "local". This will have to wait until the very last chapter of this book, where examples of "global fields" are given (see page 275).

We shall always write v_K for the unique valuation of the p-adic field K extending the valuation v_p on $\mathbb{Q}_p$, as in Theorem 2.25. By Proposition 2.27, the field K is complete for v_K. However, note that v_K may not be normalized. (We caution that other authors write v_p for all the extensions of v_p to all local number fields, and they may use v_K for the normalized valuation – be careful!).

Lemma 2.29 (and definition) *The valuation group $v_K(K^\times)$ of the p-adic field K is $\frac{1}{e}\mathbb{Z}$ for a unique integer $e = e(K) > 0$, which divides $[K : \mathbb{Q}_p]$, and is also the index $[v_K(K^\times) : v_p(\mathbb{Q}_p^\times)]$.*

Proof. The group $v_K(K^\times)$ is infinite cyclic (that is, isomorphic to $\mathbb{Z}$), and $v_p(\mathbb{Q}_p^\times) = \mathbb{Z}$ is a nontrivial subgroup, so it has a finite index that we call e. Also from our knowledge of the subgroups of an infinite cyclic group, we deduce that $v_K(K^\times)$ is generated by an element λ such that $e\lambda$ generates $\mathbb{Z}$; imposing $\lambda > 0$ gives $\lambda = \frac{1}{e}$.

The divisibility statement will follow from the material below, but we may argue directly as follows. Suppose $\alpha \in K$ verifies $v_K(\alpha) = \frac{1}{e}$. We also have $v_K(\alpha) = \frac{1}{d}v_p(a_0)$ where the minimal polynomial of α is $X^d + \cdots + a_0$. Thus

$$d = ev_p(a_0),$$

and $v_p(a_0)$ is an integer, so e divides $[\mathbb{Q}_p(\alpha) : \mathbb{Q}_p]$. □

Of course we may work with the normalized valuation $e(K)v_K$ if we wish. Still, the integer $e(K)$ is important for understanding the extension $K/\mathbb{Q}_p$ (if not K itself).

Let us at once introduce another integer attached to the p-adic field K (the reader should review Definitions 2.15 and 2.19 for this section).

Lemma 2.30 (and definition) *The residue field $\mathbb{K}$ of the p-adic field K is finite, of characteristic p. Its order is p^f, defining $f = f(K)$.*

Proof. The definitions make it clear that $\mathbb{Z}_p \subset \mathcal{O}_K$ and that, if $\mathfrak{p}$ is the unique maximal ideal of $\mathcal{O}_K$, then $\mathfrak{p} \cap \mathbb{Z}_p = p\mathbb{Z}_p$ (= the ideal of elements of positive valuation). Thus $\mathbb{K} = \mathcal{O}_K/\mathfrak{p}$ is an extension of $\mathbb{F}_p = \mathbb{Z}_p/p\mathbb{Z}_p$. It is enough to prove that $[\mathbb{K} : \mathbb{F}_p]$ is finite.

Let $a_1, \ldots, a_n \in \mathcal{O}_K$ be such that their images $\overline{a}_1, \ldots, \overline{a}_n \in \mathbb{K}$ are linearly independent over $\mathbb{F}_p$. Suppose we have a linear combination

$$f_1 a_1 + \cdots + f_n a_n = 0 \qquad (*)$$

with $f_i \in \mathbb{Q}_p$. If the coefficients are not all 0, then one of them, say f_1, has minimal p-adic valuation among the f_i. If we divide (*) by f_1, we obtain a new linear combination, this time with coefficients in $\mathbb{Z}_p$ since $v_p(f_i/f_1) \geq 0$, and with the first coefficient being a unit. Reducing mod $\mathfrak{p}$ yields a contradiction. We conclude that $a_1, \ldots, a_n$ are linearly independent over $\mathbb{Q}_p$, so $n \leq [K : \mathbb{Q}_p]$. □

Remark 2.31 The p-adic fields are among our main objects of study in the rest of this book. A lot of the material to be developed can be generalized in various ways: Some results hold for general fields equipped with a discrete valuation, some only require the field to be complete (or even "henselian", meaning that Hensel's lemma applies), sometimes one requires the residue field to be finite (or only "quasi-finite"), for some results one requires characteristic 0, or some sort of separability condition, etc. For supreme clarity, in what constitutes a first discovery of "local fields" for most readers, we stick to p-adic fields. For these, all the assumptions one could hope for actually hold, and we will not try to find the minimal hypotheses for a given result. For a thorough treatment of the general case, and a great read after this book, see [Ser79]. ■

Example 2.32 Open your web browser and look for the LMFDB, or "L-function and modular forms database". At the time of writing, the page on local number fields is available at www.lmfdb.org/LocalNumberField/.

You will see, for example, that $\mathbb{Q}_3$ has just five extensions of degree four, giving in each case the numbers e and f, as well as the minimal polynomial of a primitive element, and the Galois group (of the Galois closure, over $\mathbb{Q}_p$). ■

We embark on an initial study of the basic properties of p-adic fields, or more properly of extensions K/F where F and K are both local number fields. We set up some notation as we make our first observations.

The fields F, K are endowed with the discrete valuations v_F, v_K, both extending v_p, so that v_K extends v_F (by the uniqueness statement in Theorem 2.25). Occasionally we may work with the normalized valuation $v = e(F)v_F$ and its extension to K, which is $w = e(F)v_K$. The latter might not be normalized, of course – the study of general valuations may not be escaped. In any case, we have the corresponding absolute values $|\cdot|_F$ and $|\cdot|_K$ (which do not depend seriously on whether we normalize or not, from Lemma 2.12).

Each of F, K has its ring of integers, and recall that we denote these $\mathcal{O}_F$ and $\mathcal{O}_K$. Of course $\mathcal{O}_F \subset \mathcal{O}_K$. A first remark is that, once $\mathcal{O}_F$ is known, there is an alternative definition of $\mathcal{O}_K$:

Lemma 2.33 *The ring $\mathcal{O}_K$ consists precisely of those elements $\alpha \in K$ whose minimal polynomial* $\min(F,\alpha)$ *has coefficients in $\mathcal{O}_F$. In fact, if $P(\alpha) = 0$ for any $P \in \mathcal{O}_F[X]$ at all, then $\alpha \in \mathcal{O}_K$.*

Proof. If $P = \min(F,\alpha) = X^d + \cdots + a_0 \in \mathcal{O}_F[X]$, then $v_K(\alpha) = \frac{1}{d}v_F(a_0) \geq 0$, so $\alpha \in \mathcal{O}_K$ by definition. For the converse, we see that $\alpha \in \mathcal{O}_K$ implies that $a_0 \in \mathcal{O}_F$; however, P is irreducible and monic, so the result follows from Lemma 2.24.

For the second statement, suppose now that $P = X^d + \cdots + a_0$ is any polynomial in $\mathcal{O}_F[X]$ vanishing at α (rather than the minimal polynomial of α), and write (again) that

$$\alpha = -a_{d-1} - a_{d-2}\alpha^{-1} - \cdots - a_0\alpha^{1-d}.$$

If we had $v_K(\alpha) < 0$, we would have $v_K(a_i\alpha^{-k}) > 0$, leading to a contradiction; so $v_K(\alpha) \geq 0$ and $\alpha \in \mathcal{O}_K$ as claimed. □

Elements as in this lemma are called *integral over $\mathcal{O}_F$*. Thus the "ring of integers" of a p-adic field consists of all the elements that are integral over $\mathbb{Z}_p$.

We shall follow the notational tradition, and write $\mathfrak{p}$ for the unique maximal ideal of $\mathcal{O}_F$, comprised of the elements of positive valuation, while the unique maximal ideal of $\mathcal{O}_K$ will be written $\mathfrak{P}$. It follows that $\mathfrak{P} \cap \mathcal{O}_F = \mathfrak{p}$. The residue field of F resp. K will be denoted $\mathbb{F}$ $(= \mathcal{O}_F/\mathfrak{p})$ resp. $\mathbb{K}$ $(= \mathcal{O}_K/\mathfrak{P})$. We see $\mathbb{K}$ as an extension of $\mathbb{F}$, which is obviously finite since $\mathbb{F}$, $\mathbb{K}$ are both finite fields anyway.

Definition 2.34 Let K/F be an extension of p-adic fields. We define

$$e(K/F) = [v_K(K^\times) : v_F(F^\times)] = \frac{e(K)}{e(F)},$$

ramification index

and call it the **ramification index** of K/F. We also define

$$f(K/F) = [\mathbb{K} : \mathbb{F}] = \frac{f(K)}{f(F)},$$

inertia degree

and call it the **inertia degree** of K/F. ■

There is a fundamental relation between $e(K/F)$ and $f(K/F)$, but we need a bit of patience. We start modestly, with another interpretation of $e(K/F)$.

Lemma 2.35 *In the notation above, we have $\mathfrak{p}\mathcal{O}_K = \mathfrak{P}^e$, where $e = e(K/F)$.*

Proof. We take the opportunity to recall some classical notation: when I, J are subsets of a commutative ring R, we write IJ for the ideal generated by products ab with $a \in I$, $b \in J$. The ideal $\mathfrak{P}^e$ is understood in this notation, that is, $\mathfrak{P}^e = \mathfrak{P}\mathfrak{P}\cdots\mathfrak{P}$ with e copies of $\mathfrak{P}$ (the operation being visibly associative, brackets are not needed). Likewise for $\mathfrak{p}\mathcal{O}_K$, which is therefore the ideal of $\mathcal{O}_K$ generated by $\mathfrak{p}$. This also allows us to write principal ideals as $x\mathcal{O}_F$ and $x\mathcal{O}_K$, for example, which is less ambiguous than (x) when there are several rings around. (Of course here x is an abuse for $\{x\}$.)

Now, for the proof itself, we pick the normalized valuation v of F, and its extension w to K. Let $\pi \in F$ be a uniformizer, and let $\Pi \in K$ be such that $w(\Pi) = \frac{1}{e}$. We have $\mathfrak{p} = \pi\mathcal{O}_F$, and $\mathfrak{P} = \Pi\mathcal{O}_K$, by two applications of Proposition 2.17. But $w(\Pi^e) = 1$ so $w(\Pi^e\pi^{-1}) = 0$ and $\pi = u\Pi^e$ where u is a unit of $\mathcal{O}_K$. In the end $\mathfrak{p}\mathcal{O}_K = \pi\mathcal{O}_K = \Pi^e\mathcal{O}_K = \mathfrak{P}^e$. □

Proposition 2.36 *The ring $\mathcal{O}_K$ is a free $\mathcal{O}_F$-module of rank $n = [K : F]$.*

Recall that, for any ring R, an R-module M is said to be *free of rank n* when it is isomorphic to R^n. Recall also that the integer n is then unique, at least when R is an integral domain, for we only need to take its field of fractions F and recover n from linear algebra, as the dimension of $M \otimes_R F \cong F^n$.

To prove the proposition, we shall require a well-known result about modules over a principal ideal domain R (recall from Proposition 2.17 that $\mathcal{O}_F$ is such a ring). Namely, let M be a submodule of R^n, where R is a "PID"; then M is a free R-module of rank $\leq n$.

In fact, in our situation more is true. Indeed $\mathcal{O}_F$ is actually a euclidean domain, in the sense that long divisions may be performed: given $a, b \in \mathcal{O}_F$, with $b \neq 0$, one can find $q, r \in \mathcal{O}_F$ such that

$$a = bq + r,$$

with either $r = 0$ or $v(r) < v(b)$. (Put $r = 0$ and $q = a/b$ if $v(a) \geq v(b)$, and $q = 1$, $r = a - b$ otherwise). Any euclidean domain is a PID, but in this case, proving the above fact about modules is considerably easier. See proposition 2.1.5 in [Stea] for a very readable, algorithmic approach, which is written for the ring $\mathbb{Z}$ but works for any euclidean domain.

Finally, note that the proof to come makes use of traces, and the reader may wish to review Appendix A.

Proof of Proposition 2.36. Let v be the normalized valuation of F, and let π be a uniformizer. Let $a_1, \ldots, a_n \in K$ constitute a basis for K as an F-vector space. Multiplying by a power of π if necessary, we may assume that each $a_i \in \mathcal{O}_K$.

Let M be the $\mathcal{O}_F$-module spanned by $a_1, \ldots, a_n$, so that $M \subset \mathcal{O}_K$. By construction M is a free $\mathcal{O}_F$-module of rank n. For each $x \in \mathcal{O}_K$, we can find an integer k such that $\pi^k x \in M$ (clearing denominators of the coefficients of x written in our basis). Suppose that we could in fact find a single $d \in \mathcal{O}_F$ such that $d\mathcal{O}_K \subset M$. Then by the theory of modules over PIDs just recalled, we would know that $d\mathcal{O}_K$ is free of rank $m \leq n$, as an $\mathcal{O}_F$-module. But then the same could be said of $\mathcal{O}_K$, since $\mathcal{O}_K \cong d\mathcal{O}_K$ as modules over $\mathcal{O}_F$. The inclusion $M \subset \mathcal{O}_K$, by the same token, would give $n \leq m$, and the proposition would be proved.

So we look for d. Pick $x \in \mathcal{O}_K$ and write

$$x = \sum_{j=1}^{n} f_j a_j,$$

with $f_j \in F$. Now multiply by a fixed a_i:

$$a_i x = \sum_j a_i a_j f_j,$$

and take the trace down to F on both sides:

$$\mathrm{Tr}_{K/F}(a_i x) = \sum_j \mathrm{Tr}_{K/F}(a_i a_j) f_j.$$

Write all these equalities (when i varies) as

$$AC_1 = C_2,$$

where A is the $n \times n$ matrix $A = (\mathrm{Tr}_{K/F}(a_i a_j))_{i,j}$, while C_1 is the column containing $f_1, \ldots, f_n$ as entries, and C_2 is another column containing $\mathrm{Tr}_{K/F}(a_1 x)$, $\ldots$, $\mathrm{Tr}_{K/F}(a_n x)$. Note that C_1 has coefficients in F, and we claim that A and C_2 actually have their coefficients in $\mathcal{O}_F$. Indeed, whenever $y \in \mathcal{O}_K$, we have $\min(F, y) \in \mathcal{O}_F[X]$

by Lemma 2.33. The usual expression for the trace of an element in terms of its minimal polynomial then guarantees that $\mathrm{Tr}_{K/F}(y) \in \mathcal{O}_F$.

We shall prove that $d = \det(A) \neq 0$, so that $C_1 = A^{-1}C_2$, and the usual formula for the inverse of a matrix shows that $df_i \in \mathcal{O}_F$ for each i, so that $dx \in \mathcal{O}_F$, as we wanted.

It remains to conclude with a well-known argument: A is the matrix associated with the bilinear form $K \times K \to F$ given by $(x,y) \mapsto \mathrm{Tr}_{K/F}(xy)$. If we had $\det(A) = 0$, then there would be $y \neq 0$ such that $x \mapsto \mathrm{Tr}_{K/F}(xy)$ is identically 0; but this is absurd, since for $x = \frac{1}{y}$ we have $\mathrm{Tr}_{K/F}(xy) = \mathrm{Tr}_{K/F}(1) = n \neq 0$. This concludes the proof. □

Here is another classic from commutative algebra.

Proposition 2.37 (Nakayama's lemma) *Suppose R is a ring possessing a unique maximal ideal $\mathfrak{m}$.*

1. *Let M be a finitely generated R-module such that $M = \mathfrak{m}M$. Then $M = 0$.*
2. *Let M be an R-module with a submodule N, such that M/N is finitely generated, and $M = \mathfrak{m}M + N$. Then $M = N$.*

Proof. One obtains (2) by applying (1) to M/N, so we turn to (1). Suppose first that M is generated by a single element x. Then the map $\varphi\colon R \to M$, $a \mapsto ax$ induces an isomorphism $M \cong R/I$ where $I = \ker\varphi$. If $I \neq R$ then $I \subset \mathfrak{m}$, because I must be included in *some* maximal ideal. It follows that $M/\mathfrak{m}M \cong R/\mathfrak{m} \neq 0$, a contradiction; so $I = R$ and $M = 0$.

Now if M is spanned by $x_1, \ldots, x_n$, then the quotient $M' = M/\langle x_1, \ldots, x_{n-1}\rangle$ satisfies the hypotheses and is generated by the image of x_n, so $M' = 0$, by the case just treated. Thus we see that M can be generated by $n-1$ elements. Iterating, we conclude that M can be generated by just one element, and so $M = 0$ as proved above. □

Theorem 2.38 *Let K/F be an extension of local number fields, let $e = e(K/F)$, $f = f(K/F)$, and $n = [K : F]$. Then*

$$n = ef\,.$$

Proof. ([Neu99, II 6.8]) We write v for the normalized valuation of F, and w for its extension to K. So we have $v(F^\times) = \mathbb{Z}$ and $w(K^\times) = \frac{1}{e}\mathbb{Z}$. Pick $\pi_i \in K$ such that $w(\pi_i) = \frac{i}{e}$, for $0 \leq i < e$, and pick representatives $a_1, \ldots, a_f \in \mathcal{O}_K$ for a basis of $\mathbb{K}$ as an $\mathbb{F}$-vector space. We claim that the family $(\pi_i a_j)$ is a basis for K as a vector space over F.

Suppose we have a relation of the form

$$\sum_i \sum_j f_{ij}\pi_i a_j = 0\,,$$

with $f_{ij} \in F$. We rewrite this $\sum_i s_i\pi_i = 0$, with $s_i = \sum_j f_{ij}a_j$.

Here are initial remarks about the sum s_i, for fixed i. Suppose there is a j such that $f_{ij} \neq 0$, and pick one of the least valuation, say f_{ij_0}. Then $s_i/f_{ij_0} \in \mathcal{O}_K$ is a linear combination with coefficients in $\mathcal{O}_F$, and one of these coefficients is 1. The image of s_i/f_{ij_0} in $\mathbb{K}$ is nonzero by definition of the a_j, and in particular s_i is nonzero. What is more, s_i/f_{ij_0} is a unit of $\mathcal{O}_K$, since it does not belong to $\mathfrak{P}$, and we deduce that its valuation is 0. Thus $w(s_i) = v(f_{ij_0}) \in \mathbb{Z}$.

Now, suppose the terms $s_i\pi_i$, as i varies, are not all zero. Then two of them must have the same valuation, otherwise the valuation of the sum would be the minimum of the $w(s_i\pi_i)$, preventing it from being $+\infty$. If $e = 1$ this is already a contradiction, and otherwise say $w(s_{i_1}\pi_{i_1}) = w(s_{i_2}\pi_{i_2})$, so that $w(s_{i_1}) - w(s_{i_2}) = w(\pi_{i_2}) - w(\pi_{i_1}) = \frac{i_2 - i_1}{e} \notin \mathbb{Z}$. This contradicts our above remark, to the effect that $w(s_i) \in \mathbb{Z}$ whenever $s_i \neq 0$. That contradiction shows that $s_i\pi_i = 0$ for all i, so $s_i = 0$ for all i, and $f_{ij} = 0$ for all i,j as observed above.

Let N be the $\mathcal{O}_F$-module spanned by the $\pi_i a_j$, which we know now is a free module of rank ef. We will prove that $N = \mathcal{O}_K$, which is free of rank n by Proposition 2.36, so $n = ef$ (alternatively, $N = \mathcal{O}_K$ shows that the $\pi_i a_j$ span K as an F-vector space, while we have proved their linear independence). We also introduce the $\mathcal{O}_F$-module N_0 spanned by just the a_j. Thus

$$N = \pi_0 N_0 + \pi_1 N_0 + \cdots + \pi_{e-1} N_0 .$$

Now, since $w(\pi_1) = \frac{1}{e}$, we see that $(\pi_1) = \mathfrak{P}$, by Proposition 2.17 (applied to the normalized valuation ew, if you will). Likewise $(\pi_i) = \mathfrak{P}^i$, and π_0 is a unit. So

$$N = N_0 + \mathfrak{P}N_0 + \cdots + \mathfrak{P}^{e-1}N_0 .$$

On the other hand, we have $\mathcal{O}_K = N_0 + \mathfrak{P}$, by definition of the a_j. We obtain iteratively

$$\mathcal{O}_K = N_0 + \mathfrak{P}\mathcal{O}_K = N_0 + \mathfrak{P}(N_0 + \mathfrak{P}\mathcal{O}_K) = N_0 + \mathfrak{P}N_0 + \mathfrak{P}^2\mathcal{O}_K = \cdots = N + \mathfrak{P}^e\mathcal{O}_K .$$

But $\mathfrak{P}^e = \mathfrak{p}\mathcal{O}_K$ by Lemma 2.35. In the end $\mathcal{O}_K = N + \mathfrak{p}\mathcal{O}_K$ and we conclude that $\mathcal{O}_K = N$ from Nakayama's lemma applied with $M = \mathcal{O}_K$. Here we use Proposition 2.36 to make sure that $\mathcal{O}_K$ is finitely generated as an $\mathcal{O}_F$-module. □

We conclude the generalities on extensions of local number fields by a first inspection of the relationship between $\mathrm{Gal}(K/F)$ and $\mathrm{Gal}(\mathbb{K}/\mathbb{F})$. By Corollary 2.26, any $\sigma \in \mathrm{Gal}(K/F)$ maps $\mathcal{O}_K$ onto itself, and $\mathfrak{P}$ onto itself. Therefore, there is an induced automorphism $\overline{\sigma}$ of $\mathcal{O}_K/\mathfrak{P} = \mathbb{K}$. Since σ is the identity on F, we see that $\overline{\sigma}$ is the identity on $\mathbb{F} = \mathcal{O}_F/\mathfrak{p}$. Our conclusion is that $\sigma \mapsto \overline{\sigma}$ defines a map

$$\mathrm{Gal}(K/F) \longrightarrow \mathrm{Gal}(\mathbb{K}/\mathbb{F}),$$

which is easily seen to be a homomorphism. We shall shortly prove that it is surjective, after a quick detour.

Unramified extensions

unramified

Definition 2.39 An extension K/F of local number fields is said to be **unramified** when $e(K/F) = 1$.

By Theorem 2.38, when K/F is unramified we have $[K : F] = [\mathbb{K} : \mathbb{F}]$, and conversely. Also K/F is unramified if and only if $\mathfrak{P} = \mathfrak{p}\mathcal{O}_K$, if and only if K and F have the same valuation group.

Theorem 2.40 *Let F be a p-adic field. For any $n \geq 1$, there exists an extension L/F which is unramified and verifies $[L : F] = n$.*

Moreover, for any extension K/F such that n divides $[\mathbb{K} : \mathbb{F}]$, there are n distinct embeddings $L \to K$. In particular, the extension L/F is unique up to isomorphism, and is Galois.

Finally, the natural map $\mathrm{Gal}(L/F) \to \mathrm{Gal}(\mathbb{L}/\mathbb{F})$, where $\mathbb{L}$ is the residue field of L, is an isomorphism.

Proof. ([Bla72, Théorème V-2]) The finite field $\mathbb{F}$ has an extension $\mathbb{L}/\mathbb{F}$ of degree n, which is unique (up to isomorphism). Let $a \in \mathbb{L}$ be such that $\mathbb{L} = \mathbb{F}[a]$ (it is elementary that a exists, for example because $\mathbb{L}^\times$ is cyclic), and let $f = \min(\mathbb{F}, a)$. We have $\deg(f) = n$. Choose a polynomial $P \in \mathcal{O}_F[X]$, which is monic of degree n, such that $\overline{P} = f$. Finally, put $L = F[\alpha]$, where α is a root of P.

On the one hand $[L : F] \leq n$ since $\deg(P) = n$. On the other hand $\alpha \in \mathcal{O}_L$ by Lemma 2.33, so $\overline{\alpha}$ is a root of $\overline{P} = f$ in the residue field $\mathbb{L}$, and it follows that $f = [\mathbb{L} : \mathbb{F}] \geq n$. As a result $n \geq [L : F] = ef \geq en$ by Theorem 2.38, so we conclude that $e = 1$ and $[L : F] = n$. We have proved the existence of L/F as claimed.

Now suppose K/F is such that n divides $[\mathbb{K} : \mathbb{F}]$. By the theory of finite fields, there is a unique subfield $\mathbb{L}' \subset \mathbb{K}$ such that $[\mathbb{L}' : \mathbb{F}] = n$, and there is an $\mathbb{F}$-isomorphism between $\mathbb{L}$ and $\mathbb{L}'$. In particular, the polynomial $P \in \mathcal{O}_F$ used above is split over $\mathbb{L}'$, and we may write $\overline{P} = f = (X - a_1) \cdots (X - a_n)$ in $\mathbb{L}'[X]$. Here we note that the a_i are distinct, since f is the minimal polynomial of an element in the separable extension $\mathbb{L}/\mathbb{F}$. Hence Hensel's lemma applies, in the field K, to show us that P has n distinct roots in K, call them $\alpha_1, \ldots, \alpha_n$. The required embeddings $L = F[\alpha] \to K$ are given by $\alpha \mapsto \alpha_i$ for $i = 1, \ldots, n$.

In particular, if K/F is unramified, of degree n over F, we have $[\mathbb{K} : \mathbb{F}] = n$ and so there is an isomorphism (and in fact several isomorphisms) $L \cong K$, showing the uniqueness.

Finally, for $L = K$ we see that L/F has n distinct F-automorphisms, while $[L : F] = n$, so this extension is Galois. Any element $\sigma \in \mathrm{Gal}(\mathbb{L}/\mathbb{F})$ is determined by the image $\sigma(a)$, which is one of $a_1, \ldots, a_n$ in the notation above (for $\mathbb{L} = \mathbb{L}'$), say $\sigma(a) = a_i$. The automorphism of L mapping α to α_i then induces σ, so $\mathrm{Gal}(L/F) \to \mathrm{Gal}(\mathbb{L}/\mathbb{F})$ is surjective. It is bijective for reasons of cardinality. □

We proceed to deduce a number of corollaries. The most obvious is:

Corollary 2.41 *Suppose that L/F is an unramified extension. Then it is Galois and* $\mathrm{Gal}(L/F)$ *is cyclic, generated by a canonical element denoted* $\mathrm{Frob}_{L/F}$ *and called the* Frobenius element. *The latter is characterized by the congruence* $\mathrm{Frob}_{L/F}(x) \equiv x^q$ *mod $\mathfrak{P}$, for all $x \in \mathcal{O}_L$, where $\mathfrak{P}$ is the maximal ideal of $\mathcal{O}_L$ and q is the size of the residue field $\mathbb{L}$.* □

Next, we allow ourselves to choose algebraic closures. This will help rephrase the uniqueness statement in the theorem, based on the following elementary fact: If K/F is any Galois extension, and $\overline{F}$ is an algebraic closure of F, then all the F-homomorphisms $K \to \overline{F}$ have the same image.

Corollary 2.42 *Let F be a p-adic field. Choose an algebraic closure $\overline{F}$ of F, and an algebraic closure $\overline{\mathbb{F}}$ of the residue field $\mathbb{F}$.*

There is an inclusion preserving bijection between, on the one hand, the set of unramified extensions L/F with $L \subset \overline{F}$, and on the other hand, the set of finite extensions $\mathbb{L}/\mathbb{F}$ with $\mathbb{L} \subset \overline{\mathbb{F}}$. The correspondence takes each local number field to its residue field, and also preserves the dimensions (over F and $\mathbb{F}$ respectively).

Moreover, if $K \subset \overline{F}$ is any p-adic field with residue field $\mathbb{K}$, let K_0 be the field associated with $\mathbb{K}$ via the above bijection. Then $K_0 \subset K$, and we have $[K : K_0] = e(K/F)$. □

This corollary was also a definition, as it introduced K_0. We make a separate statement about this field, which hides the need to take algebraic closures back in the proof.

Corollary 2.43 *Let K/F be an extension of p-adic fields. There exists a unique intermediate field K_0, with $F \subset K_0 \subset K$, called the* inertia subfield *of K/F, with the following property: For any field L with $F \subset L \subset K$, the extension L/F is unramified $\Longleftrightarrow L \subset K_0$.*

One has $[K : K_0] = e(K/F)$ and $K = K_0$ if and only if K/F is unramified.

One also says that K_0/F is the *maximal, unramified extension contained in K*, but the corollary is more precise than that.

Proof. Choose algebraic closures, view K as a subfield of $\overline{F}$ and the residue field $\mathbb{K}$ as a subfield of $\overline{\mathbb{F}}$, and apply the preceding corollary. This defines $K_0 \subset K$ corresponding to $\mathbb{K}$. The subfields of $\mathbb{K}$ correspond to unramified extensions of F contained in K_0; however, by Galois theory, since K_0/F and $\mathbb{K}/\mathbb{F}$ have the same Galois group, this process exhausts the subfields of K_0. So any L with $L \subset K_0$ is unramified over F. Conversely, any unramified L defines a residue field $\mathbb{L}$ which is a subfield of $\mathbb{K}$ by construction, so $L \subset K_0$ by the preceding corollary. The uniqueness of K_0 is clear. □

We can use this to prove basic facts about unramified extensions.

Lemma 2.44 *Let L_1/F and L_2/F be two extensions of local number fields.*

1. *If $L_1 \subset L_2$ and L_2/F is unramified, then L_1/F is unramified, as is L_2/L_1.*
2. *If $L_1 \subset L_2$, if L_1/F is unramified and L_2/L_1 is unramified, then L_2/F is unramified.*
3. *If L_1 and L_2 are subfields of a common extension K/F, and if L_1/F and L_2/F are both unramified, then the compositum L_1L_2 is unramified over F.* □

Proof. Properties (1) and (2) are obvious if one recalls that K/F is unramified if and only if K and F have the same valuation group; we could have stated them earlier. For (3), however, we rely on the previous corollary, to see that $L_i \subset K_0$ for $i = 1, 2$, and so that $L_1 L_2 \subset K_0$. □

We are now in position to introduce an important subgroup.

Lemma 2.45 *Let K/F be a Galois extension of p-adic fields. Then the natural map* $\mathrm{Gal}(K/F) \to \mathrm{Gal}(\mathbb{K}/\mathbb{F})$ *is surjective. Its kernel is called the* inertia subgroup, *sometimes denoted $I(K/F)$. The fixed field of the inertia subgroup is the inertia subfield K_0.*

Proof. The extension K_0/F is unramified, so automatically Galois by Theorem 2.40, and $\mathrm{Gal}(K_0/F) \to \mathrm{Gal}(\mathbb{K}/\mathbb{F})$ is an isomorphism. Since the restriction homomorphism $r\colon \mathrm{Gal}(K/F) \to \mathrm{Gal}(K_0/F)$ is surjective with kernel $\mathrm{Gal}(K/K_0)$, the corollary will follow if we can only prove that the "natural map" $\mathrm{Gal}(K/F) \to \mathrm{Gal}(\mathbb{K}/\mathbb{F})$ factors via r.

To verify this, let $\mathcal{O}_{K_0}$ be the ring of integers of K_0, so that $\mathcal{O}_{K_0} \subset \mathcal{O}_K$. The fact that the residue field of K_0 is all of $\mathbb{K}$ implies that $\mathcal{O}_K = \mathcal{O}_{K_0} + \mathfrak{P}$. It is now clear that the action of $\sigma \in \mathrm{Gal}(K/F)$ on $\mathcal{O}_{K_0}$ determines its image in $\mathrm{Gal}(\mathbb{K}/\mathbb{F})$. □

We turn to more concrete considerations, and show that unramified extensions can be constructed as cyclotomic extensions.

Proposition 2.46 *Let F be a p-adic field with residue field $\mathbb{F}$, let N be an integer prime to p, and let $K = F(\mu_N)$ be the field obtained by adjoining to F the Nth roots of unity, forming the group μ_N. Then K/F is unramified, of degree f, the smallest integer such that $q^f \equiv 1 \bmod N$, where $q = |\mathbb{F}|$.*

The Galois group $\mathrm{Gal}(K/F)$ *is cyclic, generated by an automorphism σ with $\sigma(\zeta) = \zeta^q$, where we have picked a primitive Nth root of unity ζ.*

Finally, one has $\mathcal{O}_K = \mathcal{O}_F[\zeta]$.

Proof. The relation $\zeta^N = 1$ implies $v_K(\zeta) = 0$, so $\zeta \in \mathcal{O}_K$ (or, use Lemma 2.33). Let $P = \min(F, \zeta) \in \mathcal{O}_F[X]$, a divisor of $X^N - 1$. Its reduction $\overline{P}$ is also a divisor of $X^N - 1 \in \mathbb{F}[X]$; as we have chosen N prime to p, it follows that $\overline{P}$ does not have multiple roots. Hence, by Hensel's lemma, the polynomial $\overline{P}$ is irreducible. Since $\overline{P}$ has a root in $\mathbb{K}$, namely $\overline{\zeta}$, we see that $[\mathbb{K} : \mathbb{F}] \geq \deg(\overline{P}) = \deg(P) = [K : F] = e \cdot [\mathbb{K} : \mathbb{F}]$. It follows that $e = 1$, and K/F is unramified.

We see at the same time that $\mathbb{K} = \mathbb{F}[\overline{\zeta}]$, and moreover $\mathbb{K}$ contains all the Nth roots of unity; indeed the roots of $X^N - 1 \in \mathcal{O}_F[X]$ are in $\mathcal{O}_K$, so this polynomial factors as a product of linear terms in $\mathcal{O}_K[X]$, giving a corresponding factorization of $X^N - 1$ over $\mathbb{K}[X]$. Thus $\mathbb{K}/\mathbb{F}$ is cyclotomic (and $\overline{\zeta}$ is primitive). A finite field $\mathbb{K}$ of order q^f contains the Nth roots of unity if and only if N divides $q^f - 1$, the order of the cyclic group $\mathbb{K}^\times$, so f is as announced.

Since $\mathbb{K} = \mathbb{F}[\zeta]$, we have $\mathcal{O}_K = \mathcal{O}_F[\zeta] + \mathfrak{P}$. However, $\mathfrak{P} = \mathfrak{p}\mathcal{O}_K$ since K/F is not ramified, and we conclude by Nakayama's lemma that $\mathcal{O}_K = \mathcal{O}_F[\zeta]$.

Finally, the Galois group $\mathrm{Gal}(K/F)$ is cyclic, generated by $\sigma = \mathrm{Frob}_{K/F}$, which satisfies in particular $\sigma(\zeta) \equiv \zeta^q \bmod \mathfrak{P}$, by Corollary 2.41. However $\sigma(\zeta) = \zeta^k$ for some integer k, as ζ is primitive. Thus we have $\overline{\zeta}^{q-k} = 1$, and we have observed that $\overline{\zeta}$ is still primitive, so N divides $q - k$. It follows that $\sigma(\zeta) = \zeta^q$. □

We can now work backwards. Starting with the integer f, put $N = q^f - 1$, an integer which is prime to p, and which does not divide $q^{f'} - 1$ for $f' < f$. As a result, we see that $F[\mu_N]$ is the unique unramified extension of degree f.

Before moving on, we give a name to those extensions with the "opposite" property:

totally ramified

Definition 2.47 An extension K/F of local number fields is called **totally ramified** when $f(K/F) = 1$.

This happens precisely when K and F have the same residue field. By construction, this is the case for K and K_0, in the notation above. To summarize the discussion, if K/F is any extension of local number fields, then it can be "factorized" as a tower $F \subset K_0 \subset K$ with K_0/F unramified, and K/K_0 totally ramified. We have $f(K/F) = [K_0 : F]$ and $e(K/F) = [K : K_0]$, illustrating that $[K : F] = e(K/F)f(K/F)$.

Here is a typical example of totally ramified extension.

Proposition 2.48 *Let p be a prime number. Let $K = \mathbb{Q}_p(\mu_{p^m})$ be the field obtained by adjoining to $\mathbb{Q}_p$ the p^mth roots of unity, forming the group μ_{p^m}. Then $K/\mathbb{Q}_p$ is totally ramified, of degree $p^{m-1}(p-1)$. The Galois group $\mathrm{Gal}(K/\mathbb{Q}_p)$ is isomorphic to $(\mathbb{Z}/p^m\mathbb{Z})^\times$.*

Moreover, if ζ is a primitive root, then $\mathrm{N}_{K/\mathbb{Q}_p}(1-\zeta) = p$.

Of course this echoes Proposition 2.46, but notice how this is restricted to $F = \mathbb{Q}_p$. It remains true that $\mathcal{O}_K = \mathbb{Z}_p[\zeta]$, as follows from a general fact below (namely Proposition 2.53).

Proof. The usual facts about cyclotomic extensions remain available, of course: If ζ is a primitive root, then the homomorphism $\mathbb{Z}/p^m\mathbb{Z} \to \mu_{p^m}$ mapping k to ζ^k is an isomorphism; the primitive roots are those of the form ζ^k for $k \in (\mathbb{Z}/p^m\mathbb{Z})^\times$; the group $\mathrm{Gal}(K/\mathbb{Q}_p)$ injects into $(\mathbb{Z}/p^m\mathbb{Z})^\times$ via a map i satisfying $\sigma(\zeta) = \zeta^{i(\sigma)}$ for all σ in the Galois group.

Let $\xi = \zeta^{p^{m-1}}$. Then ξ is a primitive pth root of unity, so that

$$1 + \xi + \xi^2 + \cdots + \xi^{p-1} = 0.$$

Thus $\Phi(\zeta) = 0$ where

$$\Phi = 1 + X^{p^{m-1}} + X^{2p^{m-1}} + \cdots + X^{(p-1)p^{m-1}}.$$

Suppose we could establish that Φ (which is a cyclotomic polynomial) is irreducible over $\mathbb{Q}_p$. Then it is the minimal polynomial of ζ, and of each primitive root by the same token; the degree of $K/\mathbb{Q}_p$ is $p^{m-1}(p-1)$; and the Galois group is isomorphic to all of $(\mathbb{Z}/p^m\mathbb{Z})^\times$. We also see

$$\Phi = \prod_{\zeta' \text{ primitive}} (X-\zeta') = \prod_{\sigma\in \mathrm{Gal}(K/\mathbb{Q}_p)} (X-\sigma(\zeta)),$$

and so

$$\Phi(1) = p = \prod_{\sigma} (1-\sigma(\zeta)) = \mathrm{N}_{K/\mathbb{Q}_p}(1-\zeta).$$

Finally, by Theorem 2.25 (and the following remark), if w is the valuation of K extending v_p, we have

$$w(1-\zeta) = \frac{1}{[K:\mathbb{Q}_p]} v_p(\mathrm{N}_{K/\mathbb{Q}_p}(1-\zeta)) = \frac{1}{[K:\mathbb{Q}_p]}.$$

Thus $K/\mathbb{Q}_p$ is totally ramified.

So everything depends on proving that Φ is irreducible. The reader who knows a general version of Eisenstein's criterion will find this easy, but many will only be aware of the classical version over $\mathbb{Z}$, so we provide a direct argument. It suffices to prove that $\Phi(X+1)$ is irreducible. If $\overline{\Phi} \in \mathbb{F}_p[X]$ is the reduction of Φ mod p, then we have

$$\overline{\Phi}(X+1) = X^{p^{m-1}(p-1)}.$$

If we had $\Phi(X+1) = fg$ with $f,g \in \mathbb{Z}_p[X]$ both nonconstant, then both $\overline{f}$ and $\overline{g}$ would be (nonconstant) powers of X. Thus the constant terms of both f and g are divisible by p, and we deduce absurdly that the constant term of $\Phi(X+1)$ is divisible by p^2, but it is p. We leave it as an exercise to the reader to prove that f,g cannot be found in $\mathbb{Q}_p[X]$, either. □

Corollary 2.49 *Let p be a prime number, let $N = p^m n$ with n prime to p, and let $K = \mathbb{Q}_p(\mu_N)$. Then the inertia subfield is $K_0 = \mathbb{Q}_p(\mu_n)$. Moreover*

$$e(K) = [\mathbb{Q}_p(\mu_{p^m}) : \mathbb{Q}_p] = \varphi(p^m), \quad f(K) = [\mathbb{Q}_p(\mu_n) : \mathbb{Q}_p].$$

Proof. As $\mathbb{Q}_p(\mu_n)/\mathbb{Q}_p$ is unramified, by Proposition 2.46 the integer $f(K)$ is a multiple of $[\mathbb{Q}_p(\mu_n) : \mathbb{Q}_p]$. Likewise, since $\mathbb{Q}_p(\mu_{p^m})/\mathbb{Q}_p$ is totally ramified, by Proposition 2.48 we see that $e(K)$ is a multiple of $[\mathbb{Q}_p(\mu_{p^m}) : \mathbb{Q}_p]$. So $[K:\mathbb{Q}_p] = e(K)f(K)$ is a multiple of $[\mathbb{Q}_p(\mu_n) : \mathbb{Q}_p][\mathbb{Q}_p(\mu_{p^m}) : \mathbb{Q}_p]$. However, the field K is the composite

$$K = \mathbb{Q}_p(\mu_{p^m})\mathbb{Q}_p(\mu_n),$$

so that $[K:\mathbb{Q}_p]$ also divides the product of the degrees of these two subfields. In the end, we have the equality

$$[K:\mathbb{Q}_p] = [\mathbb{Q}_p(\mu_{p^m}) : \mathbb{Q}_p][\mathbb{Q}_p(\mu_n) : \mathbb{Q}_p] = e(K)f(K),$$

from which everything follows. □

Higher ramification groups

Definition 2.50 Let K/F be a Galois extension of local number fields. We define for $s \geq -1$ the group

$$G_s(K/F) = \{\sigma \in \mathrm{Gal}(K/F) : \sigma(a) \equiv a \bmod \mathfrak{P}^{s+1} \text{ for all } a \in \mathcal{O}_K\},$$

where as usual $\mathfrak{P}$ is the maximal ideal of $\mathcal{O}_K$.

Recall that the Galois action preserves $\mathcal{O}_K$, by Corollary 2.26, which serves to reassure us that the definition makes sense, as well as prove that $G_{-1}(K/F) = \mathrm{Gal}(K/F)$. Further, we see that $G_0(K/F)$ is the kernel of the map $\mathrm{Gal}(K/F) \to \mathrm{Gal}(\mathbb{K}/\mathbb{F})$, which we have called the inertia subgroup. It is sometimes denoted by $I(K/F)$, but we will not use this notation much.

In the following we often write G_s instead of $G_s(K/F)$, and $G = G_{-1} = \mathrm{Gal}(K/F)$.

Lemma 2.51 *For $s \geq 0$, the group $G_s(K/F)$ is a normal subgroup of* $\mathrm{Gal}(K/F)$, *and a subgroup of $G_{s-1}(K/F)$.*

Proof. It is clear that G_s is a subgroup of G_{s-1}. Let $a \in \mathcal{O}_K$, $\sigma \in G_s$, $\tau \in G$. We have $\sigma(\tau(a)) \equiv \tau(a) \bmod \mathfrak{P}^{s+1}$ by definition, and Corollary 2.26 assures us that $\tau^{-1}(\mathfrak{P}^{s+1}) = \mathfrak{P}^{s+1}$ (= the set of elements of normalized valuation $\geq s+1$). So $\tau^{-1}(\sigma(\tau(a))) \equiv a \bmod \mathfrak{P}^{s+1}$, and $\tau^{-1}\sigma\tau \in G_s$, as claimed. □

Remark 2.52 The group $G_1(K/F) \subset G_0(K/F)$ is often called the **wild inertia subgroup**; one says that K/F is wildly ramified when the latter is nontrivial, and that it is tamely ramified otherwise. For more information, see Problem 2.6 and compare with Theorem 2.58 (we will not say more about wild and tame ramification in the book).

wild inertia subgroup

We are thus in the presence of a descending chain of normal subgroups of $\mathrm{Gal}(K/F)$, looking like this:

$$\mathrm{Gal}(K/F) \supset G_0 \supset G_1 \supset G_2 \supset \cdots$$

Correspondingly we have an ascending chain

$$F \subset K_0 \subset K_1 \subset K_2 \subset \cdots$$

of fields which are Galois over F and contained in K, with $G_i = \mathrm{Gal}(K/K_i)$.

A peculiarity of the notation is that $G_s(K/F) = G_s(K/K_{s_0})$ whenever $s \geq s_0$. We will have the opportunity to use this with $s_0 = 0$, so that $G_s(K/F) = G_s(K/K_0)$ for $s \geq 0$. The groups G_s for $s > 0$ therefore contain information about the totally ramified extension K/K_0, which is why they are called the *higher ramification groups*.

The first important result is that this sequence reaches the trivial group. As a preparation for this, we prove a general fact.

Proposition 2.53 *Let K/F be an extension of p-adic fields. Then there exists an element $\zeta \in \mathcal{O}_K$ such that $\mathcal{O}_K = \mathcal{O}_F[\zeta]$.*

Moreover, if K/F is totally ramified, we may choose for ζ any element such that $(\zeta) = \mathfrak{P}$.

Note that we have already seen a particular case with Proposition 2.46. Also note that we must have $K = F[\zeta]$, and so the minimal polynomial of ζ has degree $n = [K : F]$; it follows easily that $\mathcal{O}_K$ is the free $\mathcal{O}_F$-module generated by $1, \zeta, \zeta^2, \ldots, \zeta^{n-1}$. This makes the statement of Proposition 2.36 more precise.

Proof. We resume the proof of Theorem 2.38, with all its notation. We have established that, whenever we have picked elements π_i and a_j, where $w(\pi_i) = {}^i/_e$ and the a_j are representatives for a basis of $\mathbb{K}$ as a vector space over $\mathbb{F}$, then the family of all $\pi_i a_j$ is a basis for $\mathcal{O}_K$ as a free $\mathcal{O}_F$-module. Now we choose the π_i and the a_j in a special way.

Let $\zeta \in \mathcal{O}_K$ be such that $\overline{\zeta} \in \mathbb{K}$ is a primitive element for the extension $\mathbb{K}/\mathbb{F}$, and let $f \in \mathcal{O}_K[X]$ be such that $\overline{f} = \min(\mathbb{F}, \overline{\zeta})$. Suppose we were able to choose ζ such that $w(f(\zeta)) = \frac{1}{e}$. Then we may put $\pi_i = f(\zeta)^i$, and $a_j = \zeta^j$, and we see that $\mathcal{O}_K$ is generated by the subring $\mathcal{O}_F$ and ζ, as claimed.

For any initial choice of ζ, we have $\overline{f}(\overline{\zeta}) = 0$ so $f(\zeta) \in \mathfrak{P}$, and $w(f(\zeta)) \geq \frac{1}{e}$. If $w(f(\zeta)) > \frac{1}{e}$, pick any $\Pi \in K$ such that $w(\Pi) = \frac{1}{e}$ and replace ζ by $\zeta + \Pi$: This does not change $\overline{\zeta}$, but we have from Taylor's formula

$$f(\zeta + \Pi) = f(\zeta) + f'(\zeta)\Pi \bmod \Pi^2. \qquad (*)$$

The polynomial $\overline{f'}$ does not have multiple roots, since $\mathbb{K}/\mathbb{F}$ is separable, so $\overline{f'}(\overline{\zeta}) \neq 0$, and $f'(\zeta)$ is a unit of $\mathcal{O}_K$. So (*) does show that $w(f(\zeta + \Pi)) = w(\Pi)$, and we are done.

Note that, in the totally ramified case when $\mathbb{K} = \mathbb{F}$, we could have taken $\zeta = 0$ as an initial choice, so that $f = X$, and if we follow the argument just given, we end up picking $\zeta = \Pi$. □

Corollary 2.54 *With ζ as above, for all $s \geq -1$ we have*

$$G_s(K/F) = \left\{\sigma \in \mathrm{Gal}(K/F) : \sigma(\zeta) \equiv \zeta \bmod \mathfrak{P}^{s+1}\right\}.$$

Moreover, the group $G_s(K/F)$ is trivial for s large enough.

Proof. The first (easy) statement is left to the reader. Now let $\sigma \in \mathrm{Gal}(K/F)$ be a nonidentity element. Then $x = \sigma(\zeta) - \zeta \neq 0$, and $w(x) \neq +\infty$. However, the elements of $\mathfrak{P}^{s+1}$ have valuation no less than $\frac{s+1}{e}$, so $x \notin \mathfrak{P}^{s+1}$ for large s; as a result, $\sigma \notin G_s$ for large s. There are finitely many elements in $\mathrm{Gal}(K/F)$, so there are no nonidentity elements at all in G_s when s is large enough. □

We turn to the study of the quotients G_s/G_{s+1}.

Definition 2.55 Define $U_K^{(0)} = \mathcal{O}_K^\times$, and for $s \geq 1$ define

$$U_K^{(s)} = \left\{x \in \mathcal{O}_K : x \equiv 1 \bmod \mathfrak{P}^s\right\}.$$

Notice that $U_K^{(s)} \subset U_K^{(s-1)}$, and in particular the elements of $U_K^{(s)}$ are units, and in fact one calls them *principal units*. It follows that $U_K^{(s)}$ is a group; indeed, it is obvious that the product of two elements remains in $U_K^{(s)}$, and to inspect the inverses, multiply the congruence $x \equiv 1 \bmod \mathfrak{P}^s$ by the unit x^{-1} to get $x^{-1} \equiv 1 \bmod \mathfrak{P}^s$. (Alternatively, one may argue that if $x = 1 - z$ with $z \in \mathfrak{P}^s$, then $x^{-1} = 1 + z\sum_{n=0}^{\infty} z^n$.)

Lemma 2.56 *There is an isomorphism*

$$U_K^{(0)}/U_K^{(1)} \cong \mathbb{K}^\times,$$

so that $U_K^{(0)}/U_K^{(1)}$ is an abelian group of order prime to p. Moreover, there is an isomorphism

$$U_K^{(s)}/U_K^{(s+1)} \cong \mathbb{K},$$

for $s \geq 1$, so that $U_K^{(s)}/U_K^{(s+1)}$ is an abelian group whose order is a power of p.

Proof. The first statement is obvious. For the second one, let $\Pi \in \mathcal{O}_K$ be such that $\mathfrak{P} = (\Pi)$, and consider the map $U_K^{(s)} \to \mathbb{K}$ mapping x to $(x-1)/\Pi^s$. □

Proposition 2.57 *Let $\Pi \in \mathcal{O}_K$ be such that $(\Pi) = \mathfrak{P}$. Then for $s \geq 0$ the map*

$$G_s/G_{s+1} \longrightarrow U_K^{(s)}/U_K^{(s+1)}$$

defined by $\sigma \mapsto \frac{\sigma(\Pi)}{\Pi}$ is an injective homomorphism.

Proof. Let φ be the proposed map. Let $\sigma, \tau \in \mathrm{Gal}(K/F)$, and write

$$\varphi(\sigma\tau) = \frac{\sigma(\tau(\Pi))}{\Pi} = \frac{\sigma(\tau(\Pi))}{\tau(\Pi)} \cdot \frac{\tau(\Pi)}{\Pi}.$$

We have $\tau(\Pi) \equiv \Pi \bmod \mathfrak{P}^{s+1}$ since $\tau \in G_s$, and we write this $\tau(\Pi) = u\Pi$ with $u \in U_K^{(s)}$. Thus

$$\frac{\sigma(\tau(\Pi))}{\tau(\Pi)} = \frac{\sigma(u)}{u} \cdot \frac{\sigma(\Pi)}{\Pi}.$$

Next, consider the congruence $\sigma(u) \equiv u \bmod \mathfrak{P}^{s+1}$ and multiply by the unit u^{-1} to see that $\frac{\sigma(u)}{u} \in U_K^{(s+1)}$. In the end we do have $\varphi(\sigma\tau) = \varphi(\sigma)\varphi(\tau)$, and φ is a homomorphism.

To check injectivity, we use a trick, namely that $G_s(K/F) = G_s(K/K_0)$ as already observed (since $s \geq 0$ here). So we may as well assume that $F = K_0$, that is, we may assume that the extension K/F is totally ramified. Now, Proposition 2.53 asserts that Π satisfies $\mathcal{O}_K = \mathcal{O}_F[\Pi]$. The injectivity of φ is then a consequence of Corollary 2.54. □

We summarize the information gathered in this section in the next beautiful theorem.

Theorem 2.58 *Let K/F be an extension of p-adic fields. Then* $\mathrm{Gal}(K/L)$ *is a solvable group. Moreover, the unique p-Sylow subgroup of the inertia subgroup is* $G_1(K/F)$.

Proof. We have constructed a nested sequence of normal subgroups of the Galois group $\mathrm{Gal}(K/F)$, which reaches the trivial group in finitely many steps. We have $G_{-1}/G_0 \cong \mathrm{Gal}(\mathbb{K}/\mathbb{F})$, which is a cyclic group since $\mathbb{K}$ is finite. By the last proposition, the group G_s/G_{s+1} is abelian for $s \geq 0$, so $\mathrm{Gal}(K/F)$ is solvable.

By Lemma 2.56 we see that G_0/G_1 has order prime to p, while G_s/G_{s+1} is a p-group for $s \geq 1$, so G_1 is a p-Sylow of G_0. It is in fact the only one, since it is normal. □

The reader is probably aware of Galois's theorem on polynomial equations, which can be solved by radicals if and only if appropriate Galois groups are solvable. Over $\mathbb{Q}_p$, the theorem shows that we may always do so. In other words, the roots of a polynomial with coefficients in $\mathbb{Q}_p$, even though they may not always be in $\mathbb{Q}_p$, can always be expressed using radicals. This is similar to the situation over $\mathbb{R}$, for which the worst that can happen is that we need to add $\sqrt{-1}$ in order to express the roots of a polynomial which are not real.

Problems

2.1. Let F be a field with a valuation v, and let $x, y \in F$ be such that $v(x) \neq v(y)$. Show that $v(x+y) = \min(v(x), v(y))$.

2.2. Let p be a prime number.

1. Prove that
$$v_p\binom{p^n}{k} = n - v_p(k) = v_p\left(\frac{p^n}{k}\right)$$
for $0 \leq k \leq p^n$. Here the leftmost term features a binomial coefficient, while the rightmost term involves a fraction p^n/k.
2. Show that, in $\mathbb{Q}_3$, the sequence $(2^{3^n})_{n\geq 0}$ converges to -1.

2.3. Let p be an odd prime. Show that $\mathbb{Z}_p$ has the following purely algebraic description in $\mathbb{Q}_p$:
$$\mathbb{Z}_p = \left\{a \in \mathbb{Q}_p : \exists t \in \mathbb{Q}_p, t^2 = 1 + p^3 a^4\right\}.$$

2.4. Let σ be an automorphism of the field $\mathbb{Q}_p$. Show that σ is continuous. Deduce that σ is the identity.

2.5. Let $f \in \mathbb{Z}[X_1, \ldots, X_m]$ be a multivariate polynomial. Show that the two statements below are equivalent:

1. There exist $x_1, \ldots, x_m \in \mathbb{Z}_p$ such that $f(x_1, \ldots, x_m) = 0$.
2. For each $n \geq 1$, there exist $x_1, \ldots, x_m \in \mathbb{Z}/p^n\mathbb{Z}$ such that $f(x_1, \ldots, x_m) = 0$.

In the next chapter, we shall introduce the language of inverse limits, which makes the above problem much easier to deal with. We will offer a new formulation of the same exercise (see Problem 3.4). Until then, we leave it without hint, as a challenge to the reader.

2.6. Let K/F be an extension of p-adic fields, and let $K_0 \subset K$ be the inertia subfield. The extension K/F is said to be *tamely ramified* when $[K : K_0]$ is prime to p. In particular, in this case K/K_0 is totally ramified but tamely ramified. In this problem we study such extensions. More precisely, we ask you to prove the equivalence of the next two statements, for an extension K/F of p-adic fields:

1. $K = F\left[a_1^{1/m_1}, \ldots, a_k^{1/m_k}\right]$, for some integers m_i which are all prime to p.
2. K/F is totally ramified, and tamely ramified.

Hints. (1) $\implies$ (2) *is easy if we keep Problem 1.1 in mind. For* (2) $\implies$ (1), *start with elements* $\alpha_i \in K$ *such that the* $v_K(\alpha_i)$ *generate* $v_K(K^\times)/v_F(F^\times)$*; show that these elements may be chosen so that* $\alpha_i^{m_i} \in F$ *for some integers* m_i *prime to* p *(Hensel's lemma is relevant). This gives a field* $F[\alpha_1, \ldots, \alpha_k] \subset K$ *which has the same valuation group as* K.

3 Tools from topology

In this chapter we study the interplay between algebra and topology, as illustrated by topological groups and topological fields. Of paramount importance in Galois theory are *profinite groups*, special topological groups arising when assembling together infinitely many finite groups. We will prove the fundamental theorem of Galois theory for arbitrary extensions, which involves topology even in its statement. Finally, we shall provide a proof for Proposition 2.27.

Topological groups

topological group

Definition 3.1 A **topological group** is a group G, which is also endowed with a topology, for which the multiplication map

$$G \times G \longrightarrow G, \qquad (x,y) \mapsto xy$$

as well as the "inverse" map

$$G \longrightarrow G, \qquad x \mapsto x^{-1}$$

topological isomorphism

are continuous. A **topological isomorphism** between two topological groups G and H is a group isomorphism $\varphi\colon G \to H$ which is also a homeomorphism, that is, both φ and φ^{-1} are continuous. ■

Let us discuss examples. Any group at all can be seen as a topological group with the *discrete topology*, for which every subset of G is open. Obviously this will not always be the most interesting topology, but there are cases when this is the sensible thing to do, for example when G is finite.

A very important series of examples is the following. Let $K = \mathbb{R}$ or $\mathbb{C}$ or $\mathbb{Q}_p$ or any local number field, or indeed any field with an absolute value. First, the additive group K is a topological group, as is the multiplicative group $K^\times = K \smallsetminus \{0\}$. This must be familiar over $\mathbb{R}$ or $\mathbb{C}$, and we let the reader check that the same can be said over a general K. Likewise, many statements in the next few paragraphs will require a quick, mental verification when K is not $\mathbb{R}$ or $\mathbb{C}$.

Let $\mathrm{GL}_n(K)$ denote the group of invertible $n \times n$ matrices with coefficients in K. Since K is a field, a square matrix over K is invertible precisely when its determinant is nonzero, so $\mathrm{GL}_n(K)$ is the subset of $M_n(K) = K^{n^2}$ where the map $\det\colon M_n(K) \to K$ does not vanish. Now the determinant is continuous, since it is given by a polynomial. So $\mathrm{GL}_n(K)$ is an open subset of K^{n^2}, where the latter is equipped with the product topology. We endow $\mathrm{GL}_n(K)$ with the induced topology. The multiplication on $GL_n(K)$ is then continuous, because matrix multiplication only involves polynomial expressions. What is more, the process of taking inverses, $A \mapsto A^{-1}$, is also continuous. The verification takes slightly more work, since we must know about Cramer's rule, which shows that A^{-1} only involves rational fractions in the coefficients of A. What is more, we must also know that $x \mapsto \frac{1}{x}$, as a map $K^\times \to K^\times$, is continuous (why?); indeed, this is contained in the statement that $K^\times = \mathrm{GL}_1(K)$ is a topological group, with which our running discussion started. Eventually we see that $\mathrm{GL}_n(K)$ is a topological group.

Any subgroup of a topological group can be equipped with the induced topology, and becomes a topological group in its own right. Thus all the subgroups of $\mathrm{GL}_n(K)$, for any $n \geq 1$, are examples. The group $S^1 = \{z \in \mathbb{C} : |z| = 1\} \subset \mathrm{GL}_1(\mathbb{C})$ is typical. In general a closed subgroup of $\mathrm{GL}_n(\mathbb{R})$ (for some n) is called a *linear Lie group*, and a closed subgroup of $\mathrm{GL}_n(\mathbb{Q}_p)$ is called a *linear p-adic Lie group*, although in this book we will use the phrase "Lie group" only three times. As another example, the group $\mathbb{Z}_p$ is a topological group, as a subgroup of $\mathbb{Q}_p$.

Another device to produce new examples is to take direct products, with the direct product topology. Thus $T_n = S^1 \times S^1 \times \cdots \times S^1$ (with n copies of S^1) is a topological group, called the n-dimensional compact torus. (It is indeed compact as a topological space.) The reader will prove that there is a topological isomorphism between T_n and the group of $n \times n$ diagonal matrices, with eigenvalues in S^1, with the topology induced from $\mathrm{GL}_n(\mathbb{C})$.

In this book we shall frequently have to deal with infinite products; when we get to that, we will recall in detail what the product topology is. As a teaser, let us announce that taking infinite products of discrete, indeed finite, groups will result in highly nontrivial topological groups.

When N is a normal subgroup of the topological group G, we can form G/N and give it the quotient topology – a subset U of G/N being open, by definition, if its inverse image $p^{-1}(U)$ under $p\colon G \to G/N$ is open. Typically, one will require N to be closed for the quotient to be nicely behaved: If N is not closed, then the points of G/N are not closed, and so G/N is not Hausdorff! (Recall that a topological space X is Hausdorff if for any pair of (distinct) points $x, y \in X$, we can find open sets U and V with $x \in U$, $y \in V$ and $U \cap V = \emptyset$. In a Hausdorff space, a singleton subset $\{x\}$ must be closed.) Keeping this in mind, one can otherwise prove that G/N with this topology is a topological group, as we will see in the exercises (there is certainly something to prove here).

However, for us, the interesting case will be when N has finite index in G, so that G/N is a finite group. We will only worry about whether this has the discrete topology, as expected. Let us prove a few general results.

Lemma 3.2 *Let G be a topological group, and let H be a subgroup.*

1. *If H is open, then H is closed.*
2. *If H is closed and has finite index in G, then it is open.*
3. *If G is compact, then H is open $\Longleftrightarrow$ H is closed of finite index.*

Proof. (1) The group G is the disjoint union of cosets of H, that is, sets of the form gH for $g \in G$. Each coset is open, since $x \mapsto gx$ is a homeomorphism $G \to G$, with inverse $x \mapsto g^{-1}x$. Thus any union of cosets is open, and in particular, the complement of any coset is open (being the union of all the other ones). So the complement of H is open.

(2) If we argue as above under the assumption that H is closed, we see that all the cosets are closed, and so any *finite* union of cosets is closed. If H has finite index, its complement in G is such a finite union, and so is closed.

(3) It remains to prove that, when G is compact and H is open, then H has finite index. Indeed, the space G is covered by the open sets gH, so that by compactness it is also covered by finitely many of these. However, they are disjoint and nonempty, so we cannot discard any of them; it follows that there were finitely many cosets in the first place. □

Lemma 3.3 *Let G be a topological group, let Γ be a group with the discrete topology, and let $\varphi\colon G \to \Gamma$ be a homomorphism. Then φ is continuous if and only if* $\ker(\varphi)$ *is open. When Γ is finite, this is equivalent to requiring that* $\ker(\varphi)$ *be closed.*

Proof. For any $x \in \Gamma$, the set $\varphi^{-1}(x)$ is either empty, or is a coset of $\ker(\varphi)$. If $\ker(\varphi)$ is open, we conclude that $\varphi^{-1}(x)$ is always open, and so φ is continuous. Conversely, if φ is continuous then $\ker(\varphi) = \varphi^{-1}(\{1\})$ must be open, since $\{1\}$ is open in the discrete space Γ. □

Lemma 3.4 *Let G be a topological group, and let N be a normal subgroup. The quotient topology on G/N is discrete if and only if N is open. If N has finite index, this is equivalent to requiring that N be closed.*

Proof. The quotient map $G \to G/N$ is continuous by the definition of the quotient topology, so if G/N is discrete then N is open by the previous lemma. Conversely, if N is open, then all its cosets are open, and by definition of the quotient topology we see that all the points in G/N are open. In other words, the topology on G/N is discrete. □

For example, the group $O_n(\mathbb{R})$ of $n \times n$ orthogonal matrices, with real entries, has an open-and-closed subgroup $SO_n(\mathbb{R})$, comprised of the matrices of positive determinant. In fact the determinant of a matrix in $O_n(\mathbb{R})$ can only be ± 1, so $SO_n(\mathbb{R})$ has index 2, and $O_n(\mathbb{R})/SO_n(\mathbb{R}) \cong \{\pm 1\}$ (a topological isomorphism between a quotient of two topological groups and a discrete group).

Profinite groups

We propose a definition that is not the one most commonly found in the literature. It has the advantage of being intrinsic. Alternative definitions will be afforded by Proposition 3.13.

profinite group

Definition 3.5 A **profinite group** is a topological group G which is compact and Hausdorff, and such that every open set containing the identity element of G also contains an open, normal subgroup U.

Remark 3.6 The subgroup U is then (closed and) of finite index in G, by Lemma 3.2, so G/N is finite. This is a first justification for the name "profinite", which will become more natural shortly. Note that, if U can always be found such that G/U is a p-group, or a solvable group, etc., then G is called a pro-p-group, or a pro-solvable group, etc. In this chapter we focus on profinite groups (and "profinite" is sometimes spelled "pro-finite"), but most results can be adapted to pro-p-groups, etc.

Finite groups, with the discrete topology, are profinite. To construct more examples, the first thing we can do is take infinite products of finite groups.

We recall a few things about the product topology. Let I be an indexing set, let X_i be a topological space for each $i \in I$, and let $X = \prod_i X_i$. We write $p_i \colon X \to X_i$ for the projection. The product topology is the smallest one for which the sets $p_i^{-1}(U)$ are all open, for all indices i and all open subsets $U \subset X_i$. It is designed so that a map $Y \to X$ is continuous if and only if all the compositions $Y \to X \to X_i$ are continuous. In other words, any open set U of X has the following form: select $U_{i_1}, \ldots, U_{i_n}$, open subsets of $X_{i_1}, \ldots, X_{i_n}$ respectively, and define

$$U = \{(x_i)_{i \in I} : x_{i_k} \in U_{i_k} \text{ for } 1 \leq k \leq n\}.$$

When each X_i is discrete, we see that the topology of X is in fact *generated* by sets of the form

$$\{(x_i)_{i \in I} : x_{i_0} = y\},$$

for a fixed index i_0 and a fixed $y \in X_{i_0}$. Any open subset of X can then be expressed as a union of sets of the form

$$U = \{(x_i)_{i \in I} : x_{i_k} = y_{i_k} \text{ for } 1 \leq k \leq n\},$$

where elements $y_{i_1}, \ldots, y_{i_n}$ have been selected in $X_{i_1}, \ldots, X_{i_n}$ respectively. What is more, if $z = (z_i)_{i \in I} \in X$, then any open subset of X containing z contains a U as above, where we pick $y_{i_k} = z_{i_k}$. Put differently (and somewhat more loosely): A basis of neighborhoods of $z \in X$ is provided by the sets U consisting of all the $x \in X$ which have the same coordinates as z for finitely many given indices.

A strong and important result about the product topology is Tychonoff's theorem:

Theorem 3.7 (Tychonoff) *If each X_i is compact, then so is $X = \prod_{i \in I} X_i$.*

See [Mun75, chapter 5].

It is an easy exercise, which we leave to the reader, to prove that if G_i is a topological group, for each $i \in I$, then so is $G = \prod_i G_i$. If each G_i is finite, it is trivially compact, and then G is a compact topological group by Tychonoff (it is also trivially Hausdorff). Let us check that it is in fact profinite. Indeed, by the discussion above, any open set of G containing $1 = 1_G = (1_{G_i})_{i \in I}$ contains a subset of the form

$$U = \{(g_i)_{i \in I} : g_{i_k} = 1 \text{ for } 1 \leq k \leq n\}.$$

But it is clear that this U is an open, normal subgroup of G; in fact $G/U \cong G_{i_1} \times \cdots \times G_{i_n}$.

To produce more examples, we keep the following in mind.

Lemma 3.8 *If G is a profinite group, and if H is a closed subgroup of G, then H is also profinite.*

Proof. It is clear that H is compact and Hausdorff. Any open set of H containing 1 is of the form $V \cap H$ for V open in G, by definition of the induced topology. This V contains a U which is an open, normal subgroup of G. But then $U \cap H$ is an open, normal subgroup of H. □

So here is a way to produce closed subgroups of infinite products of finite groups. It may seem surprising at first, but we will end up proving that any profinite group at all is isomorphic to one constructed in this fashion.

We will now assume that I is not just an indexing set, but a partially ordered set (poset for short), so that elements $i, j \in I$ can sometimes satisfy $i \leq j$. We assume that we are given a finite group G_i for each $i \in I$, but also a homomorphism $\varphi_{ji} : G_j \to G_i$ whenever $i \leq j$ (note the change of direction!). We require that φ_{ii} be the identity of G_i, and that $\varphi_{kj} \circ \varphi_{ji} = \varphi_{ki}$ when $i \leq j \leq k$. In this situation we define

$$\lim_{i \in I} G_i := \left\{ (x_i)_{i \in I} \in \prod_i G_i : x_i = \varphi_{ji}(x_j) \text{ for all } i \leq j \right\},$$

and call it the *limit*, or *inverse limit*, or *projective limit*, of the *system of groups* at hand. (We will see much later in this book that the "opposite" concept is called a "direct limit" or "colimit"; See Chapter 7, page 141.) Similarly, one defines inverse limits of topological spaces, rather than topological groups.

Lemma 3.9 *The limit $\lim_i G_i$ is a closed subgroup of $\prod_i G_i$. In particular, it is a profinite group.*

Proof. Let $G = \lim_i G_i$, which is clearly a group. Let $x = (x_i)_{i \in I} \in \prod G_i$ be such that $x \notin G$. Then $x_{i_0} \neq \varphi_{j_0 i_0}(x_{j_0})$ for some indices i_0, j_0. If we define then U to be the subset of elements having the same i_0th and j_0th coordinates as x, that is $U = \{(y_i) : y_{i_0} = x_{i_0} \text{ and } y_{j_0} = x_{j_0}\}$, then U is open and $U \cap G = \emptyset$. □

We have already seen an example of inverse limit in this book. Indeed, take the indexing set I to be the set of integers, and for each $n \in \mathbb{N}$ let $G_n = \mathbb{Z}/p^n\mathbb{Z}$, for some fixed prime number p; the map φ_{mn} is taken to be the natural map $\mathbb{Z}/p^m\mathbb{Z} \to \mathbb{Z}/p^n\mathbb{Z}$, which reduces mod p^n.

Proposition 3.10 *There is a topological isomorphism between* $\lim_n \mathbb{Z}/p^n\mathbb{Z}$ *and the (additive) group* $\mathbb{Z}_p$ *of p-adic integers.*

Proof. In the previous chapter we defined maps $\pi_n \colon \mathbb{Z}_p \to \mathbb{Z}/p^n\mathbb{Z}$. Together they define a map into the product of all the groups $\mathbb{Z}/p^n\mathbb{Z}$, and by construction it lands inside the inverse limit, so we have $f \colon \mathbb{Z}_p \to \lim_n \mathbb{Z}/p^n\mathbb{Z}$. That it is an isomorphism of groups can be deduced from the material of the previous chapter, and we leave this as an exercise for the reader.

Some work is needed, however, to show that f is a homeomorphism. To show that f is continuous, we must prove that for each n the map π_n is continuous, with the discrete topology on $\mathbb{Z}/p^n\mathbb{Z}$ (by definition of the product topology). By Lemma 3.3, we must prove that the kernel of this map is open, or equivalently, closed. We already knew with Lemma 2.3 that this kernel is the ideal $p^n\mathbb{Z}_p$, which is also the closed ball of radius p^{-n}, centered at 0. So f is continuous.

Next we must show that f maps open sets to open sets, and for this it suffices, as the reader will check, to establish that the image under f of any open ball centered at 0 is open. Indeed, any such ball is an ideal $p^{n_0}\mathbb{Z}_p$ for some n_0, and $f(p^{n_0}\mathbb{Z}_p)$ is the subset of $\lim \mathbb{Z}/p^n\mathbb{Z}$ of those $(x_n)_{n\in\mathbb{N}}$ such that $x_n = 0$ for $n \leq n_0$. This condition involves finitely many coordinates, and is therefore an open one. □

This affords the alternative construction of $\mathbb{Z}_p$ alluded to in the previous chapter: We could have *defined* $\mathbb{Z}_p$ to be $\lim_n \mathbb{Z}/p^n\mathbb{Z}$. Then it would have been automatically a group, and in fact, it is easy to see that $\lim_n \mathbb{Z}/p^n\mathbb{Z}$ is a ring (and that the isomorphism of the proposition is a ring isomorphism). So the arithmetic operations on $\mathbb{Z}_p$ would have been obtained "for free". Of course, as a trade-off, harder work would have been needed to prove some other properties of $\mathbb{Z}_p$.

We have decidedly learned something which was not obvious with our previous point of view:

Corollary 3.11 *The group* $\mathbb{Z}_p$ *is compact.* □

Let us say that the poset I is a *directed set* when for any $i,j \in I$, there exist k such that $i \leq k$ and $j \leq k$. Inverse limits over directed sets are sometimes better behaved. Indeed, consider an inverse limit of mere discrete topological spaces $X = \lim_{i\in I} X_i$. Any open set of X containing a given $z = (z_i)_{i\in I} \in X$ contains a set of the form $U = \{(x_i) \in X : x_{i_k} = z_{i_k}\}$ for some indices $i_1, \ldots, i_n$. Now if I is directed, we can find an index j such that $i_k \leq j$ for all k, and it follows that U contains in turn $V = V_{z,j} = \{(x_i) \in X : x_j = z_j\}$. In other words, we have an utterly simple basis of neighborhoods of a given $z \in X$: Take the set $V_{z,j}$ of elements agreeing with z in the j-coordinate, and let j run through the elements of I.

Here is an example of use.

Lemma 3.12 *Let $X = \lim_i X_i$ be an inverse limit of discrete topological spaces over a directed set, and write $p_i\colon X \to X_i$ for the projection. Let $A, B \subset X$. If for each i one has $p_i(A) \subset p_i(B)$, then $A \subset \overline{B}$.*

Here $\overline{B}$ is the closure of B, that is the smallest closed subset of X which contains B – not to be confused with the reduction modulo some ideal, as the "bar" notation sometimes means.

Proof. Let $a = (a_i)_{i \in I} \in A$, fix $j \in I$, and consider $U = \{(x_i) \in X : x_j = a_j\}$. Since $p_j(A) = p_j(B)$, there is $b = (b_i) \in B$ such that $b_j = a_j$, or in other words $b \in U$. As the open sets U obtained in this way form a basis of neighborhoods of a, we see that a is in the closure of B. □

Now, as promised, we prove:

Proposition 3.13 *The following conditions on a topological group G are equivalent:*

1. *G is a profinite group.*
2. *G is topologically isomorphic to an inverse limit $\lim_{i \in I} G_i$ of finite groups, where I is a directed set.*
3. *G is a closed subgroup of a product of finite groups.*

Proof. We must prove (1) $\Longrightarrow$ (2), all the other implications being already established. For this, we suppose that G is profinite, and take our indexing set I to be the set of all open, normal subgroups U of G. It is partially ordered by $U \leq V \Leftrightarrow V \subset U$, and it is a directed set since we always have $U \leq U \cap V$ and $V \leq U \cap V$. The finite groups we take are $G_U = G/U$, and the maps $G_V \to G_U$ are the natural ones (induced by the identity of G). We have an obvious map

$$f\colon G \longrightarrow H := \lim_{U} G/U,$$

and we will prove that it is a topological isomorphism. It is certainly continuous, for each composition $G \to G/U$ is continuous, having an open kernel. Since G is compact and H is Hausdorff, a classical argument of topology will show that f is a homeomorphism, if we can only prove that it is a bijection.

First, consider the injectivity. Perhaps surprisingly, this will follow because we assume that G is Hausdorff. Indeed, if $x \neq 1$, we see that there is an open subset of G containing 1 but not x, so there is an open, normal subgroup U such that x does not map to the identity of G/U. This means that $\ker(f) = \{1\}$.

For the surjectivity, it is enough to prove that the image of f is dense in H; indeed this image is automatically closed, since G is compact and H is Hausdorff. For this we apply the last lemma with $A = H$ and $B = f(G)$: We certainly have, for each U, that the image of H in G/U is contained in the image of $f(G)$, which is all of G/U. So $H \subset \overline{f(G)}$, and we are done. □

From now on, all inverse limits will often be tacitly taken over a directed set. The proposition shows that the groups produced as general inverse limits can also be obtained in this way.

We conclude with a proposition about $\mathrm{GL}_n(\mathbb{C})$ which will show that it is, in a sense, "incompatible" with profinite groups. This will not be used in the sequel, and is only included as an illustration, so the magnanimous reader will forgive the references, in the proof below, to material which has not been reviewed (norms on complex matrix spaces in particular).

Proposition 3.14 *There is an open subset $U \subset \mathrm{GL}_n(\mathbb{C})$ which contains $\{I\}$, but no other subgroup.*

Proof. Pick any norm $|\cdot|$ on $\mathbb{C}^n$, and the corresponding norm $\|\cdot\|$ on $M_n(\mathbb{C})$, defined by $\|M\| = \sup_{|v|=1} |Mv|$. Thus $|Mv| \leq \|M\| \cdot |v|$ for any matrix M and vector v, and $\|M_1 M_2\| \leq \|M_1\| \cdot \|M_2\|$.

First, we claim that if a subgroup $G \subset \mathrm{GL}_n(\mathbb{C})$ is bounded, in the sense of our norm, then all the eigenvalues of all the elements of G have absolute value 1. Indeed, let $A \in G$, and let v be an eigenvector, so that $Av = \lambda v$ with $\lambda \in \mathbb{C}$. Then $|\lambda v| = |\lambda| \cdot |v| \leq \|A\| \cdot |v|$, so $|\lambda| \leq C$ for some constant C. However, λ^n is an eigenvalue of $A^n \in G$ for any integer $n \in \mathbb{Z}$, so by the same token $|\lambda|^n \leq C$, for all $n \geq \mathbb{Z}$; positive values of n indicate here that $|\lambda| \leq 1$, while $|\lambda| \geq 1$ by inspecting negative values of n. In the end we must have $|\lambda| = 1$, and the claim is proved.

Now call U the open ball centered at the identity I, with radius $\frac{1}{4}$, say. We claim that if a subgroup G is contained in U, then the only eigenvalue of any element of G is 1. This time we pick an eigenvector v for some $A \in G$ and we write $|(\lambda - 1)v| = |\lambda - 1| \cdot |v| \leq \|A - I\| \cdot |v| \leq \frac{1}{4}|v|$, so $|\lambda - 1| \leq \frac{1}{4}$. Of course, we similarly have $|\lambda^n - 1| \leq \frac{1}{4}$ for all $n \in \mathbb{Z}$. Since $|\lambda| = 1$, it is now an easy exercise, left to the reader, to show that we must have $\lambda = 1$ (just write $\lambda = e^{i\theta}$ for a small value of θ, determined by the condition for $n = 1$, and examine $n\theta$).

We conclude. Suppose $G \subset U$, and let $A \in G$. We can put A in Jordan canonical form, that is, we can find P such that $P^{-1}AP$ is made of Jordan blocks along the diagonal, with 1s only on the diagonal itself. If A is not the identity, the coefficients of $P^{-1}A^nP$ will include an n somewhere, and in particular will not all remain bounded. However, $A^n \in G \subset U$ so $P^{-1}A^nP \in P^{-1}UP$, and the set $P^{-1}UP$ is clearly bounded. This contradiction shows that A can only be the identity. □

(At the end of this proof we have used the fact that being bounded for the norm $\|\cdot\|$ is equivalent to being bounded coefficient-by-coefficient. This is because all norms on $M_n(\mathbb{C})$ are equivalent, a familiar fact which we revisit later in this chapter.)

Corollary 3.15 *Let G be a profinite group, and let φ be a homomorphism $G \to \mathrm{GL}_n(\mathbb{C})$. Then φ is continuous if and only if its kernel is open. In this case the image of φ is finite.*

Proof. Suppose φ is continuous, and let U be as in the proposition. Then $\varphi^{-1}(U)$ is open, and thus contains a normal, open subgroup $V \subset G$. Now $\varphi(V)$ is a subgroup contained in U, so $\varphi(V) = \{I\}$, or in other words $V \subset \ker(\varphi)$. As a result, $\ker(\varphi)$ has finite index in G (just like V), and since it is obviously closed, it is open. Besides, φ factors via $G/\ker(\varphi)$, so it has a finite image.

Conversely, suppose $\ker(\varphi)$ is open. Then $G/\ker(\varphi)$ is finite, as just observed, and the projection $G \to G/\ker(\varphi)$ is continuous, where we endow $G/\ker(\varphi)$ with the discrete topology (Lemma 3.4). The induced map $G/\ker(\varphi) \to \mathrm{GL}_n(\mathbb{C})$ is continuous (as is any map out of a discrete space), so the composition of the two, which is φ, is also continuous. □

So the group $\mathrm{GL}_n(\mathbb{C})$ does not contain any profinite groups, except for finite groups. This is not to say, however, that homomorphisms between a profinite group G and various subgroups of $\mathrm{GL}_n(\mathbb{C})$ are not interesting to study. Rather, the corollary allows us to derive from a topological condition (the continuity of φ) an algebraic one, that the image of φ is finite.

Infinite Galois extensions

The reader has already encountered, perhaps unknowingly, another familiar example of a profinite group: As we proceed to explain, the Galois group of any algebraic extension can be naturally endowed with such a topology. As an illustration of the relevance of topological ideas in this area, we shall then extend the fundamental theorem of Galois theory – which the reader is assumed to know for finite extensions in this book – to arbitrary extensions.

So let K/F be an algebraic extension, and assume that it is Galois. We pause to recall that, if we make no assumption about the dimensions, this means that K/F is separable and normal. (When $[K : F] < \infty$, alternative definitions are available.)

Consider those intermediate fields L, that is $F \subset L \subset K$, such that L/F is finite-dimensional and Galois, and observe that they form a directed set. Indeed, write $L_1 \leq L_2$ for $L_1 \subset L_2$, and we clearly get a poset; also note that the compositum L_1L_2 is finite-dimensional and Galois over F and satisfies $L_i \leq L_1L_2$ for $i = 1, 2$.

For each such L, we can form the group $\mathrm{Gal}(L/F)$. Whenever $L_1 \leq L_2$, we have the restriction map $\mathrm{Gal}(L_2/F) \to \mathrm{Gal}(L_1/F)$, and so it makes sense to talk about the inverse limit of these groups.

Lemma 3.16 *There is an isomorphism of groups*

$$\mathrm{Gal}(K/F) \longrightarrow \varprojlim_{L} \mathrm{Gal}(L/F)$$

where the inverse limit is taken over the directed set described above.

Proof. Since K/F is assumed to be algebraic, any $x \in K$ is contained in the finite-dimensional extension $L_x = F[x]$. We shall write $\widetilde{L}_x$ for the Galois closure of L_x/F, which is finite and Galois.

Now, if we take the product of all the restriction maps $\mathrm{Gal}(K/F) \to \mathrm{Gal}(L/F)$, for all L in our directed set, then the combined map

$$\iota\colon \mathrm{Gal}(K/F) \longrightarrow \prod_L \mathrm{Gal}(L/F)$$

is injective. Indeed if $\sigma \in \mathrm{Gal}(K/F)$ is not the identity, pick $x \in K$ such that $\sigma(x) \neq x$, and observe that the restriction of σ to $\mathrm{Gal}(\widetilde{L_x}/F)$ is not the identity.

The image of ι is clearly contained in the inverse limit, and we need to prove that it is all of the inverse limit. This is almost tautological. Suppose $(\sigma_L)_L$ is a given family of automorphisms, in the limit. Any $x \in K$ is contained in some L such that L/F is finite and Galois (for example $L = \widetilde{L_x}$); let L_1 and L_2 be two such fields containing x, and let $E = L_1L_2$. Observe then that $\sigma_{L_1}(x) = \sigma_E(x) = \sigma_{L_2}(x)$: this is just rewriting that the restriction of σ_E to L_i, for $i = 1,2$, is σ_{L_i}, by definition of the inverse limit. However, this means that the element $\sigma_L(x) \in K$ does not depend on the choice of a particular L containing x. Call this element $\sigma(x)$. This defines a map $\sigma : K \to K$, and it is a simple exercice, left to the reader, to check that $\sigma \in \mathrm{Gal}(K/F)$. By construction, we have $\iota(\sigma) = (\sigma_L)_L$, and we see indeed that any element in the inverse limit is in the image of ι. □

Since $\mathrm{Gal}(K/F)$ is isomorphic to an inverse limit of finite groups, we can give it a topology by requiring the isomorphism in the lemma to be a homeomorphism. This is called the *Krull topology* on $\mathrm{Gal}(K/F)$. It is *not* just a curiosity, as we proceed to show with the "Fundamental Theorem of Galois theory", whose very statement involves the Krull topology.

It is best to start with a little combinatorial statement. Let S and T be two posets. Let $f\colon S \to T$ and $g\colon T \to S$ be two order-reversing maps ($x \leq y$ implies $f(y) \leq f(x)$ and likewise for g). Finally, suppose that $s \leq g(f(s))$ for all $s \in S$, and similarly $t \leq f(g(t))$ for all $t \in T$.

Lemma 3.17 *In the situation above:*

1. *An element $s \in S$ is in the image of g if and only if $s = g(f(s))$.*
2. *An element $t \in T$ is in the image of f if and only if $t = f(g(t))$.*
3. *There is a bijection between the image of g and the image of f. It is given by the restriction of f, and the inverse bijection is the restriction of g.*

Proof. (1) Let $s \in S$, so that $s \leq g(f(s))$ by assumption. We claim that, if we suppose that $s = g(t)$ for some $t \in T$, then $g(f(s)) \leq s$; this establishes that $s = g(f(s))$, and the nontrivial part of (1). To prove the claim, apply f to the identity $s = g(t)$ and get $f(s) = f(g(t)) \geq t$; now apply g to the inequality $f(s) \geq t$ and get $g(f(s)) \leq g(t) = s$.

(2) is of course similar. As for (3), consider the restriction $f'\colon g(T) \to f(S)$, as well as $g'\colon f(S) \to g(T)$. Then (1) shows that $g' \circ f'$ is the identity of $g(T)$, while (2) shows that $f' \circ g'$ is the identity of $f(S)$. □

In this situation we say that we have a *Galois correspondence* between S and T. The elements of S satisfying $s = g(f(s))$, or alternatively $s \in g(T)$, will be called *saturated*, as will be the elements of T satisfying $t = f(g(t))$ or alternatively $t \in f(S)$. Sometimes $g(f(s))$ is written $\overline{s}$ and called the closure of s, and likewise with the elements of T. Note that $s_1 \leq s_2$ gives $\overline{s}_1 \leq \overline{s}_2$, and $\overline{\overline{s}} = \overline{s}$.

Example 3.18 Let k be a ring, and let $n \geq 1$ be an integer. Let S be the set of subsets of $k[X_1, \ldots, X_n]$, and let T be the set of subsets of k^n. Define for $I \in S$, $V \in T$:

$$f(I) = \{x \in k^n : P(x) = 0 \text{ for all } P \in I\},$$

and

$$g(V) = \{P \in k[X_1, \ldots, X_n] : P(x) = 0 \text{ for all } x \in V\}.$$

We have a Galois correspondence. The saturated elements of T, that is the subsets of k^n that can be described as the zeros of a bunch of polynomial equations, are usually called *affine algebraic varieties*, or sometimes *Zariski closed subsets*. By the lemma, they are in bijection with the saturated subsets of $k[X_1, \ldots, X_n]$. At the very least, we can notice that these are ideals. When k is an algebraically closed field, the famous *Nullstellensatz* asserts that the saturated subsets are precisely the so-called *radical ideals* (those ideals I for which $x^n \in I$ implies $x \in I$). ■

We return to our main example of Galois correspondence, which occurs when K/F is an algebraic extension, and we let S denote the set of subgroups of $\mathrm{Gal}(K/F)$, while T will be the set of intermediate fields L (with $F \subset L \subset K$). Playing the role of f will be the map usually written $\mathcal{F}$, where $\mathcal{F}(H)$ is the fixed field of the subgroup H, that is

$$\mathcal{F}(H) = \{x \in K : \sigma(x) = x \text{ for all } \sigma \in H\}.$$

As for g, we choose the map $L \mapsto \mathrm{Gal}(K/L)$, with of course

$$\mathrm{Gal}(K/L) = \{\sigma \in \mathrm{Gal}(K/F) : \sigma(x) = x \text{ for all } x \in L\}.$$

This is indeed a correspondence in the above sense, and the challenge is to identify the saturated elements on both sides.

When K/F is finite and Galois, it is a nontrivial result, which we assume is known to the reader, that *all* subgroups are saturated, and *all* intermediate fields are saturated. This is the hard part in proving the "Fundamental Theorem", stating that there is a bijection between the subgroups of $\mathrm{Gal}(K/F)$ and the intermediate fields of K/F. We shall use this result to investigate the general case, when K/F is merely assumed to be algebraic and Galois, but not finite.

In one direction, there is no surprise.

Lemma 3.19 *Let K/F be algebraic and Galois, and let L be an intermediate field. Then*

$$L = \mathcal{F}(\mathrm{Gal}(K/L)),$$

or in other words, L is saturated.

Proof. There is always an inclusion $L \subset \mathcal{F}(\mathrm{Gal}(K/L))$ (since we have a Galois correspondence), so we show the reverse. For this we pick $x \in K$ such that $x \notin L$, and we show that x is not in $\mathcal{F}(\mathrm{Gal}(K/L))$, that is, we show the existence of $\sigma \in \mathrm{Gal}(K/L)$ not fixing x. Consider E, the Galois closure of $L[x]/L$. Then E/L is finite and Galois. Since $x \notin L$, by finite-dimensional Galois theory there exist $\tau \in \mathrm{Gal}(E/L)$ such that $\tau(x) \neq x$. Now extend τ to an element σ of $\mathrm{Gal}(K/L)$, which is possible since the extension K/L is normal. □

However, we shall see that not all subgroups of $\mathrm{Gal}(K/F)$ are saturated. In fact:

Lemma 3.20 *Let L be an intermediate field of K/F, and let $H = \mathrm{Gal}(K/L)$. Then H is closed in the Krull topology.*

Below we shall provide examples of non-closed subgroups.

Proof. Pick $\sigma \in \mathrm{Gal}(K/F)$ with $\sigma \notin H$. By definition, there exists $x \in L$ such that $\sigma(x) \neq x$. Let E be the Galois closure of $F[x]/F$, so that E/F is finite and Galois. Now, the set

$$U = \{\tau \in \mathrm{Gal}(K/F) : \tau_E = \sigma_E\}$$

is open (in the Krull topology), and $U \cap H = \emptyset$ (here τ_E is the restriction of τ to E/F, and likewise for σ_E). So the complement of H is open. □

We proceed to show the converse, that is, closed subgroups are saturated.

Lemma 3.21 *Let H be a subgroup of $\mathrm{Gal}(K/F)$, and let $\sigma \in \mathrm{Gal}(K/F)$. Then $\sigma \in \mathrm{Gal}(K/\mathcal{F}(H))$ if and only if for any finite Galois extension E/F contained in K, the restriction of σ to $\mathrm{Gal}(E/F)$ coincides with the restriction of an element of H.*

Proof. First note that for any intermediate field L, we have $\sigma \in \mathrm{Gal}(K/L)$ if and only if the restriction of σ to $\mathrm{Gal}(E/F)$, for any E as in the lemma, fixes the intersection $E \cap L$. (This assumes just that K/F is algebraic.)

Now we apply this remark with $L = \mathcal{F}(H)$. Let E be as above, and let $r\colon \mathrm{Gal}(K/F) \to \mathrm{Gal}(E/F)$ be the restriction map. If $\sigma \in \mathrm{Gal}(K/\mathcal{F}(H))$, then its restriction $\sigma' = r(\sigma)$ fixes $E \cap \mathcal{F}(H)$; in turn, this is the subfield of elements of E fixed by all the automorphisms in $H' = r(H)$. By finite-dimensional Galois theory, we see that $\sigma' \in H'$. This shows one implication. The converse is clear. □

Corollary 3.22 *The subgroup $\mathrm{Gal}(K/\mathcal{F}(H))$ is the closure, in the Krull topology, of the subgroup H.*

It follows that saturated subgroups coincide with closed subgroups.

Proof. Let $H_0 = \mathrm{Gal}(K/\mathcal{F}(H))$. We have $H \subset H_0$. The last lemma, combined with Lemma 3.12, shows that $H_0 \subset \overline{H}$, where $\overline{H}$ is the closure, in the Krull topology, of H. Moreover, by Lemma 3.20, the subgroup H_0 is closed. It follows that $H_0 = \overline{H}$. □

Summing up, we have established the following.

Theorem 3.23 (Fundamental Theorem of Galois Theory) *Let K/F be an algebraic, Galois extension. There exists a one-to-one, order-reversing correspondence between the intermediate subfields of K/F and the closed subgroups of* $\mathrm{Gal}(K/F)$*, for the Krull topology. The correspondence maps L to* $\mathrm{Gal}(K/L)$*, and H to $\mathcal{F}(H)$.*

Example 3.24 Let $F = \mathbb{F}_q$ be a finite field of cardinality q, and let $K = \overline{\mathbb{F}}_q$, an algebraic closure of $\mathbb{F}_q$. Then $\overline{\mathbb{F}}_q/\mathbb{F}_q$ is obviously normal, and it is also separable since finite fields are perfect. We proceed to describe $\mathrm{Gal}(\overline{\mathbb{F}}_q/\mathbb{F}_q)$, using the following classical facts: for any $n \geq 1$, there is just one intermediate field of dimension n over $\mathbb{F}_q$, call it $\mathbb{F}_{q^n}$; and the Galois group $\mathrm{Gal}(\mathbb{F}_{q^n}/\mathbb{F}_q)$ is cyclic of order n, generated by the Frobenius automorphism $x \mapsto x^q$. In particular we have a *canonical* isomorphism $\mathbb{Z}/n\mathbb{Z} \to \mathrm{Gal}(\mathbb{F}_{q^n}/\mathbb{F}_q)$, so

$$\mathrm{Gal}(\overline{\mathbb{F}}_q/\mathbb{F}_q) \cong \lim_n \mathbb{Z}/n\mathbb{Z},$$

where the maps involved in the inverse limit are the natural ones. In the exercises we ask you to elucidate the relationship between this group and the p-adic numbers.

To get an initial feel, though, note that the association $x \mapsto x^q$ may be seen directly as an element of $\mathrm{Gal}(\overline{\mathbb{F}}_q/\mathbb{F}_q)$, which we call again the Frobenius element and write Frob. Consider the cyclic group generated by Frob, which is clearly not finite, so $\langle \mathrm{Frob} \rangle \cong \mathbb{Z}$. Then Lemma 3.12 shows that it is dense in $\mathrm{Gal}(\overline{\mathbb{F}}_q/\mathbb{F}_q)$; alternatively, we may see this from the Galois correspondence, since the corresponding fixed field is visibly $\mathbb{F}_q$. This explains why this inverse limit is usually denoted $\widehat{\mathbb{Z}}$, to emphasize that it is a completion of $\mathbb{Z}$.

The fixed field of $\langle \mathrm{Frob}^n \rangle$ is clearly $\mathbb{F}_{q^n}$, so this intermediate subfield corresponds to the closure of $\langle \mathrm{Frob}^n \rangle$. As an exercise, prove that $\widehat{\mathbb{Z}}$ is a ring, containing $\mathbb{Z}$ as a subring. Realizing this, prove that the closure of $\langle \mathrm{Frob}^n \rangle$ is isomorphic to $n\widehat{\mathbb{Z}}$, the subgroup of elements of $\widehat{\mathbb{Z}}$ of the form nx with $x \in \widehat{\mathbb{Z}}$ (that is, the principal ideal generated by n). ■

Example 3.25 Let $F = \mathbb{Q}$, and let $K = \mathbb{Q}^{ab}$, the field obtained by adjoining all the roots of unity to the rational field (the notation $\mathbb{Q}^{ab}$ will be justified with the Kronecker–Weber theorem, much later in this book). Then K/F is normal, since it is the decomposition field of all the polynomials $X^n - 1$ for all $n \geq 1$. It is separable since F has characteristic 0, so K/F is Galois.

Let $\mathbb{Q}^{(n)} = \mathbb{Q}(e^{2i\pi/n})$. We let the reader prove that

$$\mathrm{Gal}(K/F) \cong \lim_n \mathrm{Gal}(\mathbb{Q}^{(n)}/\mathbb{Q}).$$

It is well-known that $\mathrm{Gal}(\mathbb{Q}^{(n)}/\mathbb{Q})$ is canonically isomorphic to $(\mathbb{Z}/n\mathbb{Z})^\times$: indeed to $k \in (\mathbb{Z}/n\mathbb{Z})^\times$, one associates the unique automorphism σ_k of $\mathbb{Q}^{(n)}$, which satisfies $\sigma_k(e^{2\pi i/n}) = e^{2ki\pi/n}$. So,

$$\mathrm{Gal}(K/F) \cong \lim_{n} (\mathbb{Z}/n\mathbb{Z})^\times .$$

The reader will finish the argument, and prove that this inverse limit is (topologically) isomorphic to the group $\widehat{\mathbb{Z}}^\times$ of units of the ring $\widehat{\mathbb{Z}}$, introduced in the previous example. (Here the topology on $\widehat{\mathbb{Z}}^\times$ is induced from that of $\widehat{\mathbb{Z}}$; in Problem 3.8 we will explain that one cannot always topologize a group of units in this naive manner.) ■

Locally compact fields

Recall that a topological space X is called *locally compact* if, given any $x \in X$ and any open set V with $x \in V$, we may always find an open set U with $x \in U$ such that $\overline{U}$ is compact, and $\overline{U} \subset V$. For example, a metric space in which the closed balls are compact, such as $\mathbb{R}$ or $\mathbb{C}$, is locally compact.

The first point we want to make is that $\mathbb{Q}_p$ is locally compact. In fact, we offer the following lemma about the balls in $\mathbb{Q}_p$. Recall that we use the absolute value $|x|_p = p^{-\nu_p(x)}$, and the metric $d(x,y) = |x-y|_p$.

Lemma 3.26 *Let p be a prime number, and let $r > 0$. Let*

$$B(x_0,r) = \{x \in \mathbb{Q}_p : |x-x_0|_p < r\}$$

be the open ball of radius r centered at x_0, and let

$$B_c(x_0,r) = \{x \in \mathbb{Q}_p : |x-x_0|_p \leq r\}$$

be the closed ball of radius r centered at x_0. Then

1. *If $p^{-(n+1)} < r \leq p^{-n}$, then $B(x_0,r) = B_c(x_0,p^{-(n+1)})$. In particular, any open ball is a closed ball.*
2. *If $p^{-(n+1)} \leq r < p^{-n}$, then $B_c(x_0,r) = B(x_0,p^{-n})$. In particular, any closed ball is an open ball.*
3. *All balls, open or closed, are homeomorphic, and they are compact.*

Proof. (1) and (2) are obvious, given the definitions. As for (3), it suffices now to show that all open balls (say) are homeomorphic to one another; but of course the open ball with center x_0 and radius r is mapped onto the open unit ball centered at 0 under $x \mapsto (x-x_0)/r$. This is a self-homeomorphism of $\mathbb{Q}_p$, with inverse $x \mapsto x_0 + rx$.

Finally, note that the closed unit ball, centered at 0, is nothing but $\mathbb{Z}_p$, which is compact (Corollary 3.11). □

So $\mathbb{Q}_p$ is locally compact. In the rest of this section, we collect properties that hold for any field F equipped with an absolute value, so that it is locally compact for the corresponding topology – what we call a locally compact field for short. So F may be $\mathbb{R}$, $\mathbb{C}$, or $\mathbb{Q}_p$, for example. These properties must be familiar to the reader in the case of $\mathbb{R}$ or $\mathbb{C}$, and the proofs one normally gives in undergraduate courses about this should hold in the general situation with practically no changes. Thus we allow ourselves to be brief, simply refreshing our memories of the old arguments.

norm

Definition 3.27 Let F be a field with an absolute value, and let V be an F-vector space. A **norm** on V is a map

$$V \longrightarrow \mathbb{R}, \qquad v \mapsto \|v\|,$$

such that

1. $\|v\| \geq 0$, and $\|v\| = 0 \Leftrightarrow v = 0$;
2. $\|\lambda v\| = |\lambda| \cdot \|v\|$ for $\lambda \in F$, $v \in V$;
3. $\|v+w\| \leq \|v\| + \|w\|$, for $v, w \in V$ (the **triangle identity**).

We emphasize that a norm takes its values in $\mathbb{R}$, not in F – not a typo!

Theorem 3.28 *Let F be a locally compact field. Then F is complete.*

Let V be an F-vector space of finite dimension. Then any two norms on V are equivalent. In the topology induced by any norm, compact subsets of V coincide with closed, bounded subsets. The metric induced by any norm on V is complete.

Proof. Let $(u_n)_n$ be a Cauchy sequence in F. Then it is a bounded sequence, so it lives within a closed ball. All the closed balls are homeomorphic to one another, by the argument used above for $\mathbb{Q}_p$; since F is assumed to be locally compact, there are at least some closed balls which are compact, and thus they all are. A Cauchy sequence in a compact metric space must converge. This proves the first point.

Also, a closed and bounded subset X of F is contained in a closed ball, which we know now is compact, so X is itself compact. The converse is obvious, and so the compact subsets of F are precisely the closed, bounded subsets.

Now consider the vector space F^n, for some $n \geq 1$, with the norm

$$\|(x_1, \ldots, x_n)\|_1 = |x_1| + \cdots + |x_n|.$$

We see immediately that the corresponding metric space is complete, and that the induced topology is the product topology on F^n. It follows that the compacts subsets of F^n, for this topology, are precisely the closed, bounded subsets.

If we can prove that any other norm $\|\cdot\|$ on F^n is equivalent to $\|\cdot\|_1$, then everything will be proved, clearly.

We first note that there is a $c > 0$ such that $\|x\| \leq c\|x\|_1$ for $x \in F^n$, by a direct application of the triangle identity. The "second triangle identity", as is well known, is

$$|\,\|x\| - \|y\|\,| \leq \|x - y\| \qquad (\leq c\|x-y\|_1).$$

From this we deduce that the map $f\colon F^n \to \mathbb{R}$ defined by $f(x) = \|x\|$ is continuous, where F^n is endowed with the topology coming from $\|\cdot\|_1$.

Using this initially with $\|\cdot\| = \|\cdot\|_1$, we see that the unit sphere

$$S = \{x \in F^n : \|x\|_1 = 1\}$$

is closed, since it is $f^{-1}(\{1\})$. It is also clearly bounded, so it is compact.

Now return to $\|\cdot\|$ being any norm at all. The continuity of f together with the compactness of S shows the existence of m and M such that

$$m \leq f(x) \leq M$$

for all $x \in S$; what is more, these bounds are attained, so m is not 0. We see that we have proved that for any $v \in F^n$, we have

$$m\|v\|_1 \leq \|v\| \leq M\|v\|_1$$

(try $x = v/\|v\|_1$ when $v \neq 0$). This concludes the proof. □

We finally see why Proposition 2.27 is true. Indeed, in the notation of that proposition, we can see K as an F-vector space, and the absolute value $|\cdot|_K$ extending $|\cdot|_p$ is in fact a norm. So K is complete.

Moreover, the ring of integers $\mathcal{O}_K$ is the closed unit ball of K, so it is closed and bounded, and therefore compact by the theorem. It follows that K is itself locally compact.

Problems

3.1. Let G be a topological group. Prove that G is Hausdorff if and only if $\{1\}$ is closed, where 1 is the unit of G.

3.2. Let G be a topological group, and let N be a normal subgroup. In this problem we ask you to prove that G/N is a topological group. See the comments below for why this is not obvious.

1. Let Γ be any topological space, let $f\colon G \to \Gamma$ be a continuous map with $f(gn) = f(g)$ for $g \in G$, $n \in N$. Prove that the induced map $G/N \to \Gamma$ is continuous.
2. Let U be an open subset of G, and let $X \subset G$ be any subset at all. Prove that

$$UX := \{ux : u \in U, x \in X\}$$

is open in G. Deduce that, if Ω is an open set in $G \times G$, with the property that $\Omega(N \times N) = \Omega$, then Ω is a union of open sets of the form $U \times V$, with U, V open in G, and $UN = U$, $VN = V$.

3. Use the first question to deduce the existence of a continuous isomorphism

$$\theta\colon (G \times G)/(N \times N) \to G/N \times G/N.$$

Use the second question to deduce that θ is a homeomorphism.

4. Conclude.

Comments. If $\equiv$ is an equivalence relation on the topological space X, one defines a topology on $X/\equiv$ in the obvious way. Amazingly, if $\simeq$ is another equivalence relation on Y, and we use the product relation $(\equiv,\simeq)$ on $X\times Y$, it is NOT always true that the natural map

$$(X\times Y)/(\equiv,\simeq)\longrightarrow(X/\equiv)\times(Y/\simeq)$$

is a homeomorphism. See [Mun75, chapter 2, §22, example 7] for a counter-example. This holds true, however, if Y is locally compact, see [Mun84, Theorem 20.1]. The problem shows that, in the realm of topological groups, this issue does not come up.

3.3. Let $(G_i)_{i\in I}$ be a system of groups, with maps $\varphi_{ji}\colon G_j\to G_i$ for $i\le j$, and let $G=\lim_i G_i$. Let Γ be any group, and suppose we have homomorphisms $\theta_i\colon\Gamma\to G_i$ satisfying $\varphi_{ji}\circ\theta_j=\theta_i$. Show that there is a unique homomorphism $\theta\colon\Gamma\to G$ with $p_i\circ\theta=\theta_i$, where p_i is the obvious projection $G\to G_i$, for all $i\in I$.

This is called the "universal property of inverse limits".

Show that this characterizes G; that is, if G' is a group that has maps $p_i'\colon G'\to G_i$ such that $\varphi_{ji}\circ p_j'=p_i'$, and satisfying the same "universal property", then there is a canonical isomorphism $G\cong G'$.

3.4. Let $(X_n)_{n\ge0}$ be a system of finite sets, with maps $\varphi_{mn}\colon X_m\to X_n$ when $n\le m$.

1. Suppose that each X_n is nonempty, and that each φ_{mn} is surjective. Verify that $\lim_n X_n$ is nonempty. *This should use the axiom of choice.*
2. Remove the assumption on φ_{mn}, and prove the same thing.
   Hint: introduce $Y_n=\mathrm{Im}(\varphi_{mn})$ for all m large enough.
3. Deduce a new solution to Problem 2.5 from Chapter 2. In fact, you can easily generalize to several polynomials $f_1,\dots,f_s$.

This problem is adapted from [Ser73].

3.5. Complete the statement of the fundamental theorem of Galois theory: $[L:F]$ is finite if and only if $\mathrm{Gal}(K/L)$ has finite index in $\mathrm{Gal}(K/F)$, and in this case $[L:F]=[\mathrm{Gal}(K/F):\mathrm{Gal}(L/F)]$; the extension L/F is Galois if and only if $\mathrm{Gal}(K/L)$ is normal in $\mathrm{Gal}(K/F)$. *You may assume the finite-dimensional case!*

3.6. Let G be a topological group.

1. Without assuming that the topology on G is induced from a metric, give a definition of what it means for a sequence $(g_n)_{n\ge0}$ of elements of G to be Cauchy.
2. Show that a continuous homomorphism $G\to H$ maps Cauchy sequences to Cauchy sequences.
3. Show that, when G is profinite, its Cauchy sequences converge.

3.7. Recall that we write $\widehat{\mathbb{Z}}=\lim_n\mathbb{Z}/n\mathbb{Z}$, where the poset used is $\mathbb{N}^*$ under the divisibility relation. We may see $\mathbb{Z}$ as a subset of $\widehat{\mathbb{Z}}$ (why?), so we have a topology on $\mathbb{Z}$ (usually, this is just called the profinite topology).

1. Let Γ be any profinite group. Show that, for a fixed $x\in\Gamma$, the map $\mathbb{Z}\to\Gamma$ defined by $n\mapsto x^n$ is continuous for the above topology.

2. Show that the previous map can be extended to a continuous homomorphism $\widehat{\mathbb{Z}} \to \Gamma$. *One usually writes x^n even for $n \in \widehat{\mathbb{Z}}$. To show this, you may decide to use the previous problem – or not.*
3. (Application.) Show that

$$\widehat{\mathbb{Z}} \cong \prod_p \mathbb{Z}_p .$$

4. (A generalization.) Suppose Γ is an inverse limit of finite p-groups (p is a prime). Give a meaning to x^n for $x \in \Gamma$ and $n \in \mathbb{Z}_p$.
5. (Another generalization.) Generalize questions 1 and 2, with $\mathbb{Z}$ replaced by an abstract group G, and $\widehat{\mathbb{Z}}$ replaced by $\lim_U G/U$, where U runs through the normal subgroups of finite index in G. *This inverse limit is usually written $\widehat{G}$, and is called the profinite completion of G. It is NOT always true that $G \to \widehat{G}$ is injective.*

3.8. A *topological ring* is a ring A such that the underlying abelian group is a topological group (in additive notation!), and such that the multiplication $(x,y) \mapsto xy$ is a continuous map $A \times A \to A$. *It is not assumed that the map $A^\times \to A^\times$ given by $x \mapsto x^{-1}$ is continuous for the subspace topology.*

Show that $A^\times$ may be identified with the set

$$\{(x,y) \in A \times A : xy = 1\} .$$

Deduce that $A^\times$ may be turned into a topological group. If we do assume that $x \mapsto x^{-1}$ is continuous on $A^\times$, then show that the topology just defined is the subspace topology, induced from that of A.

Later in the book, we mention the ring of "adèles", whose units form the group of "idèles", and for these, the topology described in this problem must be used.

4 The multiplicative structure of local number fields

In this chapter we describe the multiplicative group $F^\times$ when F is a local number field. The treatment parallels the usual one for the field $\mathbb{R}$ of real numbers. Recall that

$$\mathbb{R}^\times = \{\pm 1\} \times \mathbb{R}^{>0} \cong \mathbb{Z}/2\mathbb{Z} \times \mathbb{R},$$

where the logarithm is used to provide an isomorphism between $\mathbb{R}^{>0}$ and $\mathbb{R}$. Consequently, we know that $\mathbb{R}^{>0}$ is divisible, that is, each element x has an nth root for each $n \geq 1$, which we may write as $x^{\frac{1}{n}}$. Using this, one can define x^λ for any $\lambda \in \mathbb{Q}$, and then by a density argument one defines x^λ for any $\lambda \in \mathbb{R}$.

The description of $F^\times$, when F is a p-local field, will be as a product of an infinite cyclic group, a finite cyclic group, and a few copies of $\mathbb{Z}_p$. The proof uses a version of the logarithm. As a result, precise information can be obtained on nth roots of elements. In passing, we define x^λ for some $x \in F^\times$ and all $\lambda \in \mathbb{Z}_p$.

Initial observations

We let p be a prime number, and let F be a p-adic field. The usual notation will be employed, with $\mathcal{O}_F$ for the ring of integers, and $\mathfrak{p} = (\pi)$ for the maximal ideal and a uniformizer; we let $e = e(F)$ be the ramification index and write v_p for the valuation on F which extends v_p on $\mathbb{Q}_p$, while $v = ev_p$ is the normalized valuation of F. We let q denote the number of elements in the residue field $\mathbb{F}$.

We simply write $U^{(n)}$ for the subgroup of $\mathcal{O}_F^\times$ which was denoted $U_F^{(n)}$ in Definition 2.55, that is, the group of elements which are 1 modulo $\mathfrak{p}^n$.

A remark that has not been made previously is that there is an isomorphism

$$\mathbb{F} = \mathcal{O}_F/\mathfrak{p} \longrightarrow \mathfrak{p}^n/\mathfrak{p}^{n+1},$$

for all $n \geq 0$, induced by $x \mapsto \pi^n x$. As a result, the group $\mathfrak{p}^n/\mathfrak{p}^{n+1}$ is of order q, and also the group $\mathcal{O}_F/\mathfrak{p}^n$ is of order q^n.

Of course, the latter is really a ring, and $\mathcal{O}_F \to \mathcal{O}_F/\mathfrak{p}^n$ is a ring homomorphism. Correspondingly, there is a map

$$\mathcal{O}_F^\times \longrightarrow (\mathcal{O}_F/\mathfrak{p}^n)^\times ,$$

whose kernel is just $U^{(n)}$. Now, we note that this is actually surjective, so that $(\mathcal{O}_F/\mathfrak{p}^n)^\times$ can be identified with $\mathcal{O}_F^\times/U^{(n)}$. Indeed, if $y \in \mathcal{O}_F/\mathfrak{p}^n$ is a unit, then it does not belong to the ideal $\mathfrak{p}/\mathfrak{p}^n$, and any element $x \in \mathcal{O}_F$ mapping to y must be in $\mathcal{O}_F \smallsetminus \mathfrak{p}$, which we know implies that x is invertible.

We concluded the previous chapter with the observation that F is locally compact, and indeed that its closed unit ball, that is $\mathcal{O}_F$, is compact. It is in fact profinite, as we proceed to show.

Proposition 4.1 *There are topological isomorphisms*

$$\mathcal{O}_F \longrightarrow \lim_n \mathcal{O}_F/\mathfrak{p}^n,$$

and

$$\mathcal{O}_F^\times \longrightarrow \lim_n \mathcal{O}_F^\times/U^{(n)},$$

as well as

$$U^{(1)} \longrightarrow \lim_n U^{(1)}/U^{(n)}.$$

It follows that the three groups $\mathcal{O}_F$, $\mathcal{O}_F^\times$, and $U^{(1)}$ are profinite, and in particular, compact.

Proof. Each map $\mathcal{O}_F \to \mathcal{O}_F/\mathfrak{p}^n$ is continuous, since $\mathfrak{p}^n$ is open in $\mathcal{O}_F$ (cf. Lemma 3.3). It is also surjective, so that the combined map $\mathcal{O}_F \to \lim_n \mathcal{O}_F/\mathfrak{p}^n$ has a dense image (Lemma 3.12). By compactness of $\mathcal{O}_F$, it is surjective. Its kernel is the intersection of all the ideals $\mathfrak{p}^n$, which can only contain elements of infinite valuation, so it is $\{0\}$. The proposed map is thus a continuous bijection, and again by compactness, it is a homeomorphism.

By inspection, we have in fact constructed an isomorphism *of rings*. We leave it to the reader to verify that

$$\left(\lim_n \mathcal{O}_F/\mathfrak{p}^n\right)^\times \cong \lim_n \left(\mathcal{O}_F/\mathfrak{p}^n\right)^\times ,$$

while we have shown above that $(\mathcal{O}_F/\mathfrak{p}^n)^\times \cong \mathcal{O}_F^\times/U^{(n)}$. The second topological isomorphism follows easily. The third is obtained by restriction. □

Proposition 4.2 *There is a topological isomorphism*

$$F^\times \cong \mathbb{Z} \times \mathbb{Z}/(q-1)\mathbb{Z} \times U^{(1)}.$$

It follows that the topological space $F^\times$ is locally compact. However, it is not compact, and not complete.

Proof. By definition $U^{(1)}$ is the kernel of $\mathcal{O}_F^\times \to \mathbb{F}^\times$, and $\mathbb{F}^\times$ is cyclic of order $q-1$, since $\mathbb{F}$ is a finite field with q elements. The polynomial $f = X^{q-1} - 1 \in \mathcal{O}_F[X]$

splits into distinct linear factors when reduced in $\mathbb{F}[X]$, so by Hensel's lemma, f is split in $\mathcal{O}_F[X]$. In other words, the group μ_{q-1} of $(q-1)$-st roots of unity contained in F is a subgroup of $\mathcal{O}_F^\times$ mapping isomorphically onto $\mathbb{F}^\times$. It follows that $\mathcal{O}_F^\times \cong \mu_{q-1} \times U^{(1)}$. This is really a topological isomorphism, with the discrete topology on the finite group μ_{q-1}, since the bijection $\mu_{q-1} \times U^{(1)} \to \mathcal{O}_F^\times$ given by $(\omega, u) \mapsto \omega u$ is continuous, and its source is compact.

The map $f\colon \mathbb{Z} \times \mu_{q-1} \times U^{(1)} \to F^\times$ mapping (n, ω, u) to $\pi^n \omega u$ is then a bijection, by the above and (4) of Proposition 2.17. The map f is continuous when the discrete topology is employed on $\mathbb{Z}$ as a product of three obviously continuous maps. Its inverse is given by the map $F^\times \to \mathbb{Z} \times \mathcal{O}_F^\times$, $x \mapsto (v(x), \pi^{-v(x)}x)$, followed by the homeomorphism above. This is continuous, so f is a homeomorphism itself.

Since $U^{(1)}$ is profinite by the previous proposition, and the discrete topology is used on the other two factors, we see that any open set in $\mathbb{Z} \times \mathbb{Z}/(q-1)\mathbb{Z} \times U^{(1)}$ containing the identity element $(0, 0, 1)$ also contains an open, compact subgroup (of the form $\{0\} \times \{0\} \times H$ with H open in $U^{(1)}$). Using translations, we see that $\mathbb{Z} \times \mathbb{Z}/(q-1)\mathbb{Z} \times U^{(1)} \cong F^\times$ is locally compact.

If $F^\times$ were complete, it would be closed in F, but $\pi^n \to 0$ as $n \to \infty$. As a result, $F^\times$ is not compact either, of course, but it seems better to see this directly as follows. For a given $n \in \mathbb{Z}$, let $F_n^\times = \{x \in F^\times : v(x) = n\}$. Each $F_n^\times$ is open, and this family of open sets covers $F^\times$. However, they are also disjoint and nonempty, so it is not possible to extract a finite subfamily that would still cover $F^\times$. □

It remains to describe $U^{(1)}$.

Exponential and logarithm

Consider $\mathbb{Q}[[X]]$, the ring of all formal power series, that is, formal sums

$$\sum_{n \geq 0} a_n X^n,$$

with $a_n \in \mathbb{Q}$. Besides adding and multiplying, it is also possible to consider $f(g(X))$ when $f(X) \in \mathbb{Q}[[X]]$ and $g(X) \in \mathbb{Q}[[X]]$, under the condition that $g(X)$ has no constant term. Indeed, if $f(X) = \sum a_n X^n$, the various power series

$$h_N = \sum_{n=0}^{N} a_n g(X)^n,$$

defined for $N \geq 0$, all have the same kth coefficient, as soon as N is comfortably larger than k. Define this to be the kth coefficient of $f(g(X))$.

This is a purely combinatorial construction. However, there is an absolutely trivial, and absolutely crucial, fact to be noted. Suppose we have a context in which convergence makes sense, say we have $x \in \mathbb{R}$ such that $g(x)$ makes sense as a convergent series in $\mathbb{R}$, call it $y = g(x) \in \mathbb{R}$, and suppose that $f(y)$ also makes sense. Then this number $f(g(x))$ is also obtained by looking at the series $f(g(X))$ just defined, which will converge at $X = x$, toward $f(g(x))$.

This is used often as a way of obtaining identities between formal series by using well-known identities of calculus. For example, consider the series

$$\exp(X) = \sum_{n \geq 0} \frac{X^n}{n!},$$

and the series

$$\log(1+X) = \sum_{n \geq 1} (-1)^{n-1} \frac{X^n}{n}.$$

(The notation $\log(1+X)$ is purely for intuition, we do not define log of anything but $1+X$.) One would like to prove that $\exp(\log(1+X)) = 1+X$. For this, we use the fact that for all $x > -1$, we can make sense of $\log(1+x)$; for any real number y, we can make sense of $\exp(y)$, so $\exp(\log(1+x))$ has a meaning; and of course, for these x, it is well-known that $\exp(\log(1+x)) = 1+x$.

By the remark just made, the formal power series $\exp(\log(1+X))$ can be evaluated at any $x > -1$, and the result of this evaluation is $1+x$. To conclude the identity $\exp(\log(1+X)) = 1+X$ in $\mathbb{Q}[[X]]$, we use another easy fact: If $f(X)$ and $g(X)$ are two power series such that $f(x) = g(x)$ for all x in an interval, then $f(X) = g(X)$ (say, by using Taylor's formula).

Example 4.3 Consider the following classic exercise. In the algebra $\mathrm{M}_r(\mathbb{C})$, there is an exponential map that converges everywhere, defined using the power series above. Let A be a nilpotent matrix ($A^k = 0$ for some k). Prove that there exists a matrix B such that $\exp(B) = I + A$ (where I is the identity matrix). Indeed, since A is nilpotent, we can certainly make sense of $B = \log(1+A)$. By the same remark again, but this time in $\mathrm{M}_r(\mathbb{C})$ rather than in $\mathbb{R}$, we know that $\exp(B) = \exp(\log(1+A))$ can be computed by taking the formal power series $\exp(\log(1+X))$, which is just $1+X$, and evaluating at A. So $\exp(B) = I + A$. ∎

The reader will prove, using similar arguments, that $\log(1+(\exp(X)-1)) = X$, or $\log(\exp(X)) = X$ with a mild abuse of notation. Perhaps more interestingly, the reader will also establish the following identities of formal power series in two variables (part of the exercise is to formalize what this means):

$$\log((1+X)(1+Y)) = \log(1+X) + \log(1+Y),$$

$$\exp(X+Y) = \exp(X)\exp(Y).$$

Now we apply this in the context of our p-adic field F.

Proposition 4.4 *For any $x \in \mathfrak{p}$, the power series $\log(1+X)$ converges at $X = x$. As a result, there is a continuous map*

$$\log\colon U^{(1)} \longrightarrow F,$$

which is a homomorphism. Moreover, we have $\log(U^{(n)}) \subset \mathfrak{p}^n$.

For all $n > \frac{e}{p-1}$, and any $x \in \mathfrak{p}^n$, the power series $\exp(X)$ *converges at* $X = x$, *to an element of* $U^{(n)}$. *As a result, there is a continuous map*

$$\exp\colon \mathfrak{p}^n \longrightarrow U^{(n)}.$$

It follows that, for $n > \frac{e}{p-1}$, the logarithm and exponential are reciprocal isomorphisms between $U^{(n)}$ and $\mathfrak{p}^n$.

In Problem 4.1, we invite you to prove this. Notice that the only question is really one of convergence. Also note that continuity of the logarithm is actually implied by $\log(U^{(n)}) \subset \mathfrak{p}^n$, and likewise for the exponential.

This proposition affords a description of $U^{(n)}$ for n large enough. Of course there is, for all n, a topological isomorphism $\mathcal{O}_F \to \mathfrak{p}^n$ given by $x \mapsto \pi^n x$. Moreover, by Proposition 2.36, we have $\mathcal{O}_F \cong \mathbb{Z}_p^d$ where $d := [F : \mathbb{Q}_p]$. So in the end, the group $U^{(n)}$ is topologically isomorphic to $\mathbb{Z}_p^d$ for all n large enough.

To deduce something about $U^{(1)}$, we need to develop a few things. As promised in the introduction, we will define x^λ for $x \in U^{(1)}$ and $\lambda \in \mathbb{Z}_p$, and treat $U^{(1)}$ as a $\mathbb{Z}_p$-module, in multiplicative (or indeed exponential) notation. The theory of such modules will be required.

Module structures

In order to prove Proposition 2.36, we had to assume as known some material about modules over a principal ideal domain (or just over a euclidean domain). Now we will need to use more of that theory. Namely, let M be a finitely generated module over such a ring R (for example $R = \mathbb{Z}_p$). Then $M \cong R^r \oplus T$ where T is isomorphic to the *torsion submodule* of M, that is $M_{tors} = \{m \in M : \lambda m = 0 \text{ for some } \lambda \in R \smallsetminus \{0\}\}$. The integer r depends only on M and is called its *rank*. Finally, if M_0 is a submodule of M, then it is finitely generated of rank $s \leq r$, and $s = r$ occurs if and only if M/M_0 is torsion. See any standard algebra textbook (again, we point out that [Stea] contains a very readable account (see sections 2.1 and 2.2 in particular), written for the ring $\mathbb{Z}$ but really working for any euclidean domain).

There is certainly a $\mathbb{Z}_p$-module structure on $\mathfrak{p}^n$, for any n, but we also need to put one on each $U^{(n)}$ for $n \geq 1$. In fact, we do this with $U^{(1)}$, and the groups $U^{(n)}$ will appear as submodules. The reader who has worked through Problem 3.7 will see our simple arguments as a particular case of a more general phenomenon.

We first note that, if A is any abelian group in multiplicative notation, and if there is some integer N such that $a^N = 1$ for all $a \in A$, then a^λ makes sense for $\lambda \in \mathbb{Z}/N\mathbb{Z}$ (and for any $a \in A$). This turns A into a $\mathbb{Z}/N\mathbb{Z}$-module.

For example, we have proved with Lemma 2.56 that the order of the group $U^{(n)}/U^{(n+1)}$ is q, for $n \geq 1$, so the order of $U^{(1)}/U^{(n+1)}$ is q^n. Thus $U^{(1)}/U^{(n+1)}$ is a module over $\mathbb{Z}/q^n\mathbb{Z}$.

We also have

$$U^{(1)} = \lim_{\overleftarrow{n}} U^{(1)}/U^{(n+1)}, \quad \text{and} \quad \mathbb{Z}_p = \lim_{\overleftarrow{n}} \mathbb{Z}/q^n\mathbb{Z},$$

as the reader will verify. So for $x = (x_n)_{n\geq 1} \in U^{(1)}$ and $\lambda = (\lambda_n)_{n\geq 1} \in \mathbb{Z}_p$, we can define x^λ to be $(x_n^{\lambda_n})_{n\geq 1}$. *It follows that $U^{(1)}$ is a $\mathbb{Z}_p$-module.* It is immediate that each subgroup $U^{(m)} \subset U^{(1)}$ is in fact a submodule.

To understand the module structure, it is best to keep two things in mind (rather than the definition). First, when $\lambda \in \mathbb{Z} \subset \mathbb{Z}_p$, then x^λ is really just a power of x or x^{-1} in the elementary sense. (This is really part of the definition of a "module", since we must have $x^1 = x$.) Second, for fixed $x \in U^{(1)}$, the homomorphism $\lambda \mapsto x^\lambda$ is continuous. To see this, observe that when q^n divides λ, the element $x^\lambda \in U^{(n+1)}$; the claim follows easily. Since $\mathbb{Z}$ is dense in $\mathbb{Z}_p$, these two facts are clearly all we need to know. Consider the next lemma.

Lemma 4.5 *Let G and H be abelian topological groups, which are also $\mathbb{Z}_p$-modules. Suppose that $\lambda \mapsto x^\lambda$ is continuous, for any $x \in G$ or any $x \in H$. Finally, let $\varphi\colon G \to H$ be continuous. Then φ is $\mathbb{Z}_p$-linear.*

Proof. Certainly $\varphi(x^\lambda) = \varphi(x)^\lambda$ for $\lambda \in \mathbb{Z}$, since φ is a homomorphism. By continuity, this extends to $\lambda \in \mathbb{Z}_p$. □

Corollary 4.6 *The isomorphism*

$$\log\colon U^{(n)} \longrightarrow \mathfrak{p}^n$$

for $n > \frac{e}{p-1}$, from Proposition 4.4, is $\mathbb{Z}_p$-linear. In particular, we have $U^{(n)} \cong \mathbb{Z}_p^d$ as a $\mathbb{Z}_p$-module, for the same values of n.

After these arguments, which were quite formal, we can prove the following.

Proposition 4.7 *The group $U^{(1)}$ is topologically isomorphic to $\mathbb{Z}/p^a\mathbb{Z} \times \mathbb{Z}_p^d$ for some $a \geq 0$.*

Proof. Recall that $U^{(1)}/U^{(n+1)}$ has finite order q^n, so it is torsion. Since $U^{(n+1)} \cong \mathfrak{p}^{n+1} \cong \mathbb{Z}_p^d$ for n large enough, we see that $U^{(1)}$ is finitely generated as a $\mathbb{Z}_p$-module, and by the material on modules over euclidean domains recalled above, we have $U^{(1)} \cong \mathbb{Z}_p^d \times T$, where T is isomorphic to the torsion in $U^{(1)}$.

The group T is finitely generated as a $\mathbb{Z}_p$-module, and it is torsion, so in the end T is finite. At the same time, T is comprised of the roots of unity contained in $U^{(1)}$, and so it is cyclic, as a finite subgroup of $F^\times$ where F is a field.

Finally, since $U^{(n+1)}$ is torsion-free, and the order of $U^{(1)}/U^{(n+1)}$ is a power of p, it follows that the order of T is a power of p. □

Summary and first applications

The next theorem summarizes the facts discovered in this chapter.

Theorem 4.8 *The topological group $F^\times$, with the topology induced from the valuation on the p-adic field F, is topologically isomorphic to*

$$\mathbb{Z} \times \mathbb{Z}/(q-1)\mathbb{Z} \times \mathbb{Z}/p^a\mathbb{Z} \times \mathbb{Z}_p^d,$$

where $d = [F : \mathbb{Q}_p]$, the residue field $\mathbb{F}$ has q elements, and $a \geq 0$ is some integer. Moreover, under this identification:

- *the map $F^\times \to \mathbb{Z}$ corresponding to the projection to the first factor is v, the normalized valuation;*
- *the subgroup $\{0\} \times \mathbb{Z}/(q-1)\mathbb{Z} \times \mathbb{Z}/p^a\mathbb{Z} \times \mathbb{Z}_p^d$ corresponds to $\mathcal{O}_F^\times$;*
- *the subgroup $\{0\} \times \{0\} \times \mathbb{Z}/p^a\mathbb{Z} \times \mathbb{Z}_p^d$ corresponds to $U^{(1)}$;*
- *the subgroup $\{0\} \times \mathbb{Z}/(q-1)\mathbb{Z} \times \mathbb{Z}/p^a\mathbb{Z} \times \{0\}$ corresponds to the roots of unity contained in F.*

Example 4.9 We can revisit Example 2.23. Let $F = \mathbb{Q}_p$, first with p odd. By the theorem, we have

$$\mathbb{Z}_p^\times \cong \mathbb{Z}/(p-1)\mathbb{Z} \times \mathbb{Z}/p^a\mathbb{Z} \times \mathbb{Z}_p,$$

for some integer a which we will determine in an instant. Before we even figure this out, we can answer again the question: When does $x \in \mathbb{Z}_p^\times$ have a square root in $\mathbb{Z}_p^\times$? Indeed, multiplication by 2 is an isomorphism on both $\mathbb{Z}/p^a\mathbb{Z}$ and on $\mathbb{Z}_p$, since p is odd. So an element of $\mathbb{Z}/(p-1)\mathbb{Z} \times \mathbb{Z}/p^a\mathbb{Z} \times \mathbb{Z}_p$ is divisible by 2 if and only if the same can be said of its projection onto $\mathbb{Z}/(p-1)\mathbb{Z}$. The kernel of this projection is identified with $U^{(1)}$ (still by the theorem), so we conclude that $x \in \mathbb{Z}_p^\times$ has a square root if and only if its image in $\mathbb{Z}_p^\times/U^{(1)} \cong \mathbb{F}^\times$ has one. This was exactly our finding, using Hensel's lemma, when we treated Example 2.23.

To find the value of a, we simply write that $(\mathbb{Z}/p^n\mathbb{Z})^\times \cong \mathbb{Z}/p^{n-1}(p-1)\mathbb{Z} \cong \mathbb{Z}/p^{n-1}\mathbb{Z} \times \mathbb{Z}/(p-1)\mathbb{Z}$, as is well known. Taking inverse limits, we see that $\mathbb{Z}_p^\times \cong \mathbb{Z}/(p-1)\mathbb{Z} \times \mathbb{Z}_p$, so $a = 0$ (as there is no p-torsion in $\mathbb{Z}_p^\times$).

For $p = 2$, it is classical that $(\mathbb{Z}/2^n\mathbb{Z})^\times \cong \mathbb{Z}/2\mathbb{Z} \times \mathbb{Z}/2^{n-2}\mathbb{Z}$. It follows that

$$\mathbb{Z}_2^\times = U^{(1)} \cong \mathbb{Z}/2\mathbb{Z} \times \mathbb{Z}_2.$$

To investigate the squares, we need the more precise statement of Proposition 4.4: for $n > e/(p-1)$, the logarithm gives an isomorphism $U^{(n)} \cong \mathfrak{p}^n$. Here $e = 1$ and $p = 2$, so for $n \geq 2$ we have $U^{(n)} \cong \mathbb{Z}_2$. And of course $U^{(n)}/U^{(n+1)}$ has order 2.

As a result, we have $U^{(1)} = \{\pm 1\} \times U^{(2)}$ (there is no torsion in $U^{(2)} \cong \mathbb{Z}_2$, so $-1 \notin U^{(2)}$, and $|U^{(1)}/U^{(2)}| = 2$). An element of $\mathbb{Z}_2^\times = U^{(1)}$ is thus a square if and only if it belongs to $U^{(2)}$, and is a square in $U^{(2)}$. Also, the only sub-$\mathbb{Z}_2$-module ($=$ ideal) of $\mathbb{Z}_2$ having index two is $2\mathbb{Z}_2$, which we translate as: The only submodule of $U^{(2)}$ having index two is the subgroup of squares. However, $U^{(3)}$ is also a submodule of index two, so $U^{(3)}$ is the subgroup of squares in $U^{(2)}$.

Our conclusion is that the squares in $\mathbb{Z}_2^\times$ comprise precisely the subgroup $U^{(3)}$, or in other words, the group of elements which are 1 modulo 8. ■

The index of $F^{\times 2}$ in $F^\times$ can always be computed easily. In fact, we have the following general result.

Lemma 4.10 *For any integer $n \geq 1$, the subgroup $(F^\times)^n$ of the p-adic field F has finite index in $F^\times$, and in fact this index is*

$$[F^\times : (F^\times)^n] = n \cdot |\mu_n(F)| \cdot p^{dr},$$

where $n = p^r m$ with m prime to p, and $\mu_n(F)$ is the group of nth roots of unity contained in F. (As above $d = [F : \mathbb{Q}_p]$.)

Proof. By the theorem, we have $F^\times \cong \mathbb{Z} \times \mu(F) \times \mathbb{Z}_p^d$, where $\mu(F)$ is the (finite) group of all roots of unity contained in F. The index of $n\mathbb{Z}$ in $\mathbb{Z}$ is n. The map $x \mapsto x^n$ on $\mu(F)$ has $\mu_n(F)$ as its kernel, which must have the same order as the cokernel $\mu(F)/\mu(F)^n$, or in other words, the index of $\mu(F)^n$ in $\mu(F)$ is $|\mu_n(F)|$. Finally, since m is invertible in $\mathbb{Z}_p$ we have $m\mathbb{Z}_p = \mathbb{Z}_p$, so $n\mathbb{Z}_p = p^r\mathbb{Z}_p$, so the index of $n\mathbb{Z}_p^d$ in $\mathbb{Z}_p^d$ is p^{dr}. □

The next result will be used much later in this book, when dealing with class field theory.

Lemma 4.11 *Let F be a local number field.*

1. *Any subgroup of finite index in $F^\times$ is closed (or equivalently, open).*
2. *Any open subgroup of $F^\times$ contains a subgroup $U^{(n)}$ for some n. Conversely each subgroup $U^{(n)}$ is open in $F^\times$.*
3. *Any subgroup of finite index in $F^\times$ contains a subgroup of the form $\langle \pi^m \rangle \times U^{(n)}$ for some $n, m \in \mathbb{N}$.*

Proof. (1) A little general fact: If H is a subgroup of the topological group G, and if H contains a closed subgroup U of finite index in G, then H is itself closed. Indeed, H is then the union of finitely many cosets of U.

So suppose H is a subgroup of finite index n in $G = F^\times$, and put $U = (F^\times)^n$, so that $U \subset H$. The subgroup U has finite index by the previous lemma, so it remains to prove that U is closed. For this, using the theorem, we need to see that multiplication by n has a closed image on $\mathbb{Z}$ (but this is trivial since the topology used is discrete), on a finite cyclic group (ditto), and on $\mathbb{Z}_p$ (where this is automatic by compactness of $\mathbb{Z}_p$).

(2) The group $U^{(n)}$ is compact, so it is closed in $\mathcal{O}_F^\times = U^{(0)}$; the latter is also compact of course, so $U^{(n)}$ is open in $\mathcal{O}_F^\times$. Since $F^\times$ is of the form (discrete group) $\times\ \mathcal{O}_F^\times$, we see that $U^{(n)}$ is also open in $F^\times$.

For the converse, it is enough so show that any open subgroup of $U^{(1)}$ contains some $U^{(n)}$. Since in $\mathbb{Z}_p$, any neighborhood of 0 contains some ideal (p^k) for some k, we see that any open set in $\mathbb{Z}_p^d$ containing 0 must contain a submodule of the form $(p^k\mathbb{Z}_p)^d$. As the module $U^{(1)}$ is topologically isomorphic to $\mathbb{Z}/p^a\mathbb{Z} \times \mathbb{Z}_p^d$, in the end we must exactly prove that given any $k \geq 0$, we will have $U^{(n)} \subset (U^{(1)})^{p^k}$ for some n.

Indeed, let us establish that, given n, m both chosen large enough, we have $U^{(n+m)} \subset (U^{(n)})^{p^k}$. For appropriate values of n, we have seen that the logarithm gives a topological isomorphism between $U^{(n)}$ and $\mathfrak{p}^n$ (Proposition 4.4). Writing that $\mathfrak{p}^{n+m} = \pi^m \mathfrak{p}^n$, and recalling that π is divisible by p in the ring $\mathcal{O}_F$, the result is then a direct translation to multiplicative notation.

(3) Let H have finite index in $F^\times$. By (1) we know that H is open, and by (2) it contains a subgroup of the form $U^{(n)}$ for some n. The valuation group $v(H)$ must be $m_0\mathbb{Z}$ for some $m_0 > 0$, since the index of H is finite, so H contains some element $\pi^{m_0}u$ where $u \in \mathcal{O}_F^\times$ is a unit. Let k be an integer such that $u^k \in U^{(n)}$, and let $m = km_0$. Then H contains $\pi^{km_0}u^k$, and finally $\pi^m \in H$. In the end, the group H contains both $\langle \pi^m \rangle$ and $U^{(n)}$, and the product of these two subgroups is a direct product, by the theorem. □

For much later use, we note the following.

Lemma 4.12 *Let $A \subset F^\times$ be a divisible subgroup, in the sense that for any $a \in F^\times$ and any integer $n \geq 1$, there exists $b \in A$ with $b^n = a$. Then $A = \{1\}$.*

Proof. A divisible subgroup of a cyclic group (finite or not) is trivial, as we see readily. The same is true for the group $\mathbb{Z}_p$, as an element infinitely divisible by p must have infinite valuation, and so must be 0. By the theorem, the subgroup A is trivial, since its various projections on the given factors are trivial. □

The number of extensions (is finite)

Theorem 4.13 *Let F be a local number field, let $n \geq 1$ be an integer, and let $\overline{F}$ be an algebraic closure of F. There are only finitely many Galois extensions K/F, with $K \subset \overline{F}$, for which $[K : F]$ divides n.*

Proof. We start with a very special case: Suppose that F contains the nth roots of unity, and suppose that K/F is an *abelian* extension, with $[K : F]$ dividing n. We prove that there are only finitely many such Ks. Indeed, by Kummer theory these extensions are in bijection with the subgroups of $F^\times/(F^\times)^n$; but the latter is a finite group by Lemma 4.10.

Next, we remove the hypothesis about the roots of unity. Let K/F be an abelian extension of degree dividing n, let F' be obtained from F by adjoining the roots of unity, and consider the composite KF'; the situation is like this.

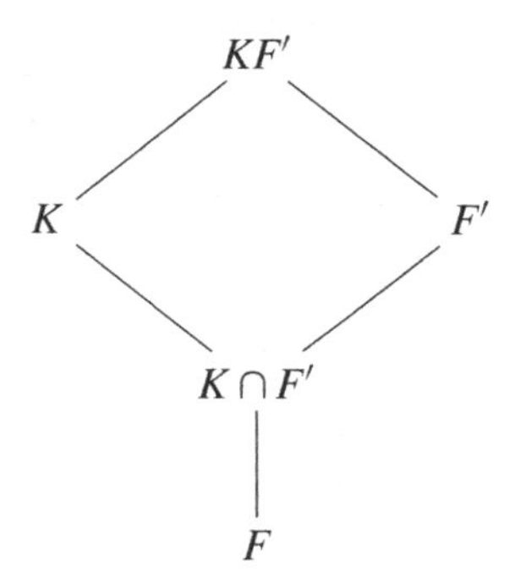

We apply Proposition 1.15. Since K/F is Galois, so is KF'/F', and we have $\mathrm{Gal}(KF'/F') \cong \mathrm{Gal}(K/K \cap F') \subset \mathrm{Gal}(K/F)$, so KF'/F' is an abelian extension of degree $[K : K \cap F']$ dividing n. By the case already treated, the field KF' is to be found among finitely many candidates, and K is one of the finitely many intermediate fields of KF'/F (recall that we are in characteristic 0, so all extensions are separable, and a finite separable extension is home to only finitely many intermediate fields). Thus our conclusion is the same: There are only finitely many possibilities for K.

Finally, we turn to the general case. Let K/F be an extension with $[K : F]$ dividing n. By Theorem 2.58, the group $\mathrm{Gal}(K/F)$ is solvable, so there is a chain of fields

$$K_0 = F \subset K_1 \subset K_2 \subset \cdots \subset K_s = K,$$

with K_{i+1}/K_i abelian, of degree $[K_{i+1} : K_i] > 1$ dividing $[K : F]$, and thus dividing n. The number s is bounded of course (by n, for example). By the cases treated, there are finitely many ways of picking K_{i+1} once K_i has been determined. In total, we find only finitely many Ks. □

Remark 4.14 One may of course rephrase this in various ways. For example, there are also finitely many extensions K/F with $[K : F] \leq n$: indeed the Galois closure of K/F, call it $\widetilde{K}/F$, has degree dividing $n!$ in this case, so there are finitely many possibilities for $\widetilde{K}$. For reasons given in the above proof, there are also finitely many possibilities for K itself.

Problems

4.1. Prove Proposition 4.4.

For a solution, see [Neu99, chapter II, section 5, propositions 5.4 and 5.5].

4.2. ***(Krasner's lemma).*** Let K/F be a finite Galois extension of p-adic fields. Let $x \in K$, and let $y \in K$ be such that $|y - x| < |y - \sigma(x)|$ for all $\sigma \in \mathrm{Gal}(K/F)$, $\sigma \neq 1$. (We use the absolute value corresponding to the valuation of K extending v_p.) Show that $x \in F(y)$.

Hint: What about $|y - \sigma(x)|$ for $\sigma \in \mathrm{Gal}(K/F(y))$?

4.3. In this problem, we let F be a field with an absolute value $|\cdot|$. No further assumptions are made.

1. Fix an integer $d \geq 1$ and a real number $M > 0$. Show that there exists $\rho > 0$ such that, whenever $P = a_0 + a_1X + \cdots + X^d$ is a monic polynomial of degree d with $|a_i| \leq M$ for all i, then each root $x \in F$ of P satisfies $|x| \leq \rho$.
2. We write $\|a_0 + a_1X + \cdots + a_dX^d\| := \max |a_i|$. Fix a monic polynomial P of degree d, which splits as a product of linear factors over F. Show that, for any given $\varepsilon > 0$, there exists $\alpha > 0$ with the following property: whenever $\|P - Q\| < \alpha$, where Q is monic of degree d, and whenever $y \in F$ is a root of Q, there exists a root x of P such that $|y - x| < \varepsilon$.

   Hint: Use the previous question and the following remark. For y, a root of Q, let x be the root of P which is closest to y. Then $|y - x|^d \leq |P(y)|$.

4.4. Let F be a local number field, and let $\overline{F}$ be an algebraic closure. Use the previous problems to show that when two irreducible polynomials $P, Q \in F[X]$ are close enough, in an appropriate sense, then their splitting fields (within $\overline{F}$) agree. Then, use the compactness of $\mathcal{O}_F$ to produce an alternative proof for Theorem 4.13.

This alternative proof is in fact classical.[1] *This problem also explains why the minimal polynomials, which the reader has observed during Example 2.32, have all their coefficients in* $\mathbb{Z}$, *rather than* $\mathbb{Z}_p$.

4.5. 1. Show Baire's theorem: In a complete metric space, a countable union of closed subsets, each of empty interior, also has empty interior; equivalently, the intersection of countably many open, dense subsets is again dense.

Hint: Let (V_n) *be the sequence of open, dense subsets, and let* W *be open. Construct a sequence of nested, closed balls* (B_n) *such that* $B_n \subset W \cap V_1 \cap \cdots \cap V_n$.

2. Let $\overline{\mathbb{Q}}_p$ be the algebraic closure of $\mathbb{Q}_p$. Show that v_p extends to $\overline{\mathbb{Q}}_p$, but that the corresponding metric is not complete.
3. Construct a field, equipped with an absolute value, which is complete and contains $\overline{\mathbb{Q}}_p$ as a dense subfield (extending its absolute value). Give a uniqueness statement. *This field is usually denoted by* $\mathbb{C}_p$.
4. Use Krasner's lemma (as in the previous problem) to show that $\mathbb{C}_p$ is algebraically closed.

The name $\mathbb{C}_p$ *is employed because one can show that* $\overline{\mathbb{Q}}_p$ *is isomorphic to* $\mathbb{C}$ *(and so* $\mathbb{C}_p$ *is a completion of* $\mathbb{C}$, *for some topology). Indeed, the classification of algebraically closed fields asserts that there is just one such field for each characteristic and each uncountable cardinality. This can be seen as the classical use of Morley's categoricity theorem, a deep result in logic; for this, see [Mar02].*[2] *But a more elementary argument is possible,*[3] *as follows: For any algebraically closed field* F, *let* F_0 *be the algebraic closure of the prime field in* F. *Pick a transcendence basis* $\{x_i : i \in S\}$, *so that* F *is an algebraic closure of the rational function field* $F(x_i : i \in S)$; *thus* F *is determined up to isomorphism by the characteristic of* F *(which determines* F_0*) and the cardinality of* S. *For* F *uncountable, the cardinality of* S *is equal to that of* F.

Finally, it is sometimes argued that the name $\mathbb{C}_p$ *is merely employed because* $\mathbb{C}_p$ *is both complete and algebraically closed, just like* $\mathbb{C}$.

[1] The author is indebted to Filippo Nuccio for pointing it out, and to Pierre Baumann for streamlining the presentation.
[2] This reference was kindly provided by Chloé Perin.
[3] An argument provided by an anonymous reviewer.

Part II

Brauer groups

In this part of the book we study *skewfields*, also known as *division algebras*, which are rings in which every nonzero element has an inverse. (Commutative skewfields are just fields.) As it turns out, the strongest results are obtained when we fix a field F and study all the skewfields whose center is F, and which are finite-dimensional over F. Perhaps surprisingly, these can be organized into an abelian group, called the *Brauer group* of F, after which Part II is titled.

We are led quickly to study algebras over F, rather than just skewfields over F, and the emphasis is on the so-called *semisimple* algebras. The theory of modules over semisimple algebras includes linear algebra (e.g. the fact that vector spaces over a field always have a basis), and we will see in passing that linear algebra over skewfields "just works" (but to convince oneself of this, it is best to think of the usual arguments over fields, and see that nothing needs to be changed).

More importantly, semisimple algebras must be understood by every student in *representation theory*. The basic facts about representations of a finite group G over the complex numbers, usually obtained through the study of characters, can also be derived from the general facts of the coming chapter. For representations over other fields (especially in positive characteristic), semisimple algebras are indispensable tools.

When K/F is a finite Galois extension, we will also define and study a group $\mathrm{Br}(K/F)$. The Brauer group $\mathrm{Br}(F)$ is the union of all the "relative" Brauer groups $\mathrm{Br}(K/F)$ (where K ranges through the fields contained in a fixed algebraic closure of F). The good news is that $\mathrm{Br}(K/F)$ can be described purely in combinatorial terms, starting from the finite group $\mathrm{Gal}(K/F)$. In fact, for each finite group G and each abelian group M with an action of G, we define a certain group $\mathrm{H}^2(G,M)$ very explicitly (it is also related to *extensions* of G). We then prove that $\mathrm{H}^2(\mathrm{Gal}(K/F),K^\times) \cong \mathrm{Br}(K/F)$, providing a powerful approach to $\mathrm{Br}(F)$.

As the notation suggests, there are also groups $\mathrm{H}^n(G,M)$ for all $n \geq 0$ (and even all $n \in \mathbb{Z}$). These are studied in the next part of the book.

This part concludes with the computation of $\mathrm{Br}(F)$ when F is a local number field. (Note that no material from Part I is needed until the last chapter of Part II.) We end up with an isomorphism $\mathrm{Br}(F) \cong \mathbb{Q}/\mathbb{Z}$, one of the highlights of the book. Class field theory, as developed in Part IV, crucially depends on this – but not on the specific understanding of skewfields over F. In a sense, in the sequel, the Brauer group itself is more important than the objects that are used to define it. For the time being, however, let us turn our attention to skewfields.

5 Skewfields, algebras, and modules

In this chapter the mains actors of Part II are introduced: skewfields and more generally algebras over a field, with special emphasis on the so-called semisimple algebras. We give several characterizations for these, some of them relying on the properties of the corresponding modules.

As some readers are perhaps not well acquainted with modules over a general ring (that is, not assumed commutative), we go over the basics. The purpose is to collect those facts that hold "just like they do in linear algebra". We caution that we put little emphasis on the definitions, which are carried over from the commutative case without change (such as the notions of submodule, quotient module, direct sum, and so on).

Skewfields and algebras

A skewfield is a "not necessarily commutative field". More precisely:

skewfield

division algebra

Definition 5.1 A ring K is called a **skewfield**, or **division algebra**, when every nonzero element of K has a (two-sided) inverse. Explicitly, for any $x \in K$, with $x \neq 0$, there must be an element $x^{-1} \in K$, easily seen to be unique, satisfying $xx^{-1} = x^{-1}x = 1$. ■

(To exclude the ring $\{0\}$ from this definition, we also require that $0 \neq 1$ in a skewfield.) If we added the requirement that K be commutative, we would have a field; and fields are particular skewfields. To discuss more examples, it will be convenient to use the language of algebras:

center

Definition 5.2 Let A be a ring. The **center** of A is

$$\mathcal{Z}(A) = \{z \in A : xz = zx \text{ for all } x \in A\}.$$

algebra

Let F be a commutative ring. We say that A is an **algebra** when there is a homomorphism $\iota\colon F \longrightarrow A$ such that $\iota(F) \subset \mathcal{Z}(A)$. ■

Our use of the letter F betrays our intention to mostly deal with the case when F is a field (in this case K is a vector space over F, which is often useful). Still, in this chapter we give some definitions in general – notice that any ring at all is an algebra over $\mathbb{Z}$, for instance. Also, in the vast majority of our examples, the map ι is injective, and in fact F is a subring of A.

The reader should prove the following.

Lemma 5.3 *The center of a ring is a commutative ring. The center of a skewfield is a field. Every skewfield K is an algebra over its center $F = \mathcal{Z}(K)$.* □

Some very important examples of skewfields (which are actually not fields) are constructed using the following proposition.

Proposition 5.4 *Let F be a field, and let $a,b \in F^\times$. There exists a unique algebra over F, written $\mathbb{H}_{a,b}$ or $(a,b)_F$, which is four-dimensional as an F-vector space, with a basis $1,i,j,k$ (where 1 is the unit of the algebra) satisfying*

$$i^2 = a1, \qquad j^2 = b1, \qquad ij = -ji = k.$$

An element $x1 + yi + zj + tk$, with $x,y,z,t \in F$, is invertible if and only if $x^2 - ay^2 - bz^2 + abt^2 \neq 0$.

This algebra is called the *quaternion algebra* with parameters a,b, the letter $\mathbb{H}$ being for Hamilton. In practice one does not write $x1$ when $x \in F$, but simply x, so F is viewed as a subring of $\mathbb{H}_{a,b}$. The rules are then $i^2 = a$ and $j^2 = b$, for example.

Proof. Uniqueness is immediate: The required properties of the multiplication tell us how to multiply any two elements, so any algebra satisfying these must be isomorphic to the one we are about to construct.

To prove the existence, we could start from the formulae, express the product of two generic elements, and then face the daunting task of proving all the properties for an algebra (think of associativity!). It will prove a little less computational, and more interesting, to reason as follows. Let us imagine that $\mathbb{H}_{a,b}$ exists. To any $q \in \mathbb{H}_{a,b}$, let us associate the multiplication map $L_q \colon \mathbb{H}_{a,b} \to \mathbb{H}_{a,b}$ with $L_q(x) = qx$. Then $q \mapsto L_q$ gives an isomorphism between $\mathbb{H}_{a,b}$ and an algebra of linear operators on a four-dimensional vector space, and if we take a basis, say $1,i,j,k$, we even have an isomorphism between $\mathbb{H}_{a,b}$ and a subalgebra of $\mathrm{M}_4(F)$ (the 4×4 matrices with entries in F). If $q = x1 + yi + zj + tk$, we find that the matrix of L_q is

$$\begin{pmatrix} x & ya & zb & -abt \\ y & x & tb & -bz \\ z & -at & x & ay \\ t & -z & y & x \end{pmatrix}. \tag{*}$$

Having made this analysis, we now *define* $\mathbb{H}_{a,b}$ to be the set of matrices of the form (*), for $x,y,z,t \in F$. We call i the matrix obtained for $x = z = t = 0$ and $y = 1$, and similarly for j and k, and we allow ourselves to write 1 for the identity matrix.

It is clear that $\mathbb{H}_{a,b}$ is a four-dimensional vector space over F, with basis $1, i, j, k$; one computes the products $i^2 = a1, j^2 = b1$ and $ij = -ji = k$. It follows that $\mathbb{H}_{a,b}$ is stable under matrix multiplication, and is thus an F-algebra satisfying our requirements.

For any "quaternion" $q = x1+yi+zj+tk \in \mathbb{H}_{a,b}$, its *conjugate* is $\overline{q} = x1-yi-zj-tk$. We compute $q\overline{q} = \overline{q}q = (x^2 - ay^2 - bz^2 + abt^2)1$, an element of $F \subset \mathbb{H}_{a,b}$. If $q\overline{q} = 0$, certainly q cannot be invertible, for we would deduce $\overline{q} = 0$ and thus $q = 0$. On the other hand, if $f = q\overline{q} \neq 0$, then the element $\frac{1}{f}\overline{q}$ (which makes sense since $\mathbb{H}_{a,b}$ is an F-algebra, of course) is an inverse for q. □

Example 5.5 Take $F = \mathbb{R}$ and $a = b = -1$. In this case we write simply $\mathbb{H}$ for $\mathbb{H}_{-1,-1} = (-1,-1)_{\mathbb{R}}$, and call it the algebra of (ordinary) quaternions. Since the expression $x^2 + y^2 + z^2 + t^2$ is always > 0 when one of x, y, z, t is nonzero, we see that $\mathbb{H}$ is a skewfield, our first genuine example. Of course $(-1,-1)_{\mathbb{Q}}$ is also a skewfield, but $(-1,-1)_{\mathbb{C}}$ is not. ■

As it happens, when we attempt a study of skewfields, the object which we understand the best is the following.

Definition 5.6 Let F be a field. We write $\mathrm{Br}(F)$ for the set of isomorphism classes of skewfields whose center is precisely F, and which are finite-dimensional over F. We call it the **Brauer group** of F. ■

Brauer group

Example 5.7 When F is algebraically closed, there is only one element in $\mathrm{Br}(F)$, namely the class of F. Indeed, if $x \in K$, where K is a skewfield as above, then consider $F(x)$, the smallest skewfield containing F and x; then $F(x)$ is actually a field, since its elements are rational fractions in x with coefficients in F, and x commutes with the elements of F. Moreover $F(x)$ is finite-dimensional over F because K is. So $F(x)/F$ is algebraic and thus $F(x) = F$, and $K = F$. ■

The results we are about to obtain include:

- $\mathrm{Br}(F)$ is actually an abelian group. Proving this will require us to temporarily give an alternative definition of $\mathrm{Br}(F)$, involving F-algebras, before ultimately proving that the two definitions agree.
- $\mathrm{Br}(F)$ is trivial (that is, contains only the class of F) when F is a finite field. This is known as *Wedderburn's theorem*, which says, more plainly, that a finite skewfield must be commutative. Our proof will definitely be different from the elementary ones that are often given in undergraduate courses.
- $\mathrm{Br}(\mathbb{R})$ contains just two elements, the classes of $\mathbb{R}$ and $\mathbb{H}$.
- When F is a local number field, the group $\mathrm{Br}(F)$ is isomorphic to $\mathbb{Q}/\mathbb{Z}$. This is much harder, and is one of the most important results in this book.

We conclude this section by introducing a class of algebras that is important for motivational reasons. Indeed, while we are essentially compelled to study algebras in general in order to understand skewfields, they are certainly interesting in their own right. In this part of the book we shall prove many results that are normally part of a course on the *representation theory of groups*, simply because our theorems on algebras can be applied in the following special case.

Let G be a group (in multiplicative notation), and let F be a commutative ring. We let $F[G]$ denote the free F-module over the set G, whose elements are formal sums

$$\sum_{g\in G} \alpha_g g, \tag{*}$$

with $\alpha_g \in F$, and only finitely many of the coefficients α_g are nonzero. Alternatively, we could define $F[G]$ to be the F-module of functions $\alpha\colon G \to F$, which take nonzero values for only finitely many different elements of G, and view (*) as a suggestive way of writing the function α with $\alpha(g) = \alpha_g$.

We define a multiplication on $F[G]$ by the formula

$$\left(\sum_{g\in G} \alpha_g g\right)\cdot\left(\sum_{h\in G} \beta_h h\right) = \sum_{\sigma\in G}\left(\sum_{g,h \text{ such that } gh=\sigma} \alpha_g\beta_h\right)\sigma\,.$$

One checks all the axioms for a ring. Elements of G can be seen as elements of $F[G]$ in the obvious way, and the identity element $1 \in G$ also serves as the unit for the ring $F[G]$; also, the multiplication on $F[G]$ extends the multiplication on G. Of course, the previous sentence determines the multiplication entirely. The definition is arrived at by requiring distributivity – note that the product above is also equal to

$$\sum_{g,h\in G} \alpha_g\beta_h gh\,,$$

an actual (as opposed to formal) sum.

With the "function" point of view, the product of α and β is the function $G \to F$ defined by:

$$(\alpha\beta)(\sigma) = \left(\sum_{g,h \text{ such that } gh=\sigma} \alpha(g)\beta(h)\right) = \sum_{g\in G} \alpha(g)\beta(g^{-1}\sigma)\,,$$

sometimes called the *convolution product* of α and β.

We can identity F with the set of elements of the form $f1$ with $f \in F$, and we see that $F[G]$ is actually an F-algebra. It is called the *group algebra* of G (over F).

Example 5.8 When G is infinite cyclic, generated by an element T (so G is isomorphic to $\mathbb{Z}$, but in multiplicative notation), the group algebra $F[G]$ is isomorphic to the ring $F[T,T^{-1}]$ of Laurent polynomials. When G is cyclic of order n, the group algebra is $F[T]/(T^n - 1)$, which by the Chinese Remainder theorem splits as a direct sum of rings corresponding to the factorization of $T^n - 1$ over F. We shall see that, at least when G is finite, the algebra $F[G]$ is never a skewfield, unless G is trivial and F is a field. ■

Modules and endomorphisms

Let A be a ring. We recall the familiar definition of an A-module (so as to compare it with another definition below): An A-module is an abelian group V together with a map $A \times V \to V$, written $(a,v) \mapsto a \cdot v$, which is bilinear, and satisfies

$$a \cdot (b \cdot v) = (ab) \cdot v.$$

Also, it is required that $1 \cdot v = v$, where 1 is the unit of A.

Of course when A is a field, an A-module is just an A-vector space. We note that, when A is an algebra over the ring F, and so is equipped with $\iota\colon F \to A$, then an A-module can also be seen as an F-module, and we shall write $f \cdot v$ for this structure, rather than $\iota(f) \cdot v$, when $f \in F$, $v \in V$.

In fact, it is good to think of modules over an F-algebra A as F-modules "with an extra structure". For example, let $A = F[T]$ be a polynomial ring over the field F; then an A-module is precisely an F-vector space V with a distinguished F-linear map $V \to V$, given by $v \mapsto T \cdot v$. Likewise, if F is a field and $A = F[G]$ is the group algebra of the group G, then an A-module is an F-vector space V endowed with an action of the group G by F-linear maps; or in other words, there is a group homomorphism $G \to \mathrm{GL}(V)$. This is sometimes called a *representation* of G over F.

We assume that the reader is familiar with the operations of direct sums and direct products, and with the notions of submodule and quotient module. (Recall that finite direct sums are the same as finite direct products, and in this book this is essentially all we need.)

The proper name for modules, actually, ought to be *left modules*. There is also a notion of *right module* over the ring A, and that is an abelian group V with a map $V \times A \to V$, written $(v,a) \mapsto v \cdot a$ or v^a, which is bilinear and satisfies

$$(v \cdot a) \cdot b = v \cdot (ab) \quad \text{or} \quad (v^a)^b = v^{ab}.$$

Also required is $v \cdot 1 = v$.

In order to give examples of right modules, we show that they are left modules in disguise.

opposite ring

Definition 5.9 Let A be a ring, with multiplication written $a \cdot b$ for $a, b \in A$. The **opposite ring**, denoted by A^{op}, is the same underlying group endowed with the new multiplication $a \star b$, defined by

$$a \star b = b \cdot a.$$

One checks that the opposite ring is indeed a ring. ∎

Example 5.10 A ring A is commutative if and only if $A = A^{op}$. The ring $\mathrm{M}_n(F)$ of matrices with entries in the field F is not commutative for $n \geq 2$, so $\mathrm{M}_n(F) \neq \mathrm{M}_n(F)^{op}$; however, these two rings are isomorphic. Indeed the map $M \mapsto M^t$ (the transpose of M) satisfies $(MN)^t = N^t M^t$, and so it gives an isomorphism $\mathrm{M}_n(F) \to \mathrm{M}_n(F)^{op}$. A little more subtle is the fact that, when K is a skewfield, trying to play the same trick only gives us an isomorphism $M_n(K) \to M_n(K^{op})^{op}$. Since we always have $(A^{op})^{op} = A$, we can apply this to K^{op}, showing that there is an isomorphism $M_n(K^{op}) \to M_n(K)^{op}$, which is not the identity of the underlying sets. ∎

The connection with right modules is given by:

Lemma 5.11 *A right module over A is exactly the same thing as a left module over A^{op}.*

Proof. Given a right A-module V, define $a \cdot v := v \cdot a$ for $a \in A$, $v \in V$. We have

$$a \cdot (b \cdot v) = (v \cdot b) \cdot a = v \cdot (ba) = (ba) \cdot v = (a \star b) \cdot v,$$

where $\star$ is the multiplication of A^{op}. This turns V into a left A^{op}-module. The same trick converts a left A^{op}-module into a right $(A^{op})^{op} = A$-module. □

So it seems that we can avoid talking about right modules entirely, and indeed in this book they will be seldom mentioned – at the cost of having the "op" symbol here and there. At the same time, the concept is natural, and most examples are usually variants of this one: let R be a ring and $A \subset R$ be a subring, then R is a right A-module, using right multiplication. (Of course, it is also a left A-module using left multiplication.)

Finally, we will need to talk about bimodules:

bimodule

Definition 5.12 Let A and B be rings. An A–B-**bimodule** is an abelian group V which is simultaneously an A-module, with the operation written $a \cdot v$ for $a \in A$, $v \in V$ and a B-module, with the operation written $b \star v$ for $b \in B$, and such that the two operations commute, that is

$$a \cdot (b \star v) = b \star (a \cdot v).$$

∎

Example 5.13 In the case when A is a subring of R, then R is an A–A^{op}-bimodule. The actions are $a \cdot r = ar$ and $a \star r = ra$. Generally speaking, an A–B^{op}-bimodule is simultaneously a left A-module and a right B-module, with obvious compatibility. ∎

We point out that, in practice, operations are often written without a symbol at all, that is av or bv, and the requirement is $abv = bav$. We also warn the reader that some authors would speak of an A–B-bimodule for the objects which, in this book, are called A–B^{op}-bimodules.

We turn to the study of linear maps.

Definition 5.14 Let V and W be two A-modules. We write $\mathrm{Hom}_A(V, W)$ for the set of all group homomorphisms $\varphi\colon V \to W$ which are A-linear, in the sense that $\varphi(a \cdot m) = a \cdot \varphi(m)$ for $a \in A$, $m \in V$. We also write $\mathrm{End}_A(V) = \mathrm{Hom}_A(V, V)$, and *endomorphisms* call its elements **endomorphisms** of V. Two modules V and W are called isomorphic when there is a group isomorphism $\varphi\colon V \to W$ which is A-linear (φ^{-1} is then automatically A-linear).

The first time one encounters noncommutative rings, one is usually not aware that $\mathrm{Hom}_A(V, W)$ may not be an A-module itself. The only thing one can in fact claim is this:

Lemma 5.15 *The set* $\mathrm{Hom}_A(V, W)$ *is an abelian group. If A is an F-algebra over the (commutative) ring F, then* $\mathrm{Hom}_A(V, W)$ *is an F-module.*

Proof. The first sentence is left for the reader to prove. If $\varphi \in \mathrm{Hom}_A(V, W)$ and $f \in F$, let $f\varphi$ be the map $V \to W$ defined by $(f\varphi)(m) = f \cdot \varphi(m)$. Then for any $a \in A$ we have

$$(f\varphi)(a \cdot m) = f \cdot (\varphi(a \cdot m)) = f \cdot (a \cdot \varphi(m)) = (fa) \cdot \varphi(m) = (af) \cdot \varphi(m) = a \cdot (f\varphi)(m).$$

Note how we have used $fa = af$, since the elements of F commute with the elements of A, by definition of an F-algebra. So $f\varphi$ is A-linear, i.e., it is an element of $\mathrm{Hom}_A(V, W)$. The association $(f, \varphi) \mapsto f\varphi$ turns $\mathrm{Hom}_A(V, W)$ into an F-module. □

For example, the module $\mathrm{End}_A(V)$ always contains the identity, and so also contains all maps of the form $m \mapsto f \cdot m$ for $f \in F$, called scalar transformations. However, a map of the form $m \mapsto a \cdot m$ for $a \in A$ may very well fail to be A-linear, when A is not commutative.

One can also put a ring structure on $\mathrm{End}_A(V)$. For $\varphi_1, \varphi_2 \in \mathrm{End}_A(V)$, consider the composition $\varphi_1 \circ \varphi_2$, satisfying $\varphi_1 \circ \varphi_2(x) = \varphi_1(\varphi_2(x))$. One checks easily that $\mathrm{End}_A(V)$ becomes a ring, and it is also an F-module if A is an F-algebra; we leave it to the reader to verify that $\mathrm{End}_A(V)$, in this case, is really an F-algebra itself.

Example 5.16 As a simple example of use of endomorphism rings, consider the case when A is an F-algebra. Then every A-module is in particular an F-module. Moreover, for each $a \in A$ the map $L_a\colon A \to A$ defined by $L_a(x) = a \cdot x$ is F-linear, so belongs to $\mathrm{End}_F(V)$. The association $L\colon A \to \mathrm{End}_F(V)$ mapping a to L_a is then an F-algebra homomorphism, as a consequence of the definitions. (We have used it already, during the proof of Proposition 5.4.) Conversely, an F-module V for which there exists such a map $L\colon A \to \mathrm{End}_F(V)$ can be turned into an A-module, with $a \cdot v = L_a(v)$. This makes our initial description of A-modules as "F-modules with extra structure" more precise.

Example 5.17 Another exercise in definitions is this: When V is an A-module and $B = \mathrm{End}_A(V)$, then V is naturally an A–B-bimodule. In fact, B is in some sense the largest ring for which this is true. The B-module structure is simply given by $\varphi \cdot v = \varphi(v)$, of course.

Let us study a particular case. Recall that for $n \geq 1$ the notation A^n denotes the direct product of n copies of A, which is an A-module with $a \cdot (a_1, \ldots, a_n) = (aa_1, \ldots, aa_n)$. We recall this merely to justify our use of the notation A^1 to denote A itself, but considered as an A-module. Having a distinction between the ring A and the module A^1 will clarify things. Now, we trust that after reading this chapter, the reader will be convinced that the module A^1, called the *regular module*, is central to the study of all the others.

The importance of the next lemma can hardly be overstated.

Lemma 5.18 *For $a \in A$, define $\varphi_a \colon A^1 \to A^1$ by $\varphi_a(x) = xa$. Then $\varphi_a \in \mathrm{End}_A(A^1)$, and conversely any element of $\mathrm{End}_A(A^1)$ is of the form φ_a for a unique $a \in A$.*

The bijection $A \to \mathrm{End}_A(A^1)$ given by $a \mapsto \varphi_a$ is an isomorphism of abelian groups, and of F-modules if A is an F-algebra, but it satisfies

$$\varphi_{ab} = \varphi_b \circ \varphi_a$$

for $a, b \in A$. In other words, there is an isomorphism of rings

$$\mathrm{End}_A\left(A^1\right) \cong A^{op}\,.$$

Proof. If $\varphi \in \mathrm{End}_A(A^1)$, we have $\varphi(x) = \varphi(x1) = x\varphi(1)$, so if we put $a = \varphi(1)$, we have $\varphi = \varphi_a$. The existence of the bijection follows easily. And we note

$$\varphi_{ab}(x) = xab = (xa)b = \varphi_b(\varphi_a(x))\,,$$

as announced. □

We deduce:

Lemma 5.19 *Let V be an A-module, and let $B = \mathrm{End}_A(V)$. Consider V^n, the direct product of n-copies of V, for some $n \geq 1$. Then*

$$\mathrm{End}_A\left(V^n\right) \cong \mathrm{M}_n(B)\,,$$

where the right-hand side is the algebra of $n \times n$-matrices with entries in B. This is an isomorphism of F-algebras when A is an F-algebra. Similarly there is an isomorphism

$$\mathrm{Hom}_A\left(V^n, V^m\right) \cong \mathrm{M}_{n,m}(B)\,,$$

where the right-hand side is the algebra of matrices with entries in B, having n columns and m rows.

Proof. We prove just the first case, for simplicity.

If v is an element of V^n, and M is a matrix in $\mathrm{M}_n(B)$, we can make sense of Mv: We see elements of V^n as column-matrices with entries in V, and we employ the usual matrix multiplication, using that V is a B-module. We have immediately $M(av) = a(Mv)$, since the actions of A and B commute. So M defines an element of $\mathrm{End}_A(V^n)$.

and we have a map $\mathrm{M}_n(B) \to \mathrm{End}_A(V^n)$. It is a ring homomorphism, simply because $(M_1M_2)v = M_1(M_2v)$ (associativity of the matrix product).

To construct an inverse, let $\iota_k\colon V \to V^n$ be the inclusion via the k-th coordinate, and let $p_k\colon V^n \to V$ be the kth projection, for $1 \le k \le n$. To $\varphi \in \mathrm{End}_A(V^n)$, we associate the matrix $(p_i \circ \varphi \circ \iota_j)_{ij}$. This gives a map $\mathrm{End}_A(V^n) \to \mathrm{M}_n(B)$, which is an inverse for that of the previous paragraph. □

It is wise to test these results against what we know in the following familiar cases.

Example 5.20 Suppose $A = F$ is a field. Lemma 5.18 says that the linear maps of the vector space F^1 are the scalar transformations $x \mapsto xf$, with $f \in F$. Lemma 5.19, on the other hand, shows that the endomorphisms of F^n form an algebra isomorphic to $\mathrm{M}_n(F)$, and so the same can be said of any n-dimensional vector space over F (which, by linear algebra, is isomorphic to F^n). Incidentally, let C be the space of $n \times 1$-column-matrices. Then the proof of Lemma 5.19 shows directly that any F-linear endomorphism of C is given by multiplication by a matrix on the left. ■

Example 5.21 Now suppose F is a field and take $A = \mathrm{M}_n(F)$. Combining Lemma 5.18 and Example 5.10, we have $\mathrm{End}_A(A^1) \cong A^{op} \cong \mathrm{M}_n(F)$. Here is a remark about the module A^1: When we multiply two matrices M and N, what we do is multiply M by each of the columns of N, and place the resulting columns side-by-side; this remark shows that the module A^1 is in fact isomorphic to C^n, where C is the space of $n \times 1$-matrices as above, viewed as an A-module.

So Lemma 5.19 applies to $A^1 = C^n$, yielding

$$A = \mathrm{M}_n(F) \cong A^{op} \cong \mathrm{End}_A\left(C^n\right) \cong \mathrm{M}_n(B),$$

with $B = \mathrm{End}_A(C)$. We certainly expect that B is just F here, and indeed, we can see this trivially as follows: B is an F-algebra, so an F-vector space, and a dimension count on both sides of the isomorphism $\mathrm{M}_n(F) \cong \mathrm{M}_n(B)$ shows that $B = F$.

We can also see this more directly. An A-linear endomorphism of C is in particular F-linear. We have revisited the fact, in the course of the previous example, that the F-linear maps of C are given by matrix multiplication on the left. Therefore, any $\varphi \in \mathrm{End}_A(C) \subset \mathrm{End}_F(C)$ is in fact given by $x \mapsto ax$ with $a \in A = \mathrm{M}_n(F)$. The A-linearity of φ expresses the fact that a is actually in the center of $\mathrm{M}_n(F)$. It is a traditional exercise to show that this center consists of scalar matrices, that is, matrices of the form fI, where I is the identity matrix, and $f \in F$. The dimension count just presented provides an unusual solution to this exercise. ■

Semisimplicity

Our objective is the study of skewfields, and we have argued that the study of algebras is then forced upon us. In truth, it will suffice to restrict attention to the so-called semisimple algebras, which we define below. We start with the related topic of semisimple modules. In this section, A is a ring.

simple

Definition 5.22 A nonzero A-module V is called **simple** if it does not have any submodules, beside $\{0\}$ and V itself.

We often say that a submodule of V is *proper* when it is neither $\{0\}$ nor V.

Example 5.23 Suppose $A = F$ is a field. Then any one-dimensional vector space over F is obviously simple. More generally, suppose that A is an F-algebra, where F is a field. Any A-module is an F-vector space, so again, an A-module V which is one-dimensional over F is simple. Consider, for instance, the case $A = F[G]$ where G is a group, and recall that an $F[G]$-module is a vector space over F together with an F-linear action of G. Then F itself can be endowed with a trivial action (elements of G acting as the identity); the result is called the *trivial G-module*, and it is simple.

On the other hand, with the same A, consider the module A^1, and assume that G is finite for simplicity. Consider $v = \sum_{g\in G} g \in A^1$. Then the module spanned by v is isomorphic to the trivial one, and in particular it has dimension 1 over F. As soon as G has order > 1, this shows that A^1 is not simple.

Also, when $A = \mathbb{Z}$, the module A^1 has infinitely many proper submodules.

semisimple

Definition 5.24 An A-module V is called **semisimple** when it is (isomorphic to) a direct sum of simple modules.

Example 5.25 When $A = F$ is a field, the familiar fact that every F-vector space V possesses a basis means that V is semisimple.

Another classic example is the following. Let F be a field, let G be a finite group, and let $A = F[G]$. Crucially, *assume that the characteristic of F does not divide the order of G*. Suppose we are given an A-module V with a submodule W. Linear algebra provides us with a projection map $p\colon V \to W$, far from unique, which is F-linear, surjective, and satisfies $p \circ p = p$. Now the trick is to use the average of several projectors derived from p, namely define $\pi : V \to W$ by

$$\pi(v) = \frac{1}{|G|} \sum_{g\in G} g \cdot p\left(g^{-1} \cdot v\right),$$

an expression that makes sense because the order of G is invertible in F by assumption. One checks that π maps V into W and is the identity on W, so it is a projection much like p is, but this time π is G-equivariant, or in other words is A-linear. If $W' = \ker(\pi)$, then one has $V = W \oplus W'$, and W' is an A-module.

If we assume that V is finitely generated as an A-module (that is, V is a quotient of A^n for some $n \geq 1$), then it is finite-dimensional over F. This allows us to argue as follows. If V is not simple, let W_1 be a proper submodule, which is itself simple (there must be one, for reasons of dimensions). Then $V = W_1 \oplus W_1'$. If W_1' is not simple, write it as $W_1' = W_2 \oplus W_2'$ where W_2 is simple, by the same token, so that $V = W_1 \oplus W_2 \oplus W_2'$; and keep going with W_2' and so on. This game stops after finitely many steps, again for reasons of dimension, and we see that V is semisimple. We have shown that *all finitely generated $F[G]$-modules are semisimple*, under our hypotheses.

To give an example of a module that is not semisimple, let A be the algebra of 2×2-matrices with coefficients in the field F, such that the bottom-left coefficient is 0. Let V be the vector space of 2×1-matrices, viewed as an A-module via matrix multiplication. We leave it as an exercise to the reader to prove that V is not semisimple. ▪

The next result is the celebrated "Schur's lemma".

Proposition 5.26 *Suppose that V is a simple A-module. Then* $\mathrm{End}_A(V)$ *is a skew-field.*

Moreover, suppose U and V are two simple A-modules, which are not isomorphic. Then $\mathrm{Hom}_A(U,V) = \{0\}$.

Proof. Let $\varphi \in \mathrm{End}_A(V)$ be nonzero; we must prove that φ has an inverse. Consider $\ker(\varphi)$, which is a submodule of V: By simplicity, one has either $\ker(\varphi) = \{0\}$ or $\ker(\varphi) = V$. But the latter implies $\varphi = 0$, which is not the case, so φ is injective. Likewise, the image of φ is either V or $\{0\}$, but the latter is absurd. So φ is bijective, and its inverse is automatically A-linear, so $\varphi^{-1} \in \mathrm{End}_A(V)$ exists.

A similar argument, starting from any $\varphi\colon U \to V$, shows that φ is an isomorphism if it is nonzero. □

We add, as a way of describing the skewfield $\mathrm{End}_A(V)$ in many cases:

Lemma 5.27 *Let A be an F-algebra over the field F, and let V be a simple A-module which is finite-dimensional over F. Then the skewfield $K = \mathrm{End}_A(V)$ is also finite-dimensional over F. As a result, if F is algebraically closed, then we simply have $K = F$.*

In courses on the representation theory of groups, the emphasis is often on finite-dimensional modules over $A = \mathbb{C}[G]$, and Schur's lemma is often stated in the form "when V is simple, $\mathrm{End}_{\mathbb{C}[G]}(V) \cong \mathbb{C}$".

Proof. The algebra $\mathrm{End}_A(V)$ consists of F-linear maps of a finite-dimensional vector space, so it is isomorphic to an algebra of matrices, and it must be finite-dimensional itself. Now argue as in Example 5.7. □

It is not difficult to go from simple modules to semisimple modules:

Proposition 5.28 *Suppose the A-module V is a direct sum of finitely many simple A-modules. Then* $\mathrm{End}_A(V)$ *is isomorphic to an algebra of the form*

$$\mathrm{M}_{n_1}(K_1) \times \cdots \times \mathrm{M}_{n_k}(K_k)$$

where each K_i is a skewfield. In fact $K_i = \mathrm{End}_A(S)$ for some simple submodule $S \subset V$.

Proof. We can write V in the form

$$V = V_1 \oplus \cdots \oplus V_k$$

where each V_i is a direct sum of simple modules which are all isomorphic to one another (V_i is then called *isotypical*). In other words $V_i \cong S_i^{n_i}$ for some simple A-module S_i. We can also arrange things so that S_i is not isomorphic to S_j for $i \neq j$.

The second part of Schur's lemma tells us that a map $V_i \to V_j$ must be zero if $i \neq j$. If follows easily that

$$\mathrm{End}_A(V) = \mathrm{End}_A(V_1) \times \cdots \times \mathrm{End}_A(V_k) .$$

On the other hand $\mathrm{End}_A(V_i) \cong \mathrm{M}_{n_i}(K_i)$ where $K_i = \mathrm{End}_A(S_i)$, by Lemma 5.19. Finally, the first part of Schur's lemma asserts that K_i is a skewfield. □

Corollary 5.29 *Let A be a ring such that the module A^1 is the direct sum of finitely many simple A-modules. Then A is isomorphic to a product of matrix algebras, over various skewfields.*

Proof. The proposition applies to $\mathrm{End}_A(A^1)$, showing that it is the product of matrix algebras $\mathrm{M}_{n_i}(K_i)$. However, $A \cong \mathrm{End}_A(A^1)^{op}$ (Lemma 5.18) and $\mathrm{M}_{n_i}(K_i)^{op} \cong \mathrm{M}_{n_i}(K_i^{op})$ (Example 5.10). □

Example 5.30 Suppose $A = F[G]$ where G is a finite group, and the characteristic of the field F does not divide the order of G, as in Example 5.25. We have seen that every finitely generated module is semisimple in this case, and in particular A^1 satisfies the hypotheses of the corollary. Thus $F[G]$ is a product of matrix algebras!

The skewfields K_i involved in this decomposition, if we go back to the more precise statement of Proposition 5.28, are of the form $\mathrm{End}_A(S)$ where $S \subset A^1$ is simple. Since S must be finite-dimensional, Lemma 5.27 applies, so K_i is finite-dimensional over F. If $F = \mathbb{R}$ for example, our study of the Brauer group will lead us to conclude that K_i is either $\mathbb{R}$, $\mathbb{C}$ or $\mathbb{H}$. If F is algebraically closed, say $F = \mathbb{C}$, we see that $\mathbb{C}[G]$ is a product of algebras $\mathrm{M}_{n_i}(\mathbb{C})$. Making this explicit is, in general, a challenge. ■

This result is a first indication of the importance of the module A^1, since it constrains strongly the algebra A. This will be reinforced after the following proposition.

Proposition 5.31 *Let U be a semisimple module.*

1. *If $V \subset U$ is a submodule, then there exists another submodule W such that $U = V \oplus W$. In fact, if $U = \oplus_{i \in I_0} S_i$ with S_i simple, then we can take $W = \oplus_{i \in I} S_i$ for some appropriate subset $I \subset I_0$.*
2. *Every submodule of U is semisimple.*
3. *Every quotient module of U is semisimple.*

Proof. We conduct the proof assuming that U is a direct sum of *finitely many* simple modules, for convenience (this is all we use in the sequel; the reader will generalize the argument using Zorn's lemma).

So suppose $U = S_1 \oplus \cdots \oplus S_n$ where S_i is simple, and let us use the notation $S_I = \oplus_{i \in I} S_i$ for $I \subset \{1, \ldots, n\}$. Now let I be maximal with the property $S_I \cap V = \{0\}$, and let us prove that $U = V \oplus S_I$.

The sum is obviously direct. Suppose $V + S_I$ is not all of U, then there exists an index j such that S_j is not contained in $V + S_I$. Since V_j is simple, this implies $S_j \cap (V + S_I) = \{0\}$. In turn, if $J = I \cup \{j\}$, then $S_J \cap V = \{0\}$, contradicting the maximality of I. This is absurd, and $V + S_I = U$.

So (1) is proved and $W = S_I$ is semisimple. We see that $U/V \cong S_I$ is semisimple, proving (3). But $V \cong U/W$ is semisimple by (3), so we have (2). □

Corollary 5.32 *The following statements are equivalent:*

1. *The module A^1 is semisimple.*
2. *Every A-module is semisimple.*

Proof. Any module is a quotient of a direct sum of copies of A^1, which is semisimple if A^1 is. □

(Note that we are chiefly concerned with finitely generated modules, and with the case when A^1 is the sum of finitely many simple modules. In fact, we are going to specialize rapidly to the situation when A is a finite-dimensional algebra over a field F.)

semisimple ring

Definition 5.33 A ring A satisfying the equivalent conditions of the corollary is called **semisimple**. ■

By Corollary 5.29, a semisimple algebra which is finite-dimensional over a field F must be a product of matrix algebras, over various skewfields containing F. In fact, the converse is not hard to see: Products of matrix algebras are semisimple. We shall establish this formally below (see Proposition 5.47), but we encourage the reader to think about this now, as an exercise.

Here is yet another virtue of A^1:

Corollary 5.34 *Suppose*

$$A^1 \cong S_1 \oplus \cdots \oplus S_n,$$

with S_i simple. Then any simple A-module S is isomorphic to some S_i. In particular, there are only finitely many isomorphism classes of simple modules, in this case.

Proof. Take $s \in S$, which is nonzero; the A-module generated by s must be all of S by simplicity, hence we have a surjective map $p\colon A^1 \to S$. Apply the proposition with $U = A^1$ and $V = \ker(p)$ to get

$$A^1 = \ker(p) \oplus S_{i_1} \oplus \cdots \oplus S_{i_k}.$$

It follows that S is isomorphic to $A^1/\ker(p) \cong S_{i_1} \oplus \cdots \oplus S_{i_k}$, and the latter can only be simple if $k = 1$. □

We conclude this section with a uniqueness statement, which the curious reader may have been wondering about. This can be widely generalized (to become the Krull–Schmidt theorem), but this version will suffice for our purposes.

Proposition 5.35 *Let A be an algebra over a field F. Suppose*

$$V = S_1 \oplus \cdots \oplus S_n,$$

with each S_i simple, and suppose

$$V' = S'_1 \oplus \cdots \oplus S'_m,$$

with each S'_i simple. Finally, assume that V and V' are finite-dimensional over F.

If there is an isomorphism $V \cong V'$, then $n = m$, and after renumbering if necessary, we have $S_i \cong S'_i$ for each index i.

Proof. First we treat the case when there is a single simple module S such that $S_i \cong S'_i \cong S$ for all i. Then $V \cong S^n$ and $V' \cong S^m$, and we must prove that $n = m$. With our hypotheses, this follows by a dimension count.

In general, write $V = V_1 \oplus \cdots \oplus V_k$ where each V_i is "isotypical" as in the proof of Proposition 5.28, and in such a way that for $i \neq j$ the modules V_i and V_j correspond to non-isomorphic simple modules. Likewise, write $V' = V'_1 \oplus \cdots \oplus V'_\ell$.

Suppose no S'_i were isomorphic to S_1. Then the isomorphism $\varphi\colon V \to V'$ would satisfy $S_1 \subset \ker(\varphi)$, by Schur's lemma, and this is absurd. Say that, after renumbering, we have $S_1 \cong S'_1$, and that V_1 and V'_1 are the isotypical summands containing S_1 and S'_1 respectively. We have $\varphi(V_1) \subset V'_1$, again by Schur's lemma; this holds for φ^{-1} as well, so in the end $\varphi(V_1) = V'_1$.

The argument is then finished by applying an immediate induction (either on n, or on the dimensions), since V/V_1 and V'/V'_1 are isomorphic, and noting that V_1 and V'_1 are as in the first case. □

The radical

We give a characterization of semisimple algebras in terms of ideals. This will motivate the definition of "simple" algebras, which are particular semisimple algebras.

It is time for us to stick to the case when A is a finite-dimensional algebra over a field F. The arguments below can be generalized in various ways, sometimes trivially, sometimes with ingenuity, but on the whole the cost of this generality would be too much, given our purposes.

The notion of an ideal is of course known to the reader (we have seen many ideals in previous chapters). However, since A is not assumed to be commutative, we must point out that a left ideal is a subgroup of A which is stable under left multiplication, or in other words, a submodule of A^1; a right ideal is a subgroup of A stable under right multiplication; and a two-sided ideal is both a left and a right ideal. In the rest of this section, we use "ideal" for "left ideal", as these will come up most frequently, but please note that this convention is not universal.

An ideal M of A is called *maximal* if an inclusion $M \subset I$, where I is another ideal, can only happen for $I = M$ and $I = A$; moreover, a maximal ideal is required to be $\neq A$. We point out that M is maximal if and only if the module A^1/M is simple. Indeed, the ideals I such that $M \subset I \subset A$ are in one-to-one correspondence with the submodules of A/M, under $I \mapsto I/M$. (Besides, simple modules are nonzero by definition.)

A first connection between modules and ideals is given by a simple definition:

annihilator

Definition 5.36 If $v \in V$, where V is an A-module, the **annihilator** of v is

$$\operatorname{ann}(v) = \{a \in A : a \cdot v = 0\}.$$

It is an ideal of A.

Lemma 5.37 *Let S be a simple A-module, and $s \in S$. Then the ideal* $\operatorname{ann}(s)$ *is maximal.*

Conversely, let M be a maximal ideal. Then there exists a simple module S such that $M = \operatorname{ann}(s)$ for some $s \in S$.

Proof. Let $s \in S$ with $s \neq 0$. The module generated by s of all of S, hence a surjective map $A^1 \to S$, $a \mapsto as$; its kernel is $\operatorname{ann}(s)$, so $S \cong A^1/\operatorname{ann}(s)$. It follows that $M = \operatorname{ann}(s)$ is maximal.

Conversely, if M is maximal, let $S = A^1/M$, which is simple, and let s be the image of $1 \in A$. Then $\operatorname{ann}(s) = M$. □

This lemma justifies what is in the next definition.

radical

Definition 5.38 The intersection of all the maximal ideals of A is called the **radical** of A, and is written $\operatorname{rad}(A)$. It consists of the elements $a \in A$ such that $a \cdot s = 0$ for any $s \in S$ where S is any simple module. It follows that $\operatorname{rad}(A)$ is a two-sided ideal.

Lemma 5.39 *Let A be a finite-dimensional algebra over the field F. Then A is semisimple* $\iff \operatorname{rad}(A) = 0$.

Proof. Suppose that A is semisimple, so A^1 is a direct sum of simple modules. The elements of $\operatorname{rad}(A)$ act trivially (that is, as 0) on any simple module, hence on any semisimple module, and in particular on A^1. However, the element $1 \in A^1$ is such that $a1 = a$ for any $a \in A$, of course, so for $a \in \operatorname{rad}(A)$ we draw $a = 0$. Thus $\operatorname{rad}(A) = 0$.

We do better than prove the converse, and establish that $A^1/\operatorname{rad}(A)$ is always semisimple. By finite-dimensionality of A^1, we can find finitely many maximal ideals $M_1, \ldots, M_n$ such that $\operatorname{rad}(A) = M_1 \cap M_2 \cap \cdots \cap M_n$. It follows that $A^1/\operatorname{rad}(A)$ injects into $A^1/M_1 \oplus \cdots \oplus A^1/M_n$, and each A^1/M_i is simple. So $A^1/\operatorname{rad}(A)$ is identified with a submodule of a semisimple module, which implies that it is itself semisimple. □

There exists another characterization of the radical which is useful. When I and J are both ideals of A, we define IJ to be the ideal generated by all products ab with

$a \in I$ and $b \in J$. More generally, when V is an A-module, we write IV for the submodule of V generated by all elements of the form $a \cdot v$ with $a \in I$, $v \in V$. The product of I with itself k times is written I^k. Finally, an ideal N is said to be *nilpotent* if $N^k = 0$ for some $k \geq 1$. Nilpotent ideals are almost trivially contained in the radical.

Lemma 5.40 *Let N be a nilpotent ideal, and let S be a simple module. Then we have $NS = 0$. As a result, $N \subset \mathrm{rad}(A)$.*

Proof. If NS were nonzero, we would have $NS = S$ by simplicity of S. Then $N^k S = S$ for any $k \geq 1$, by an immediate induction. However, for k large enough, we have $N^k = 0$, so $S = 0$. This is absurd. □

This lemma has a converse, which uses the fact that A is finite-dimensional over F.

Lemma 5.41 *Let N be an ideal such that $N \subset \mathrm{rad}(A)$. Then N is nilpotent.*

Proof. Let V be an A-module which is finite-dimensional over F. We argue that $N^k V = 0$ for k large enough.

This is clear when V is simple (or zero), since the elements of $\mathrm{rad}(A)$ act trivially on simple modules, by definition. If V is not simple, and nonzero, let $W \subset V$ be a submodule which is maximal, that is, such that $S = V/W$ is simple (W must exist for dimension reasons). We have $NS = 0$. This translates as $NV \subset W$, and in particular $\dim_F NV < \dim_F V$ as soon as V is nonzero.

Applying this to $N^k V$ instead of V, we see that $\dim_F N^{k+1} V < \dim_F N^k V$ as long as $N^k V \neq 0$. This cannot go on forever, so $N^k V = 0$ for large k.

Finally, when $V = N$, which is indeed finite-dimensional, we see that $N^k N = N^{k+1} = 0$ for k large, so N is nilpotent. □

Corollary 5.42 *The radical* $\mathrm{rad}(A)$ *is nilpotent, and is in fact the largest nilpotent ideal of A. The algebra A is semisimple if and only if it does not contain any nilpotent ideal.*

Example 5.43 Let us examine $A = F[G]$, where F is a field and G a finite group, now assuming that $p = \mathrm{car}(F)$ divides the order of G. Let $g \in G$ have order p; then $a = g - 1 \in A$ satisfies $a^p = g^p - 1 = 0$, so A has nilpotent elements. If we suppose further that we were able to find g in the center of G (which is always possible when G is a p-group), then the two-sided ideal generated by $g - 1$ is nilpotent, and we conclude that $F[G]$ is not semisimple. In fact, one can show, but it is a little bit more difficult, that $F[G]$ is never semisimple when the characteristic divides the order of G. ■

It is natural, at this point, to introduce another class of algebras.

simple ring

Definition 5.44 The ring A is called **simple** when it does not possess any two-sided ideal, beside $\{0\}$ and A itself. ■

In particular, when A is simple we must have $\mathrm{rad}(A) = 0$ (since $\mathrm{rad}(A)$ is a two-sided ideal, and $1 \notin \mathrm{rad}(A)$), and we see that simple algebras are semisimple.

Natural examples of simple algebras are provided by skewfields.

Lemma 5.45 *If A is simple, then there exists a skewfield K and an integer $n \geq 1$ such that $A \cong \mathrm{M}_n(K)$.*

Proof. Since A must be semisimple, it is isomorphic to an algebra of the form

$$\mathrm{M}_{n_1}(K_1) \times \cdots \times \mathrm{M}_{n_k}(K_k),$$

where K_i is a skewfield (Corollary 5.29). However if $k > 1$, then the set of elements whose first coordinate in this decomposition is 0 forms a proper two-sided ideal, which is a contradiction. So $k = 1$. □

The reader is bound to be wondering about the converse, and has long awaited the converse to Corollary 5.29, so we turn to this now.

Matrix algebras

In the previous sections, we have established that a semisimple algebra, such as $\mathbb{C}[G]$ where G is a finite group, is a product of matrix algebras, and also that a simple algebra is a single matrix algebra. We have encouraged the reader to think about the elementary converse: Matrix algebras are always simple, and products of matrix algebras are always semisimple. It is time we gave a formal proof of this.

Proposition 5.46 *Let $A = \mathrm{M}_n(K)$ where K is a skewfield, and $n \geq 1$ is an integer. Then A is simple.*

Moreover, let C denote the set of $n \times 1$-matrices with coefficients in K, viewed as an A-module. Then C is simple, and indeed, it is up to isomorphism the only simple A-module.

Suppose that K is finite-dimensional over its center F. Then $\mathrm{End}_A(C) \cong K^{op}$. Finally, if $A \cong \mathrm{M}_m(K')$ where K' is a skewfield, then $n = m$ and K' is isomorphic to K.

The hypothesis on K is here to make sure that A is itself finite-dimensional over F. This makes the proof easier, and besides, we have restricted attention to these algebras already. However, F does not enter the statement anywhere else.

Proof. Let I be a two-sided ideal of A. Suppose it contains a nonzero matrix $M = (a_{ij})$, with $a_{i_0j_0} \neq 0$. Multiply M on the left by the matrix $P = (p_{ij})$ whose only nonzero entry is $p_{i_1i_0} = 1$, for some arbitrary index i_1; then multiply on the right by the matrix $Q = (q_{ij})$ whose only nonzero entry is $q_{j_0j_1} = a_{i_0j_0}^{-1}$, for some index j_1. Then $PMQ \in I$, and its only nonzero entry is a 1 at the position (i_1,j_1). Clearly, then, $I = \mathrm{M}_n(K)$. This proves that A is simple.

Proving that C is a simple module can be done in the same way, and is left as an exercise. Moreover, we recover this fact below. Since $A^1 = C^n$, we use Corollary 5.34 to draw that any simple A-module must be isomorphic to C.

Suppose K is finite-dimensional over F. Consider $L = \mathrm{End}_A(C)$. Then we can observe that L contains a subring isomorphic to K^{op}, since for any $x \in K$ we can consider the right multiplication by x on each coordinate of C, and this is A-linear. Since $A^1 = C^n$, we have $A \cong \mathrm{End}_A(A^1)^{op} \cong \mathrm{M}_n(L^{op}) \cong \mathrm{M}_n(K)$; counting dimensions over F, we see that $L = K^{op}$, as desired.

Moreover, this shows in particular that $\mathrm{End}_A(C)$ is a skewfield, and the argument did not use *a priori* that C is simple, so we can recover this, as follows: The algebra A is semisimple, so C is semisimple, and if C were the sum of more than one simple module, then $\mathrm{End}_A(C)$ would not be a skewfield, by Proposition 5.28.

Finally, suppose that $A = \mathrm{M}_n(K) \cong \mathrm{M}_m(K')$ for a skewfield K'. From the previous point, both K and K' are isomorphic to L^{op}, the endomorphism skewfield of the unique simple A-module. So $\mathrm{M}_n(K) \cong \mathrm{M}_m(K)$, and we count dimensions to see that $n = m$. □

Proposition 5.47 *Let $A = \mathrm{M}_{n_1}(K_1) \times \cdots \times \mathrm{M}_{n_k}(K_k)$, where each K_i is a skewfield. Then A is semisimple. Moreover,*

$$A^1 = S_1^{n_1} \oplus \cdots \oplus S_k^{n_k},$$

where each S_i is a simple module (described in the proof), the various S_i are not pairwise isomorphic, and each simple A-module is isomorphic to some S_i.

Finally, suppose that each K_i is a finite-dimensional algebra over a field F. Then $\mathrm{End}_A(S_i) \cong K_i^{op}$. The decomposition of A as a product of matrix algebras is unique.

Proof. We give two arguments showing that A is semisimple. Consider the projection $p_i \colon A \to \mathrm{M}_{n_i}(K_i)$, which is a surjective algebra homomorphism. The image $p_i(\mathrm{rad}(A))$ is a two-sided ideal of the simple algebra $\mathrm{M}_{n_i}(K_i)$. If this image is the whole matrix algebra, then there is an $a \in \mathrm{rad}(A)$ mapping to the identity matrix; this element cannot be nilpotent, whereas the ideal $\mathrm{rad}(A)$ is nilpotent, so we have a contradiction. We conclude that $p_i(\mathrm{rad}(A)) = 0$ for all i, so $\mathrm{rad}(A) = 0$ and A is semisimple.

However, we can also argue as follows. If we let S_i be the unique simple $\mathrm{M}_{n_i}(K_i)$-module (cf. previous proposition), then we may see it as an A-module via p_i, and it is clearly simple as an A-module. So A^1 splits as a direct sum of simple modules as announced, and A is semisimple by definition.

We must prove that S_i is not isomorphic to S_j when $i \neq j$. For this we may consider

$$\mathrm{ann}(S_i) := \bigcap_{s \in S_i} \mathrm{ann}(s) = \{a \in A : a \cdot s = 0 \text{ for all } s \in S_i\}.$$

If $S_i \cong S_j$ then $\mathrm{ann}(S_i) = \mathrm{ann}(S_j)$, clearly. However, $\mathrm{ann}(S_i)$ is precisely comprised of the elements $a \in A$ satisfying $p_i(a) = 0$, so we recover i from S_i.

In the event that each K_i is finite-dimensional over a field F, we apply the previous proposition to $A_i = \mathrm{M}_{n_i}(K_i)$. We have $\mathrm{End}_A(S_i) = \mathrm{End}_{A_i}(S_i)$, clearly, so $K_i^{op} \cong \mathrm{End}_A(S_i)$.

The uniqueness statement means this. The integer k is the number of simple A-modules up to isomorphism, so it is determined by A. The integer n_i is the number of times that S_i appears when A^1 is expressed as a direct sum of simple modules, and this is well-defined by Proposition 5.35. Finally K_i is the opposite of the endomorphism skewfield of S_i. □

As a summary of the study of semisimple algebras undertaken in this chapter, we state the following theorem.

Theorem 5.48 *Let A be a finite-dimensional algebra over a field F. The following statements are equivalent.*

1. *A is semisimple, that is, the module A^1 is semisimple.*
2. *A is isomorphic to a product of matrix algebras $\mathrm{M}_{n_1}(K_1) \times \cdots \times \mathrm{M}_{n_k}(K_k)$, where each K_i is a skewfield containing F in its center.*
3. *$\mathrm{rad}(A) = 0$.*
4. *A does not have any nilpotent ideals.*

Moreover, the following conditions are also equivalent to one another.

1. *A is simple, that is, it does not have any two-sided ideals apart from $\{0\}$ and A itself.*
2. *A is isomorphic to a matrix algebra $\mathrm{M}_n(K)$ where K is a skewfield.*
3. *A is semisimple, and has only one simple module, up to isomorphism.*

When these three conditions hold, one has $K \cong \mathrm{End}_A(V)^{op}$, where V is the unique simple A-module, and $\mathcal{Z}(K) \cong \mathcal{Z}(A)$.

All this was established before, except that we have used a simple remark: The center of $\mathrm{M}_n(K)$ coincides with the center of K, and the center of a product of algebras is the product of their centers. Exercise!

Problems

5.1. Assuming the description of $\mathrm{Br}(\mathbb{R})$ given in this chapter, show that a skewfield containing $\mathbb{R}$ in its center, and finite-dimensional over $\mathbb{R}$, is isomorphic to either $\mathbb{R}$, $\mathbb{C}$, or $\mathbb{H}$.

5.2. Let F be a field.

1. Show that $(a,b)_F \cong (b,a)_F$, and that $(u^2a, v^2b)_F \cong (a,b)_F$. Here $a,b,u,v \in F^\times$, and the isomorphisms are as F-algebras.
2. Show that $\mathrm{M}_2(F) \cong (1,b)_F$ for any $b \in F^\times$.

A quaternion algebra over F is usually called split *when it is isomorphic to $\mathrm{M}_2(F)$; in the next chapter we shall rather say "Brauer equivalent to F". This problem shows* *that $(a,b)_F$ is split when a or b is a square, and that any quaternion algebra splits over a quadratic extension of F.*

5.3. Let F be a field, and let $A = (a,b)_F$. Show the equivalence of the conditions below:

1. The algebra A is split (as in the previous problem).
2. The algebra A is not a skewfield.
3. There exists $q \in A$, with $q \neq 0$, such that $q\overline{q} = 0$. *Here $\overline{q}$ is as defined in the proof of Proposition 5.4.*
4. The element $b \in F^\times$ is the norm of some element in $F[\sqrt{a}]$.

Hints: (1) $\Longrightarrow$ (2) $\Longrightarrow$ (3) *is very easy. To show* (3) $\Longrightarrow$ (4), *notice that the equation $q\overline{q} = 0$ directly expresses b as a quotient of two norms from $F[\sqrt{a}]$. For* (4) $\Longrightarrow$ (1), *write $b^{-1} = r^2 - as^2$ for $r, s \in F$, and put $u = rj + sk \in A$ and $v = (1+a)i + (1-a)ui \in A$. Then $1, u, v, uv$ is a basis for A, showing that $A \cong (1, 4a^2)$. A complete solution is in [GS06, proposition 1.1.7].*

This problem should be compared with Proposition 1.28 and Problem 1.3.

5.4. *(Skewfield of dimension* 9 *over* $\mathbb{Q}$*.).*

1. Let $L = \mathbb{Q}\left[\cos(\frac{2\pi}{7})\right]$. Show that a possible basis for L as a $\mathbb{Q}$-vector space is given by
$$u = \theta + \theta^6, \qquad v = \theta^2 + \theta^5, \qquad w = \theta^3 + \theta^4,$$
where $\theta = \exp(\frac{2i\pi}{7})$. Also, show that there is a generator σ for $\mathrm{Gal}(L/\mathbb{Q})$ which permutes u, v, w cyclically.
2. Write $\mathrm{N} = \mathrm{N}_{L/\mathbb{Q}}$. Find an explicit formula for $\mathrm{N}(xu + yv + zw)$, where $x, y, z \in \mathbb{Q}$. Deduce that, if x, y, z are integers which are not all even, then $\mathrm{N}(xu + yv + zw)$ is an odd integer.
3. We define K to be the free L-vector space, on the right, on three symbols $1, a, a^2$; in other words:
$$K = \left\{ 1\xi + a\eta + a^2\zeta \ : \ \xi, \eta, \zeta \in L \right\}.$$
Show that there is a ring structure on K, such that: 1 is the identity; a^2 is the square of a; the product extends that of $L \cong \{1\xi : \xi \in L\} \subset K$, and multiplying by elements of L on the right agrees with the vector space structure; one has $\xi a = a\sigma(\xi)$ for $\xi \in L$, where σ is as in question 1; finally, one has $a^3 = 2$.
4. Show that K is a skewfield.
Hints: For $\alpha = xu + yv + zw$, consider the L-linear endomorphism of K given by left multiplication by α; it is enough to show that its determinant is $\neq 0$ when $\alpha \neq 0$. Things can be arranged so that this determinant is $\mathrm{N}(\xi) + 2\mathrm{N}(\eta) + 4\mathrm{N}(\zeta) +$ *(some even integer), so that question 2 is relevant (multiplying by a^{-1} if necessary).*
5. Show that the center of K is $\mathbb{Q}$.
Hint: For $x \in K \smallsetminus \mathbb{Q}$, what is $[\mathbb{Q}(x) : \mathbb{Q}]$?

This example is taken from [Bla72, proposition I-10]. Hints are provided there to prove that K is not isomorphic to K^{op}.

5.5. Let G be a finite group, and let $A = \mathbb{C}[G]$.

1. Let S be a simple A-module. *Usually this is called an irreducible, complex representation of G.* Show that, if A^1 is written as a direct sum of simple modules, then the number of occurrences of S in this decomposition is $\dim_{\mathbb{C}} S$. Use this to write the order of G as a sum of squares of integers.
2. For each group G of order 8, find all the simple modules, and write A as a product of matrix algebras.
   Here we ask for the types of matrix algebras that occur, which is deduced from the dimensions of the simple modules. Asking for an explicit isomorphism would make a much harder question!
   Some hints: There are five groups of order 8. *One of them is Q_8, the subgroup generated by i,j,k in $\mathbb{H}^\times$. Then $\mathbb{H}$ itself gives an A-module – some care is needed, in order to make sure that the action of G is really $\mathbb{C}$-linear.*
3. Turn $\mathbb{H}$ into an $\mathbb{R}[Q_8]$-module, and show that $\mathrm{End}_{\mathbb{R}[Q_8]}(\mathbb{H}) \cong \mathbb{H}$. Write $\mathbb{R}[Q_8]$ as a product of matrix algebras.

5.6. Let F be a field of characteristic p, and let G be a p-group. Use Lemma 1.30 to describe all the simple $F[G]$-modules, and the annihilators of their elements. Deduce that $\mathrm{rad}(F[G])$ has codimension 1, and is the unique maximal ideal in $F[G]$ (this holds whether "ideal" is taken to mean left ideal, right ideal, or two-sided ideal).

5.7. An A-module V is called *indecomposable* if it cannot be written $V = U \oplus W$ with U and W both nonzero. Let F be an algebraically closed field, and let G be a cyclic group (finite or not). Describe all the indecomposable $F[G]$-modules.

Hint: Jordan. In particular, when G is a cyclic group of finite order n, one finds n distinct indecomposable modules. A solution is given in [Alp86, chapter II, section 4].

5.8. Let A be a finite-dimensional algebra over a field, and let V be an A-module. Define $\mathrm{rad}(V) := \mathrm{rad}(A) \cdot V$. Show that $\mathrm{rad}(V)$ is the smallest submodule of V such that $V/\mathrm{rad}(V)$ is semisimple. You may want to show first that $A/\mathrm{rad}(A)$ is a semisimple algebra.

When $A = \mathbb{F}_p[C_{p^r}]$ and $V = A^1$, describe $\mathrm{rad}(V)$, $\mathrm{rad}(\mathrm{rad}(V))$, and so on.

Hint: In a certain basis for V, the action is given by a single Jordan block.

5.9. Let K/F be a field extension. When V is an $F[G]$-module, for some finite group G, then $K \otimes_F V$ is a $K[G]$-module (why?). In this problem, all modules are finite-dimensional.

1. Write V^G for the fixed points of G in V. Show that $(K \otimes_F V)^G = K \otimes_F \left(V^G\right)$.
   Hint: Fixed points are computed by Gaussian elimination.
2. When V and W are two $F[G]$-modules, we see $\mathrm{Hom}_F(V, W)$ as an $F[G]$-module itself, by $(\sigma \cdot f)(v) = \sigma \cdot f(\sigma^{-1} \cdot v)$. Describe $(\mathrm{Hom}_F(V, W))^G$ and $K \otimes_F \mathrm{Hom}_F(V, W)$.
3. Show that, if $K \otimes_F V$ and $K \otimes_F W$ are isomorphic as $K[G]$-modules, then V and W are isomorphic as $F[G]$-modules.
   Hints and comments: When F is infinite, there is an easy argument along the following lines. Let $\varphi\colon K \otimes V \to K \otimes W$ be $K[G]$-linear. Then the previous

questions show that one can write $\varphi = \sum_i k_i \otimes \varphi_i$, *where* $k_i \in K$, $\varphi_i\colon V \to W$ *is* $F[G]$*-linear, and* $k_i \otimes \varphi_i$ *maps* $x \otimes v$ *to* $k_i x \otimes \varphi_i(v)$. *Now, using a determinant argument, one can replace the* k_i'*s by elements of* F, *in such a way that the resulting map* $V \to W$ *is invertible.*

Other easy cases present themselves when G *is cyclic (use the classification of modules over* $F[X]$*), or when* V *and* W *are semisimple (use Schur's lemma).*

For the general case, it seems best to argue as follows. One reduces to the case when K/F *is finite, say* $d = [K : F]$. *Then* $K \otimes V$ *may be seen as an* $F[G]$*-module, and it isomorphic to* $V^{\oplus d}$ *(direct sum of* d *copies of* V*). Likewise for* W, *so we must draw* $V \cong W$ *from* $V^{\oplus d} \cong W^{\oplus d}$. *This is a job for the Krull–Schmidt theorem, a generalization of Proposition 5.35. See [CR06, Theorem 14.5]; also, there is an easy proof in [Alp86, chapter II], written for algebraically closed fields, but the reader will adapt the argument.*

4. *Application: the normal basis theorem.* Suppose K/F is finite and Galois, and put $G = \mathrm{Gal}(K/F)$. Use the previous question to show that K, as an $F[G]$-module, is isomorphic to the regular module $F[G]^1$.

6 Central simple algebras

In this chapter we will define central, simple algebras over the field F, and use these to redefine the Brauer group $\mathrm{Br}(F)$. We prove classical results such as the Skolem–Noether theorem, or Wedderburn's theorem, and systematically translate what we obtain in terms of the Brauer group.

Conventions for this chapter. When we mention an algebra A over F, it is understood that F is a field, and that A is finite-dimensional over F. Its dimension will be written $[A : F]$ (a traditional notation for field extensions). Also, any A-module we mention is implicitly assumed finitely generated (and so finite-dimensional over F). Of course, this may be repeated for emphasis.

The Brauer group revisited

central F-algebra

Definition 6.1 Let F be a field. We say that the ring A is a **central F-algebra** when $\mathcal{Z}(A) \cong F$, and when $[A : F]$ is finite.

A central F-algebra is of course an F-algebra, but the original condition $F \subset \mathcal{Z}(A)$ is replaced by $F = \mathcal{Z}(A)$.

The next lemma is obvious from the material of the previous chapter, but it is useful to have it right here; see Theorem 5.48 for more. Since this theorem is also a summary, for the benefit of those readers who decided to skip Chapter 5, we point out that Lemma 5.45 and Proposition 5.46 are relevant.

Lemma 6.2 *Let A be an F-algebra. The following are equivalent.*

1. *A is a central F-algebra, and is simple.*
2. *$A \cong \mathrm{M}_n(K)$ where K is a skewfield which is finite-dimensional over F, with $\mathcal{Z}(K) = F$, and $n \geq 1$ is some integer.*

Moreover, when these conditions hold, the skewfield K is uniquely determined, as well as the integer n. □

Fixing K and letting n vary, we obtain a "Brauer class" of algebras. More precisely:

Brauer equivalent

Definition 6.3 Let A and B be central, simple F-algebras. We say that A and B are **Brauer equivalent** when there is a skewfield K, with $\mathcal{Z}(K) = F$, such that $A \cong \mathrm{M}_n(K)$ and $B \cong \mathrm{M}_m(K)$ for some integers n, m. ■

Brauer equivalence is an equivalence relation, on the set of isomorphism classes of central, simple F-algebras, and we may speak of the *Brauer class* of A. Note that there is just one (isomorphism class of) skewfield belonging to a given Brauer class.

Brauer group

Definition 6.4 The **Brauer group** of F is the set of all Brauer classes of all central, simple F-algebras. We write it $\mathrm{Br}(F)$. ■

Of course, on the surface this seems to clash with Definition 5.6, but a moment's thought reveals that the two definitions are equivalent, by the remarks just made. (Officially, though, we switch to this new definition.)

It is still not obvious that the Brauer group is actually a group! However, the reformulation in terms of central algebras will make this easy to establish, as well as the other properties announced in the introduction to the previous chapter (that is, right after Definition 5.6).

Tensor products

We assume that the reader is familiar with tensor products over fields, so that $V \otimes_F W$ makes sense for V, W two vector spaces over F. (Possible references are [Lan02, chapter XVI] or [CR06, section 12]. The basics are also provided in the Appendix.) We will now define the tensor product of two algebras, and as an exercise to brush up on tensor products, the reader will provide the proof for the following lemma.

Lemma 6.5 *Let A and B be algebras over F. On the vector space $A \otimes_F B$, there exists a unique structure of F-algebra, where the product satisfies*

$$(a_1 \otimes b_1)(a_2 \otimes b_2) = a_1 a_2 \otimes b_1 b_2$$

for $a_i \in A$, $b_i \in B$. □

We will always implicitly equip $A \otimes_F B$ with this product. Also note that whenever F is obvious, we may use unadorned tensor products, such as $A \otimes B$.

Example 6.6 As another exercise, the reader will show that the abelian group V is an $A \otimes B$-module if and only if it is an A–B-bimodule. Even bimodules are just plain modules! ■

The next two lemmas will be used constantly.

Lemma 6.7 *Let n, p be positive integers. There is an isomorphism of F-algebras:*

$$\mathrm{M}_n(F) \otimes_F \mathrm{M}_p(F) \cong \mathrm{M}_{np}(F).$$

Proof. Let V be an n-dimensional vector space, and let W be a p-dimensional vector space, so that $V \otimes W$ has dimension np. To $\varphi \in \mathrm{End}_F(V)$ and $\psi \in \mathrm{End}_F(W)$, we associate $T(\varphi, \psi) \in \mathrm{End}_F(V \otimes W)$, which is characterized by $T(\varphi, \psi)(v \otimes w) = \varphi(v) \otimes \psi(w)$. This induces

$$\mathrm{End}_F(V) \otimes \mathrm{End}_F(W) \longrightarrow \mathrm{End}_F(V \otimes W), \tag{*}$$

mapping $\varphi \otimes \psi$ to $T(\varphi, \psi)$. We have used the basic properties of tensor products here. The definition of (*) makes it clear that it is a homomorphism of F-algebras.

Now, in order to show that (*) is an isomorphism, we pick bases for V and W, so that $\mathrm{End}_F(V) \cong \mathrm{M}_n(F)$ and $\mathrm{End}_F(W) \cong \mathrm{M}_p(F)$; we obtain automatically a basis for $V \otimes W$ and an isomorphism $\mathrm{End}_F(V \otimes W) \cong \mathrm{M}_{np}(F)$, using some bijection ν between $\{1, \ldots, n\} \times \{1, \ldots, p\}$ and $\{1, \ldots, np\}$. The map (*) translates into a homomorphism,

$$f\colon \mathrm{M}_n(F) \otimes_F \mathrm{M}_p(F) \longrightarrow \mathrm{M}_{np}(F).$$

Let $(e_{ij})_{ij}$ be the canonical basis for $\mathrm{M}_n(F)$, and likewise with $(\varepsilon_{k\ell})_{k\ell}$ for $\mathrm{M}_p(F)$ and $(E_{rs})_{rs}$ for $\mathrm{M}_{np}(F)$. Looking at the definitions, we see that $f(e_{ij} \otimes \varepsilon_{k\ell}) = E_{\nu(i,k),\nu(j,\ell)}$. This formula would have made it painful for us to check that f is a homomorphism of algebras, but luckily we knew that; on the other hand, it is clear that f is an isomorphism. $\square$

Lemma 6.8 *Let A be an F algebra, and let n be a positive integer. There is an isomorphism of F-algebras:*

$$A \otimes_F \mathrm{M}_n(F) \cong \mathrm{M}_n(A).$$

Proof. Let $V = A^1$, the regular A-module, and let $W = F^n$. We can and we do identify $V \otimes_F W$ with A^n; the A-module structure on $V \otimes_F W$ is then given by $a \cdot (v \otimes w) = (a \cdot v) \otimes w$. Arguing as in the previous lemma, we define an algebra homomorphism,

$$\mathrm{End}_A(V) \otimes_F \mathrm{End}_F(W) \longrightarrow \mathrm{End}_A(V \otimes W). \tag{*}$$

In fact, this is *precisely* the map from the previous lemma, restricted to $\mathrm{End}_A(V)$, a subalgebra of $\mathrm{End}_F(V)$. So we know that (*) is injective. We now count dimensions to show that it is an isomorphism.

Since $\mathrm{End}_A(V) = \mathrm{End}_A(A^1) \cong A^{op}$, on the left-hand side the dimension is $n^2 \cdot \dim_F(A)$. On the other hand, since $\mathrm{End}_A(V \otimes W) = \mathrm{End}_A(A^n) \cong \mathrm{M}_n(A^{op})$ (by Proposition 5.19), we see that the dimension on the right-hand side is also $n^2 \cdot \dim_F A$. So (*) is an isomorphism, and we have just described its source and target as $A^{op} \otimes \mathrm{M}_n(F)$ and $\mathrm{M}_n(A^{op})$. This proves the result with A^{op} replacing A. $\square$

Here is a little application. Recall that we have defined, at the beginning of this chapter, what it means for two central, simple algebras to be Brauer equivalent. We have now an alternative definition:

Lemma 6.9 *The central, simple F-algebras A and B are Brauer equivalent $\iff$ there exist integers n and m such that $\mathrm{M}_n(A) \cong \mathrm{M}_m(B)$.*

Proof. Suppose n and m exist. Write $A \cong \mathrm{M}_p(K)$ for a skewfield K, and $B \cong \mathrm{M}_q(L)$ for another skewfield L; we want to show that $K \cong L$. Indeed, $\mathrm{M}_n(A)$ is isomorphic to each of:

$$\mathrm{M}_n(F) \otimes A \cong \mathrm{M}_n(F) \otimes \mathrm{M}_p(K) \cong \mathrm{M}_n(F) \otimes \mathrm{M}_p(F) \otimes K \cong \mathrm{M}_{np}(F) \otimes K \cong \mathrm{M}_{np}(K)\,.$$

Likewise, we find $\mathrm{M}_m(B) \cong \mathrm{M}_{mq}(L) \cong \mathrm{M}_{np}(K)$. We deduce $K \cong L$, and $np = mq$ (Lemma 6.2).

For the converse, suppose $K = L$, and pick a common multiple $np = mq$ of p and q. The computation above shows that $\mathrm{M}_n(A) \cong \mathrm{M}_m(B) \cong \mathrm{M}_{np}(K)$. □

We embark on a proof that a tensor product of central, simple algebras is again simple. The main technical point is this:

Proposition 6.10 *Let K be a skewfield such that $F \subset \mathcal{Z}(K)$ (the center of K), and let A be an F-algebra. Then any two-sided ideal of $K \otimes_F A$ is generated by a two-sided ideal of $\mathcal{Z}(K) \otimes_F A$.*

Proof. Let n be the dimension of A as an F-vector space (which is supposed finite, as in the rest of this chapter), and let $e_1, \ldots, e_n$ be a basis. Of course $K \otimes_F A$ is a K-module, with $\lambda \cdot (x \otimes a) = (\lambda x) \otimes a$ (in fact, you could call it a vector space, if you feel comfortable about linear algebra over skewfields – after all, the previous chapter has shown that all usual results extend to this setting). The K-submodule of $K \otimes A$ generated by e_i will be written $K \otimes e_i$, so that

$$K \otimes A = \bigoplus_{1 \le i \le n} K \otimes e_i\,.$$

Let I be a two-sided ideal of $K \otimes A$. It is also a K-submodule, so in particular there is a set E such that

$$K \otimes A = I \oplus \bigoplus_{i \in E} K \otimes e_i\,,$$

by Proposition 5.31 (note that each $K \otimes e_i$ is isomorphic to K^1, and so is simple). For $q \notin E$ let us write

$$1 \otimes e_q = \varepsilon_q + \sum_{i \in E} x_{q,i} \otimes e_i\,,$$

where $\varepsilon_q \in I$ and $x_{q,i} \in K$; this can be done *uniquely*. The elements ε_q generate I as a K-module (this is obvious if we think of I as a quotient of $K \otimes A$, as the direct sum decomposition allows us to do). We seek to prove that $\varepsilon_q \in \mathcal{Z}(K) \otimes A$.

Indeed, fix $y \in K$ and let $\sigma(x) = y^{-1}xy$ for $x \in K$, and then extend this to an automorphism $\widetilde{\sigma}$ of $K \otimes A$ satisfying $\widetilde{\sigma}(x \otimes a) = \sigma(x) \otimes a$. In fact, $\widetilde{\sigma}$ is conjugation by $y \otimes 1 \in K \otimes A$, so that $\widetilde{\sigma}(I) \subset I$, as I is assumed to be a two-sided ideal. We compute

$$1 \otimes e_q = \widetilde{\sigma}(1 \otimes e_q) = \widetilde{\sigma}(\varepsilon_q) + \sum_{i \in E} \sigma(x_{q,i}) \otimes e_i .$$

Since $\widetilde{\sigma}(\varepsilon_q) \in I$, we see by uniqueness that $\widetilde{\sigma}(\varepsilon_q) = \varepsilon_q$, and more importantly $\sigma(x_{q,i}) = x_{q,i}$ for all q, i. In other words, we have $x_{q,i} \in \mathcal{Z}(K)$. This concludes the proof. □

Theorem 6.11 *Let A and B be simple F-algebras, and assume that one of them is central. Then $A \otimes_F B$ is simple.*

Proof. First assume that $A = K$ is a skewfield with $\mathcal{Z}(K) = F$. By the proposition, any two-sided ideal I of $K \otimes B$ is generated by a two-sided ideal J of $F \otimes_F B \cong B$. Since B is simple, we have either $J = 0$, implying immediately that $I = 0$; or alternatively $J = B$, implying that I contains $F \otimes_F B$, and in particular contains invertible elements, so $I = K \otimes B$. The algebra $K \otimes B$ is indeed simple.

In general, if the center of A is F, then $A \cong \mathrm{M}_n(K)$ where K is a skewfield such that $\mathcal{Z}(K) = F$. Write also $B \cong \mathrm{M}_m(L)$ where L is a skewfield. In this situation the algebra $A \otimes_F B$ is isomorphic to

$$\mathrm{M}_n(K) \otimes \mathrm{M}_m(L) \cong \mathrm{M}_n(F) \otimes K \otimes \mathrm{M}_m(F) \otimes L \cong M_{nm}(F) \otimes K \otimes L \cong K \otimes \mathrm{M}_{nm}(L) .$$

By the particular case treated, and since $\mathrm{M}_{nm}(L)$ is a simple F-algebra, we see that $A \otimes B$ is simple. □

Example 6.12 Let $A = \mathbb{H} = (-1,-1)_{\mathbb{R}}$, the algebra of quaternions over the reals. It is a skewfield, so a simple, central algebra over its center, and this center is $\mathbb{R}$ (exercise). Hence $\mathbb{H} \otimes_{\mathbb{R}} \mathbb{H}$ must be a simple algebra over $\mathbb{R}$. Therefore $\mathbb{H} \otimes_{\mathbb{R}} \mathbb{H} \cong \mathrm{M}_n(K)$ for some skewfield containing $\mathbb{R}$ in its center. As $\mathbb{H} \otimes_{\mathbb{R}} \mathbb{H}$ is of dimension 4 over $\mathbb{R}$, there can be only two possibilities: either it is a skewfield itself (and $n = 1$), or $n = 2$ and $K = \mathbb{R}$.

We can rule out the first possibility. Indeed $\mathbb{H} \otimes_{\mathbb{R}} \mathbb{H}$ contains $\mathbb{H} \otimes_{\mathbb{R}} \mathbb{C}$, and the latter is $(-1,-1)_{\mathbb{C}}$. We have already observed that this algebra contains zero divisors, and so cannot be contained in a skewfield. We conclude:

$$\mathbb{H} \otimes_{\mathbb{R}} \mathbb{H} \cong \mathrm{M}_2(\mathbb{R}) .$$

We give a more general explanation for this in just a few paragraphs. ■

Example 6.13 Let $\overline{F}$ be the algebraic closure of F. When A is a simple, central F-algebra, we see that $A \otimes_F \overline{F}$ is a simple $\overline{F}$-algebra, and in particular, is isomorphic to $M_n(K)$ for some skewfield containing $\overline{F}$ in its center, and finite-dimensional over it. However, this implies $K = \overline{F}$, as in Example 5.7.

There is actually a converse. Suppose that A is an F-algebra such that $A \otimes_F \overline{F} \cong M_n(\overline{F})$ (or, what is equivalent now, such that $A \otimes_F \overline{F}$ is simple). If I is a two-sided ideal of A, then $I \otimes_F \overline{F}$ is a two-sided ideal of $A \otimes \overline{F}$, so by a dimension count we see that $I = \{0\}$ or A, and that A is simple.

So an F-algebra is simple if and only if $A \otimes_F \overline{F} \cong M_n(\overline{F})$ for some n. One says that A is a *form* of $M_n(\overline{F})$ over F. This is sometimes taken as the definition of a simple algebra. The whole approach is called "proceeding by Galois descent", which certainly sounds good.

Example 6.14 Here is an example to show that one of A or B really needs to be central for the theorem to hold. Indeed, consider $A = B = \mathbb{C}$ and $F = \mathbb{R}$. As $\mathbb{C} \cong \mathbb{R}[X]/(X^2+1)$, we see easily that $\mathbb{C} \otimes_{\mathbb{R}} \mathbb{C} \cong \mathbb{C}[X]/(X^2+1)$, and this is isomorphic to $\mathbb{C} \times \mathbb{C}$ by the Chinese Remainder theorem, as X^2+1 splits over $\mathbb{C}$. Thus $\mathbb{C} \otimes_{\mathbb{R}} \mathbb{C}$ is not simple.

Definition 6.15 Let A be an F-algebra, and let B be a subalgebra of A. The **commutant** of B in A, also called the centralizer of B in A, is

commutant

$$\mathcal{Z}_A(B) := B' := \{a \in A : ab = ba \text{ for all } b \in B\}.$$

The commutant of B is another subalgebra of A.

The center of A, for example, is $\mathcal{Z}(A) = \mathcal{Z}_A(A)$.

Lemma 6.16 *Let A_1 and A_2 be F-algebras, and let B_1 resp. B_2 be a subalgebra of A_1 resp. A_2. Then*

$$\mathcal{Z}_{A_1 \otimes_F A_2}(B_1 \otimes_F B_2) = \mathcal{Z}_{A_1}(B_1) \otimes_F \mathcal{Z}_{A_2}(B_2),$$

or more informally $(B_1 \otimes B_2)' = B_1' \otimes B_2'$. In particular

$$\mathcal{Z}(A_1 \otimes A_2) = \mathcal{Z}(A_1) \otimes \mathcal{Z}(A_2).$$

Proof. We certainly have $B_1' \otimes B_2' \subset (B_1 \otimes B_2)'$. For the converse, let $a = \sum x_i \otimes y_i$ be an element of $(B_1 \otimes B_2)'$, with $x_i \in A_1$, $y_i \in A_2$. Writing that a commutes with $b_1 \otimes 1$ for any $b_1 \in B_1$, we have

$$\sum_i (b_1 x_i - x_i b_1) \otimes y_i = 0.$$

Now suppose we had chosen the elements y_i linearly independent over F, as is clearly possible. Then the last equation implies $x_i \in B_1'$, and we have shown $(B_1 \otimes B_2)' \subset B_1' \otimes A_2$.

Similarly, we can establish that $(B_1 \otimes B_2)' \subset A_1 \otimes B_2'$. So the proof will be complete if we can be convinced that

$$(B_1' \otimes A_2) \cap (A_1 \otimes B_2') = B_1' \otimes B_2'.$$

To see this, pick a basis of B_1' resp. B_2' and complete it into a basis of A_1 resp. A_2; the identity is then clear. □

We have now proved:

Corollary 6.17 *Whenever A and B are central, simple algebras over F, so is $A \otimes_F B$.* □

The group law

The tensor product operation will become the group law on $\mathrm{Br}(F)$. From now on let us write $[A]$ for the Brauer class of A.

Lemma 6.18 *Let A and B be central, simple algebras over F. The Brauer class $[A \otimes_F B]$ only depends on $[A]$ and $[B]$.*

Proof. There is just one skewfield K, with center F, in the Brauer class $[A]$ (up to isomorphism). Likewise, the class $[B]$ defines, up to isomorphism, a unique skewfield L with center F. The algebra $A \otimes B$ is Brauer equivalent to $K \otimes L$, since isomorphisms $A \cong \mathrm{M}_n(K)$ and $B \cong \mathrm{M}_m(L)$ imply

$$A \otimes B \cong \mathrm{M}_n(F) \otimes K \otimes \mathrm{M}_m(F) \otimes L \cong \mathrm{M}_{nm}(F) \otimes K \otimes L \cong \mathrm{M}_{nm}(K \otimes L)\,.$$

Thus $[A \otimes B] = [K \otimes L]$ depends only on $[A]$ and $[B]$. □

Definition 6.19 We define a composition law on $\mathrm{Br}(F)$ by setting $[A] \cdot [B] = [A \otimes_F B]$. □

By the lemma, this is well defined. Our task is to show that $\mathrm{Br}(F)$ is then an abelian group. Indeed, the basic properties of the tensor product show that the composition law is associative and commutative. There is a neutral element, which is $[F]$: indeed $[A] \cdot [F] = [A \otimes_F F] = [A]$.

Slightly more difficult is the existence of inverses. The next theorem, a classic, says just that.

Theorem 6.20 *Let K be a skewfield, let $F = \mathcal{Z}(K)$, and let $n = [K : F]$, assumed finite. Then there is an isomorphism*

$$K \otimes_F K^{op} \cong \mathrm{M}_n(F)\,.$$

In $\mathrm{Br}(F)$ this may be written $[K] \cdot [K^{op}] = [F]$. Any Brauer class is of the form $[K]$ for such a K, so the theorem proves the existence of inverses.

Proof. The skewfield K acts on itself by left and right multiplication, so it is both a left K-module and a right K-module, and the two operations commute. The right action can be seen as a structure of K^{op}-module (Lemma 5.11), and the commutativity means that K is in fact a K–K^{op}-bimodule, or equivalently a $K \otimes K^{op}$-module (Example 6.6).

Each element of $K \otimes K^{op}$ thus gives an F-linear map of the n-dimensional vector space K, hence an algebra homomorphism

$$f\colon K \otimes K^{op} \longrightarrow \mathrm{M}_n(F).$$

The dimensions match, so to show that f is an isomorphism, it is enough to show injectivity. This is automatic, however: The algebra $K \otimes K^{op}$ is simple by Theorem 6.11, so the two-sided ideal $\ker(f)$, which does not contain $1 \otimes 1$, must be $\{0\}$. □

Corollary 6.21 *The Brauer group is indeed a group, and* $[A]^{-1} = [A^{op}]$.

Proof. If A is Brauer equivalent to K, then A^{op} is Brauer equivalent to K^{op}, so $[A][A^{op}] = [K][K^{op}] = [K \otimes K^{op}] = [F]$ by the theorem. □

Definition 6.22 For any $n \geq 1$, we define $\mathrm{Br}_n(F)$ to be the n-torsion subgroup of $\mathrm{Br}(F)$, that is

$$\mathrm{Br}_n(F) = \{x \in \mathrm{Br}(F) : x^n = 1\},$$

where we write $1 = [F]$ for the identity of $\mathrm{Br}(F)$. ■

Example 6.23 Let $\mathbb{H}_{a,b} = (a,b)_F$ be a quaternion algebra over F. Then $\mathbb{H}_{a,b}$ is always a central, simple F-algebra: Indeed, Problem 5.3 shows that such an algebra is either a skewfield or a matrix algebra, and the center is easily worked out. By a dimension count, either $\mathbb{H}_{a,b} \cong \mathrm{M}_2(F)$, or $\mathbb{H}_{a,b}$ is a skewfield. Now, there is an operation of conjugation on $\mathbb{H}_{a,b}$, written $q \mapsto \overline{q}$, and defined in the proof of Proposition 5.4. The reader will check that $\overline{q_1 q_2} = \overline{q}_2 \overline{q}_1$. It follows that the conjugation is an isomorphism $\mathbb{H}_{a,b} \cong \mathbb{H}_{a,b}^{op}$. As a result, we have $\mathbb{H}_{a,b} \otimes \mathbb{H}_{a,b} \cong \mathrm{M}_4(F)$ by the last theorem, or equivalently $[\mathbb{H}_{a,b}]^2 = 1$. So, quaternion algebras define elements of $\mathrm{Br}_2(F)$.

There is a converse to this, given by a particular case of the beautiful (and difficult!) *Merkurjev–Suslin theorem*: For any field F, the subgroup $\mathrm{Br}_2(F)$ is actually generated by quaternion algebras. A proof of this is given in [GS06]. ■

Lemma 6.24 *Let E/F be any field extension. The operation* $[A] \mapsto [A \otimes_F E]$, *from* $\mathrm{Br}(F)$ *to* $\mathrm{Br}(E)$, *is well defined, and is a group homomorphism.*

Proof. When A is a central, simple F-algebra, then $A \otimes_F E$ is simple by Theorem 6.11, and its center is $F \otimes_F E \cong E$ by Lemma 6.16. So $A \otimes_F E$ defines a class in $\mathrm{Br}(E)$.

If K is the unique skewfield in $[A]$, then $A \cong \mathrm{M}_n(K)$ for some n, so $A \otimes E \cong \mathrm{M}_n(F) \otimes K \otimes E \cong \mathrm{M}_n(K \otimes E)$, and $A \otimes_F E$ is Brauer equivalent to $K \otimes_F E$. As a result, the class $[A \otimes_F E]$ depends only on $[A]$.

Finally, we write $(A \otimes_F E) \otimes_E (B \otimes_F E) \cong A \otimes_F (E \otimes_E (B \otimes_F E)) \cong A \otimes_F (B \otimes_F E) \cong (A \otimes_F B) \otimes_F E$. Taking Brauer classes, and if we write $\iota\colon \mathrm{Br}(F) \to \mathrm{Br}(E)$ the operation just defined, we have $\iota([A]) \cdot \iota([B]) = \iota([A \otimes_F B]) = \iota([A] \cdot [B])$, so ι is a group homomorphism. □

relative Brauer group

Definition 6.25 We write $\iota_{E/F}\colon \mathrm{Br}(F) \to \mathrm{Br}(E)$ the homomorphism defined in the lemma. Its kernel is written $\mathrm{Br}(E/F)$ and is called the **relative Brauer group** of the extension E/F.

For example $\mathrm{Br}(F) = \mathrm{Br}(\overline{F}/F)$, where $\overline{F}$ is the algebraic closure of F, since $\mathrm{Br}(\overline{F})$ is trivial. Ultimately, we shall show that there are powerful tools to study $\mathrm{Br}(E/F)$ when E/F is finite and Galois, and that $\mathrm{Br}(F) = \mathrm{Br}(\overline{F}/F)$ can be recovered from all these.

The fundamental theorems

Theorem 6.26 (the commutant theorem) *Let A be a central, simple F-algebra, and let B be a simple subalgebra. Write $B' = \mathcal{Z}_A(B)$, the commutant of B. Then one has*

1. *B' is simple.*
2. *$[B : F][B' : F] = [A : F]$.*
3. *$\mathcal{Z}_A(B') = B$, that is $B'' = B$.*

Proof. We first notice that if we can prove that (1) and (2) always hold, then so does (3). Indeed, it is clear that $B \subset B''$, and it is enough to prove $[B'' : F] = [B : F]$; however, by (1) applied to B we see that B' is simple, and by (2) applied to B' we see that $[B' : F][B'' : F] = [A : F]$, implying the desired identity.

As a particular case (to which we will easily reduce things below), we begin with $A = \mathrm{M}_n(F)$. Let $V = F^n$, the unique simple A-module up to isomorphism. When V is considered as a B-module, it splits as a direct sum $V = W^q$ where W is the unique simple B-module. In this situation we have $A = \mathrm{End}_F(V)$ and $B' = \mathrm{End}_B(V)$, so $B' = \mathrm{End}_B(W^q) \cong \mathrm{M}_q(L)$ where $L = \mathrm{End}_B(W)$ (Lemma 5.19). However if K is the skewfield such that $B \cong \mathrm{M}_p(K)$, then $L \cong K^{op}$ is another skewfield (Proposition 5.46). We have already established that $B' \cong M_q(K^{op})$ is simple.

Next we count dimensions. We have $[B : F] = p^2 r$ where $r = [K : F]$, while $[B' : F] = q^2 r$. The product of these two numbers is $(pqr)^2$, so to prove (2) we need to see that $pqr = n$. And indeed $n = \dim_F(V) = q \dim_F(W) = q \dim_F(K^p) = pq[K : F] = pqr$.

To prove the general case, we consider $A \otimes A^{op}$, which we know is Brauer equivalent to F, so we have proved (1) and (2) for simple sub-algebra of $A \otimes A^{op}$, such as $B \otimes_F F \cong B$. The commutant of $B \otimes F$ is $B' \otimes A^{op}$ by Lemma 6.16, which is therefore simple; we deduce that B' is simple (for a two-sided ideal I of B' gives birth to a two-sided ideal $I \otimes A^{op}$ of $B' \otimes A^{op}$). Counting dimensions, we obtain

$$[B : F][B' : F] = [B \otimes F : F]\frac{[B' \otimes A^{op} : F]}{[A^{op} : F]} = \frac{[A \otimes A^{op} : F]}{[A^{op} : F]} = [A : F],$$

again using the particular case. □

Corollary 6.27 *Let A be a central, simple F-algebra. Let B be a simple and commutative subalgebra, and let B' be its commutant. Then B is the center of B'. Moreover, the three assertions below are equivalent:*

1. *B is a maximal commutative subalgebra of A.*
2. $B = B'$.
3. $[A : F] = [B : F]^2$.

Proof. By assumption $B \subset B'$. Statement (3) of the theorem gives immediately that B is the center of B'.

Any $x \in B'$ yields a commutative algebra $B[x]$ containing B, so (1) implies (2); any commutative algebra containing B is contained in B', so (2) implies (1). By the theorem, the equivalence of (2) and (3) is clear. □

The next corollary is one of the most well-known facts about skewfields. It can be proved directly by induction on the dimension.

Corollary 6.28 *Let K be a skewfield, let $F = \mathcal{Z}(K)$, and suppose $[K : F]$ is finite. Then the integer $[K : F]$ is a square.*

Likewise, if A is any central, simple F-algebra, then $[A : F]$ is a square.

Proof. Pick a field E such that $F \subset E \subset K$, and which is maximal with this property (by finite-dimensionality, there must be one). Then E is simple and is a maximal commutative subalgebra of K, so the corollary applies. Then $[K : F] = [E : F]^2$.

For the general case, we consider $A \cong \mathrm{M}_n(K)$ for some n, and see immediately that $[A : F] = n^2[K : F]$. □

degree

The integer $\sqrt{[A : F]}$, when A is a central, simple F-algebra, is sometimes called the **degree** of A, although we shall not use this term often in the rest of the book.

Theorem 6.29 (Skolem–Noether) *Let A be a central, simple algebra over F. Let B_1 and B_2 be simple sub-algebras of A, and let $f\colon B_1 \longrightarrow B_2$ be an isomorphism of F-algebras. Then f extends to an inner automorphism of A, that is, there exists $a \in A^\times$ such that $f(b) = aba^{-1}$ for all $b \in B_1$.*

Proof. Let V be the unique simple A-module, and let $K = \mathrm{End}_A(V)$. As will become routine to us, we view A as a subalgebra of $\mathrm{End}_F(V)$, since the A-module structure yields a homomorphism $A \to \mathrm{End}_F(V)$ whose kernel, a two-sided ideal not containing 1, must be $\{0\}$. By definition, the skewfield K is also a subalgebra of $\mathrm{End}_F(V)$, and indeed it is the commutant of A, that is $K = A'$. By the last part of the commutant theorem, we have also $A = K'$.

The vector space V is an $A \otimes K$-module, or an A–K-bimodule (Example 5.17), so for $a \in A$, $\lambda \in K$, $v \in V$ we can make sense of av and λv, and $a\lambda v = \lambda a v$.

Now let V_1 be the F-vector space V, seen as a $B_1 \otimes K$-module by restriction, that is, equipped with $(b \otimes \lambda) \cdot v = b\lambda v$. On the other hand, we let V_2 be the same vector space, also seen as a $B_1 \otimes K$-module, but this time via $(b \otimes \lambda) \cdot v = f(b)\lambda v$.

The center of K is F, since that of A is F and $A \cong M_n(K^{op})$ for some n. Therefore, the tensor product $B_1 \otimes_F K$ is a simple algebra, by Theorem 6.11. In particular, the modules V_1 and V_2 must be isomorphic (they have the same dimension). Let $\varphi\colon V_1 \to V_2$ be an isomorphism of $B_1 \otimes K$-modules, so

$$\varphi(b\lambda v) = f(b)\lambda\varphi(v),$$

with the notation as above. For $b = 1$, this reads $\varphi(\lambda v) = \lambda\varphi(v)$. Since $V_1 = V_2 = V$ as F-vector spaces, if we view φ as an element of the algebra $\mathrm{End}_F(V)$, then we have just proved that it commutes with all $\lambda \in K$. In other words $\varphi \in K' = A$. Let us write $a = \varphi$ and go back to

$$ab\lambda v = f(b)\lambda av$$

for all $v \in V$, or $ab\lambda = f(b)\lambda a$. For $\lambda = 1$ we have $ab = f(b)a$, and $f(b) = aba^{-1}$. □

Example 6.30 In particular, for $B_1 = B_2 = A$, we see that any automorphism of A is inner. The map $A^\times \to \mathrm{Aut}(A)$, taking a to the conjugation by a, has $\mathcal{Z}(A)^\times = F^\times$ as its kernel, so in the end

$$\mathrm{Aut}(A) \cong A^\times/F^\times.$$

For $A = M_n(F)$, we see that $\mathrm{Aut}(M_n(F)) \cong \mathrm{PGL}_n(F)$. ■

Example 6.31 Suppose $F \subset E \subset A$ where E/F is a Galois extension of fields. Applying the Skolem–Noether theorem with $B_1 = B_2 = E$ and an automorphism $\sigma \in \mathrm{Gal}(E/F)$, we see that there is an element $a_\sigma \in A$ such that conjugation by a_σ induces σ on E. The elements a_σ will be crucial in the sequel, ultimately allowing us to reconstruct the algebra A from a well-chosen Galois extension E/F contained in it. ■

Theorem 6.32 (Wedderburn) *Every finite skewfield is a field.*

Proof. Let $F = \mathcal{Z}(K)$. All the maximal subfields E with $F \subset E \subset K$ have the same dimension of F, namely the square root of $[K : F]$, by Corollary 6.27. Hence they all have the same cardinality. As is well known from the theory of finite fields, they must all be isomorphic. Finally, by Skolem–Noether, we see that all the maximal subfields are conjugate. It must also be pointed out that any $x \in K$ belongs to the subfield $F(x)$, so is contained in some maximal subfield.

Fix such an E. The group $K^\times$ is then the union of all the conjugate subgroups $xE^\times x^{-1}$. By the lemma below, this implies $K^\times = E^\times$, so that K is commutative. □

Lemma 6.33 *Let G be a finite group, and let H be a subgroup such that the union of the conjugates of H covers G (or equivalently, such that H intersects all the conjugacy classes of G). Then $H = G$.*

Proof. Let X be a subset of G of minimal cardinality such that G is the union of the xHx^{-1} for $x \in X$. Since H is contained in its own normalizer, of course, the number of distinct conjugates of H is no greater than $[G : H]$, so $|X| \leq [G : H]$. Let us write

$$|G| = \left| \bigcup_{x \in X} xHx^{-1} \right| \leq [G : H]\,|H| = |G|\,.$$

Thus we must have equality throughout. As the various sets xHx^{-1} are not disjoint (they all contain 1), this is only possible if there is only one of them, and $|G| = |H|$. □

Splitting fields

Definition 6.34 Let A be a central, simple F-algebra, and let E/F be a field extension. We say that E is a **splitting field** for A when $A \otimes_F E$ is Brauer equivalent to E. Equivalently, the field E is a splitting field for A, or for its Brauer class $[A]$, when $[A] \in \mathrm{Br}(E/F) = \ker(\mathrm{Br}(F) \to \mathrm{Br}(E))$.

splitting field

For example, the algebraic closure $\overline{F}$ of F is a splitting field for all algebras, since $\mathrm{Br}(\overline{F})$ is trivial. We shall be concerned with finding splitting fields which are finite-dimensional over F, and we shall find them as sub-algebras of A.

Proposition 6.35 *Let A be a central, simple F-algebra. Suppose $E \subset A$ is a field, which is also a maximal commutative subalgebra of A. Then E is a splitting field for A.*

Proof. Let V be the unique simple A-module, let $L = \mathrm{End}_A(V)$, and view V as an $A \otimes_F L$-module. This affords a homomorphism $A \otimes L \to \mathrm{End}_F(V)$, which must be injective since $A \otimes L$ is simple. By counting dimensions, we see that it is an isomorphism, and we identify $A \otimes L$ with $\mathrm{End}_F(V)$.

We can also view V as an $E \otimes L$-module, and in particular as an E-vector space. We note that the commutant of $E \otimes_F L$ in $A \otimes_F L = \mathrm{End}_F(V)$ is in fact $E \otimes_F F \cong E$, using Lemma 6.16, Corollary 6.27 (exploiting the maximality of E), and the fact that $\mathcal{Z}(L) = \mathcal{Z}(A) = F$. By the commutant theorem, we draw that $E \otimes L$ is the commutant of E in $\mathrm{End}_F(V)$, or in other words that $E \otimes_F L \cong \mathrm{End}_E(V)$. (The reader can check that the dimension of V over E is r where $[L : F] = r^2$, so that $\mathrm{End}_E(V) \cong \mathrm{M}_r(E)$.)

We have shown that E is a splitting field for the skewfield L. However $[A] = [L^{op}]$ (by Theorem 5.48, say), so $[A]^{-1}$ belongs to the subgroup $\mathrm{Br}(E/F)$ of $\mathrm{Br}(F)$, and as a result, so does $[A]$. □

Corollary 6.36 *Any central, simple F-algebra has a splitting field which is finite-dimensional over F.*

Proof. By definition, splitting fields only depend on the Brauer class, so we may as well assume that we are dealing with a skewfield K whose center is F. Pick a field E with $F \subset E \subset K$ which is maximal for this property (there is one, by finite-dimensionality of K over F). The proposition applies, showing that E is a splitting field for K. □

There is a partial converse to the proposition, which will allow us to prove that the splitting field just constructed has in fact the minimal possible dimension over F. But first, we need a lemma.

Lemma 6.37 *Let A be a central, simple F-algebra, which is a subalgebra of $\mathrm{M}_n(F)$ for some n. Let A' be its commutant. Then A' is also a central, simple F-algebra, and $[A'] = [A]^{-1}$ in $\mathrm{Br}(F)$.*

Proof. Let V be the unique, simple $\mathrm{M}_n(F)$-module, and write $V \cong W^q$ as A-modules, where W is the unique simple A-module. The algebra A' is $\mathrm{End}_A(V)$; and so $A' = \mathrm{End}_A(W^q) \cong \mathrm{M}_q(L)$ where $L = \mathrm{End}_A(V)$. However $\mathcal{Z}(L) = F$ and A is Brauer equivalent to L^{op}, as we know from Theorem 5.48. It follows that A' is indeed a central, simple F-algebra which is Brauer equivalent to A^{op}. □

Proposition 6.38 *Let A be a central, simple F-algebra. If E is a splitting field for A, then A is Brauer equivalent to an algebra B, of which E is a maximal commutative subalgebra.*

Proof. By assumption $A \otimes_F E \cong \mathrm{End}_E(V)$, for some E-vector space V. We can view V as an A-module, and consider $B = \mathrm{End}_A(V)$, which is the commutant of A in $\mathrm{End}_F(V)$. Thus B is a simple F-algebra by the commutant theorem, and it contains E. We count dimensions briefly: if $[A : F] = n^2$ and $[E : F] = q$, then we see that $\dim_E(V) = n$; by the commutant theorem we also have $[B : F][A : F] = \dim_F(V)^2 = n^2q^2$, so $[B : F] = q^2$, and thus E is a maximal commutative subalgebra of B by Corollary 6.27.

By the lemma $[B] = [A]^{-1}$, so A is Brauer equivalent to B^{op}, and E is a maximal commutative subalgebra of B^{op}, so we are done. □

Corollary 6.39 *Let K be a skewfield of dimension n^2 over $F = \mathcal{Z}(K)$. Then K has a splitting field of dimension n over F, and the dimension of any splitting field for K is a multiple of n.*

Proof. In the proof of the last corollary we have exhibited a splitting field, which had dimension n. By the proposition applied to $A = K$, the dimension of any splitting field is the square root of $[B : F]$, where B is Brauer equivalent to K. If $B \cong \mathrm{M}_p(K)$, then $[B : F] = p^2n^2$. □

Separability

We want to show that any element of the Brauer group of F has a splitting field E such that E/F is finite and Galois. Of course, if E is a splitting field for A, and if $E \subset E'$, then E' is also a splitting field for A, and thus the problem is reduced to finding E such that E/F is finite and separable (then we put E' = the Galois closure of E/F).

Most of the time, we work over a field F of characteristic 0, with the occasional remark about finite fields, so separability is automatic (and we have established the existence of splitting fields E with E/F finite). Thus the reader may have little motivation for the proof in the general case. Luckily, however, the pattern of proof is interesting, and will be reused later: After a little lemma which is specific to separable extensions, we use an inductive argument which is very general.

Lemma 6.40 *Let K be a skewfield of finite-dimension over its center $F = \mathcal{Z}(K)$. Assume that K is not a field. Then there exists $\alpha \in K$ such that $F[\alpha]/F$ is separable, and $\alpha \notin F$.*

Proof. Start with any $y \in K \smallsetminus F$. The minimal polynomial of y is of the form $f(X^{p^k}) \in F[X]$, where p is the positive characteristic of F (there being nothing to prove if the characteristic is 0), and where $f \in F[X]$ is a separable polynomial. So y^{p^k} is separable over F. If y^{p^k} is not in F, we are of course done. Otherwise, let u be a power of y such that $u \notin F$ but $u^p \in F$.

Consider the F-linear map σ of K given by $\sigma(z) = u^{-1}zu$. Then σ is not the identity since $u \notin \mathcal{Z}(K)$, but $\sigma^p = 1$, and $(\sigma - 1)^p = \sigma^p - 1 = 0$. Let $z \in K$ be such that $\sigma(z) \neq z$, and let $r \geq 1$ be such that $w := (\sigma - 1)^r(z) \neq 0$ but $(\sigma - 1)(w) = 0$. Then let $v = (\sigma - 1)^{r-1}(z)$ so that $(\sigma - 1)(v) = w$. So w is fixed by σ, and thus w^{-1} is also fixed, while $\sigma(v) = v + w$. It follows that $x = w^{-1}v$ satisfies $\sigma(x) = x + 1$.

We replace y by x and try our luck again. The minimal polynomial of x is of the form $g(X^{p^s})$ where g is separable (and irreducible), so x^{p^s} is separable over F. Can it be that $x^{p^s} \in F$, this time? This only happens when g is of degree 1, so that the minimal polynomial of x is $\mu = X^{p^s} - a$ for some $a \in F$. However, the extension $F[x]/F$ possesses an automorphism given by σ, mapping x to $x+1$, so $x+1$ has the same minimal polynomial as x; but $\mu(x+1) = \mu(x) = 0$ implies $1 = 0$. We are done, with $\alpha = x^{p^s}$. □

Here is the inductive argument mentioned above. Aside from the references to the lemma just proved, it has nothing to do with separability *per se*. Later we shall use it with "separable" replaced by "unramified", when F is a local number field.

Proposition 6.41 *Let K be a skewfield, of finite-dimension over its center $F = \mathcal{Z}(K)$. There exists a field E with $F \subset E \subset K$ which is maximal, and such that E/F is separable.*

Proof. We proceed by induction on $n = [K : F]$. There is nothing to prove if $n = 1$, so let us assume that the result holds whenever $[K : F] < n$.

Let $E_0 \subset K$ be a field obtained from the previous lemma, so that E_0/F is separable, while E_0 is strictly larger than F. We let E_0' denote the commutant of E_0 in K, so that E_0 is the center of E_0' by Corollary 6.27. We can apply the induction hypothesis to the skewfield E_0', so that there exists E with $E_0 \subset E \subset E_0'$, which is a maximal subfield of E_0', and such that E/E_0 is separable.

By transitivity, the extension E/F is separable (as E/E_0 and E_0/F both are). We show by a dimension count that E is a maximal subfield of K. Maximality in E_0' implies $[E_0' : E_0] = [E : E_0]^2$ (Corollary 6.27). On the other hand $[K : F] = [E_0 : F][E_0' : F]$ by the commutant theorem. Thus

$$[E : F]^2 = [E : E_0]^2[E_0 : F]^2 = [E_0' : E_0][E_0 : F]^2 = [E_0' : F][E_0 : F] = [K : F].$$

By Corollary 6.27 yet again, we see that E is maximal, and we are done. □

We collect what we know about splitting fields in a theorem. We have established that any $x \in \mathrm{Br}(F)$ lies in some $\mathrm{Br}(E/F)$, for some extension E/F which is finite and Galois. However, we can be more economical, based on an obvious remark: If E/F is isomorphic to E'/F, as F-extensions, and if E is a splitting field for an algebra A, then E' is also a splitting field. This allows us to look for E only among the subfields of an algebraic closure $\overline{F}$, if we want. Here is the promised summary.

Theorem 6.42 *The Brauer group* $\mathrm{Br}(F)$ *is the union of all its subgroups of the form* $\mathrm{Br}(E/F)$ *with* E/F *finite and Galois. Moreover, let* $\overline{F}$ *be an algebraic closure of* F*. Then* $\mathrm{Br}(F)$ *is also the union of its subgroups of the form* $\mathrm{Br}(E/F)$ *with* $F \subset E \subset \overline{F}$ *and* E/F *finite and Galois.*

7 Combinatorial constructions

We adopt the same conventions as in the previous chapter: All algebras are finite-dimensional over a field.

Introduction

Our purpose is the study of the group $\mathrm{Br}(E/F)$, where E/F is a finite Galois extension. How much information do we need to construct an algebra A whose Brauer class is in $\mathrm{Br}(E/F)$?

Since we work up to Brauer equivalence, we know from Proposition 6.38 that we may assume that E is a maximal, commutative subalgebra of A. Say $[E : F] = n$, then $[A : F] = n^2$ by Corollary 6.27, and so $[A : E] = n$.

There are some elements of A which we know must exist. Indeed, let $G = \mathrm{Gal}(E/F)$. For each $\sigma \in G$, we know (as in Example 6.31) that the Skolem–Noether theorem provides us with an element a_σ such that $a_\sigma x a_\sigma^{-1} = \sigma(x)$ for all $x \in E$. Of course a_σ is not unique; in fact, let us collect a few easy properties of these elements.

Proposition 7.1 *Choose an element a_σ for each $\sigma \in G$ as above. Then:*

1. *If $a'_\sigma \in A$ is another element such that conjugation by a'_σ induces σ on E, then there exists $e \in E^\times$ such that $a'_\sigma = e a_\sigma$.*
2. *There exist elements $c(\sigma, \tau) \in E^\times$ such that*

$$a_\sigma a_\tau = c(\sigma, \tau)\, a_{\sigma\tau},$$

for all $\sigma, \tau \in G$.

3. *Each $a_\sigma \in A^\times$.*
4. *The various elements a_σ, for $\sigma \in G$, form a basis for A as an E-vector space.*

Proof. (1) The element $a'_\sigma a_\sigma^{-1}$ is in the commutant of E, which is E itself by Corollary 6.27.

(2) The elements $a_\sigma a_\tau$ and $a_{\sigma\tau}$ both induce, by conjugation, the same automorphism $\sigma\tau$ of E, so (2) follows from (1).

(3) We have $a_\sigma a_{\sigma^{-1}} = c(\sigma,\sigma^{-1})a_1$ by (2), and $a_1 \in E^\times$ by (1), so a_σ is invertible.

(4) For reasons of dimensions, it is enough to prove that these elements are linearly independent. Suppose

$$\sum_{\sigma\in G} e_\sigma a_\sigma = 0, \tag{*}$$

with $e_\sigma \in E$. Assume that not all the coefficients are 0. Suppose there are in fact two different nonzero coefficients, say $e_\tau \neq 0$, $e_{\tau'} \neq 0$ for $\tau \neq \tau'$. Pick $x \in E$ such that $\tau(x) \neq \tau'(x)$. Now multiply (*) by $\tau(x)$ on the left, and by x on the right, to obtain

$$\sum_{\sigma\in G} e_\sigma(\tau(x) - \sigma(x))a_\sigma, \tag{**}$$

since $a_\sigma x = \sigma(x)a_\sigma$. The relation (**) has fewer nonzero coefficients than (*), since the coefficient of a_σ is now 0, but it still has nonzero coefficients, for example the coefficient of $a_{\tau'}$ is $\neq 0$. Using this trick several times, we arrive at a relation with only one nonzero coefficient. However, (3) shows that this is absurd. □

We see that A has a basis in bijection with the elements of G. Indeed as an E-vector space, we can identify A with $E[G]$, the group algebra of G, although the multiplication is different: elements of E and elements of the form a_σ are multiplied using $a_\sigma e = \sigma(e)a_\sigma$, and the a_σ are multiplied according to $a_\sigma a_\tau = c(\sigma,\tau)a_{\sigma\tau}$. As a result, we can reconstruct A as soon as we know the numbers $c(\sigma,\tau) \in E^\times$.

We can anticipate several minor issues. An arbitrary collection of such numbers will not define an associative multiplication, for a start; but also, different choices may very well lead to isomorphic algebras. To give a first hint of the right way to go about this, let us state the following lemma, whose proof is immediate from the proposition.

Lemma 7.2 *Let Γ be the subgroup of $A^\times$ generated by $E^\times$ and all the elements a_σ for $\sigma \in G$. Then Γ does not depend on the particular choices made for the a_σ, and in fact, it is the normalizer of $E^\times$ in $A^\times$. Moreover, there is an isomorphism $\Gamma/E^\times \cong G$, also independent of choices.* □

We shall say, starting with the next section, that Γ is an "extension of G by $E^\times$". Ultimately, it is this extension that matters, that is, the group Γ together with its homomorphism to G with kernel $E^\times$.

Our agenda for the chapter is now clear. We turn our attention to group extensions, and prove that they are controlled by numbers $c(\sigma,\tau)$ as above, this time in a fairly straightforward fashion: We discover that c must be a "cocycle", and that different cocycles give rise to equivalent extensions precisely when they differ by a "coboundary" (definitions below, of course). Then we shall prove that an extension of $\mathrm{Gal}(E/F)$ by $E^\times$ defines an algebra, called a *crossed product algebra*, which is in $\mathrm{Br}(E/F)$. It will be obvious that any algebra in $\mathrm{Br}(E/F)$ is constructed in this way (we have essentially made that analysis already).

There is more: Group extensions form a group, and we will identify this group, in our case, with $\mathrm{Br}(E/F)$. Things become very computable when $\mathrm{Gal}(E/F)$ is cyclic, and perhaps we should highlight now one of the most important results of the chapter: when E/F is a finite Galois extension with $\mathrm{Gal}(E/F)$ cyclic, we have $\mathrm{Br}(E/F) \cong F^\times / \mathrm{N}_{E/F}(E^\times)$. (See Corollary 7.22, and subsequent discussion.)

Group extensions

Here is a preliminary definition.

extension

Definition 7.3 Let G, Γ, and N be groups. We say that Γ is an **extension** of G by N when there is an injective homomorphism $\iota \colon N \to \Gamma$ and a surjective homomorphism $\pi : \Gamma \to G$ such that $\iota(N) = \ker(\pi)$. ■

(Some authors call this an extension of N by G.)

It is traditional to present the situation in a diagram, as follows:

$$1 \longrightarrow N \xrightarrow{\ \iota\ } \Gamma \xrightarrow{\ \pi\ } G \longrightarrow 1\,.$$

Here, 1 stands for the trivial group; all arrows are group homomorphisms (so the two arrows without a label are determined); and the image of each arrow is the kernel of the "next" one (that is, the one immediately to the right). Such a diagram is called a *short exact sequence*. We will see many more exact sequences in the next part of this book.

Example 7.4 At the beginning of this chapter we defined a group extension

$$1 \longrightarrow E^\times \xrightarrow{\ \iota\ } \Gamma \xrightarrow{\ \pi\ } \mathrm{Gal}(E/F) \longrightarrow 1\,.$$

■

We go back to the general case. The group $\iota(N)$ is normal in Γ, being the kernel of a homomorphism. Now, let us assume that N is *abelian*, and so also $\iota(N)$. The conjugation action of Γ on $\iota(N)$ then factors through $\Gamma/\iota(N)$, which we identify with G using π; more concretely, we have argued that for $\sigma \in G$ and $x \in \iota(N)$, we define an action by

$$\sigma \cdot x := \widetilde{\sigma} x \widetilde{\sigma}^{-1}\,,$$

where $\widetilde{\sigma} \in \Gamma$ is any element such that $\pi(\widetilde{\sigma}) = \sigma$.

Of course one has $\sigma \cdot (x_1 x_2) = (\sigma \cdot x_1)(\sigma \cdot x_2)$. Being an abelian group endowed with a group action by group automorphisms, the group $\iota(N)$ is really a $\mathbb{Z}[G]$-module! We switch to an additive notation for N, and reformulate a definition.

extension of G by M

Definition 7.5 Let G, Γ be groups (usually written multiplicatively), and let M be a $\mathbb{Z}[G]$-module (usually written additively). We say that Γ is an **extension of** G **by** M when there is an injective homomorphism of groups $\iota \colon M \to \Gamma$ and a surjective homomorphism of groups $\pi: \Gamma \to G$ such that $\iota(M) = \ker(\pi)$, and such that ι is a map of G-modules, when we endow $\iota(M)$ with the action induced by conjugation. ■

We usually depict this as follows:

$$0 \longrightarrow M \xrightarrow{\iota} \Gamma \xrightarrow{\pi} G \longrightarrow 1.$$

Here, the trivial group is written additively on the left.

Example 7.6 If we go back to the previous example, then $E^\times$ has a natural $\mathrm{Gal}(E/F)$-action, of course. This action also coincides with the action induced by Γ by conjugation, by construction, and this is obviously a key point. So Γ is really an extension of the group $\mathrm{Gal}(E/F)$ by the $\mathbb{Z}[\mathrm{Gal}(E/F)]$-module $E^\times$. Of course here the module is not written additively! In the abstract case, the additive notation will help us considerably. ■

Much as in the introduction to this chapter, but in a slightly easier setting, we analyze a given Γ which is an extension of G by M, and try to see how complicated it can be. For each $\sigma \in G$, let $s(\sigma) \in \Gamma$ be an element such that $\pi(s(\sigma)) = \sigma$. One says that the map $s: G \to \Gamma$ is a section for π. In general there is no reason for us to be able to find such an s which is a group homomorphism rather than a mere map of sets. In fact, we put $c(\sigma,\tau) = \iota^{-1}\left(s(\sigma)s(\tau)s(\sigma\tau)^{-1}\right)$, thus defining $c: G \times G \to M$, which measures the defect (s is a homomorphism exactly when c vanishes identically). Let us show that c controls everything.

We define a bijection of sets

$$\Phi: M \times G \longrightarrow \Gamma$$

by $\Phi(m,\sigma) = \iota(m)s(\sigma)$ (its inverse is $g \mapsto (\iota^{-1}(gs(\pi(g))^{-1}), \pi(g))$, an expression we shall not use). Next we *transport* the group structure of Γ to $M \times G$, by requiring Φ to be an isomorphism of groups. Thus we have a multiplication map

$$(M \times G) \times (M \times G) \longrightarrow M \times G$$
$$(m_1,\sigma),(m_2,\tau) \mapsto (\mu(m_1,\sigma,m_2,\tau), \nu(m_1,\sigma,m_2,\tau)).$$

The notation μ, ν will be short-lived. Write that Φ is a homomorphism, so that

$$\iota(m_1)s(\sigma)\iota(m_2)s(\tau) = \iota(\mu(m_1,\sigma,m_2,\tau))s(\nu(m_1,\sigma,m_2,\tau)).$$

We rework the left-hand side using that $s(\sigma)\iota(m_2) = s(\sigma)\iota(m_2)s(\sigma)^{-1}s(\sigma) = \sigma\cdot\iota(m_2)\,s(\sigma) = \iota(\sigma\cdot m_2)s(\sigma)$, using the notation for the action of G on M and on $\iota(M)$. What is more, we have $s(\sigma)s(\tau) = \iota(c(\sigma,\tau))s(\sigma\tau)$. We have reached the expression

$$\iota(m_1 + \sigma \cdot m_2 + c(\sigma,\tau))s(\sigma\tau) = \iota(\mu(m_1,\sigma,m_2,\tau))s(\nu(m_1,\sigma,m_2,\tau)).$$

Applying Φ^{-1}, we find $\mu(m_1,\sigma,m_2,\tau) = m_1 + \sigma \cdot m_2 + c(\sigma,\tau)$, and $\nu(m_1,\sigma,m_2,\tau) = \sigma\tau$.

To summarize, the group law on $M \times G$, defined so that Φ is an isomorphism, is

$$(m_1, \sigma)(m_2, \tau) = (m_1 + \sigma \cdot m_2 + c(\sigma, \tau), \sigma\tau), \tag{7.1}$$

where $c(\sigma, \tau) = \iota^{-1}\left(s(\sigma)s(\tau)s(\sigma\tau)^{-1}\right)$. In same vein, and assuming that we had chosen $s(1) = 1$ (as we may), the reader will check that

$$(m, \sigma)^{-1} = \left(-\sigma^{-1} \cdot m - c\left(\sigma^{-1}, \sigma\right), \sigma^{-1}\right). \tag{7.2}$$

It is the general shape of equations (7.1) and (7.2) which we are interested in. For example, they definitely show that c "controls everything", as was the plan.

Example 7.7 Coming back one last time to the group Γ defined in the introduction, which is an extension of $\mathrm{Gal}(E/F)$ by $E^\times$, we see that the choice of an element a_σ for each $\sigma \in \mathrm{Gal}(E/F)$ was tantamount to giving a section $s\colon \mathrm{Gal}(E/F) \to \Gamma$. The expression $c(\sigma, \tau)$ appearing in Proposition 7.1 coincides with the one being used in the ongoing discussion. ■

It is time we tried to work backwards.

Proposition 7.8 *Let G be a group, let M be a $\mathbb{Z}[G]$-module, and let $c\colon G \times G \to M$ be a map (of sets). Let $\Gamma = M \times G$ be equipped with the composition law $\Gamma \times \Gamma \to \Gamma$ defined by equation (7.1), and with the map* inv$\colon \Gamma \to \Gamma$ *defined by equation (7.2). The following assertions are equivalent.*

1. *Γ is a group for this composition law, with inverses given by the map* inv, *and it is an extension of G by M fitting inside the short exact sequence*

$$0 \longrightarrow M \xrightarrow{m \mapsto (m,1)} \Gamma \xrightarrow{(m,\sigma) \mapsto \sigma} G \longrightarrow 1.$$

2. *For all $\sigma \in G$ one has $c(\sigma, 1) = c(1, \sigma) = 0$, and for all $\sigma, \tau, \rho \in G$ one has*

$$\sigma \cdot c(\tau, \rho) + c(\sigma, \tau\rho) = c(\sigma\tau, \rho) + c(\sigma, \tau).$$

Proof. When (1) holds, the identity of Γ must be $(0, 1)$ (from the exact sequence). Spelling out that $(m, \sigma)(0, 1) = (0, 1)(m, \sigma) = (m, \sigma)$ for all m, σ, or even just for $m = 0$, shows that $c(\sigma, 1) = c(1, \sigma) = 0$ for all σ. On the other hand, writing out the associativity law

$$[(m_1, \sigma)(m_2, \tau)](m_3, \rho) = (m_1, \sigma)[(m_2, \tau)(m_3, \rho)]$$

for all $\sigma, \tau, \rho \in G$ and all $m_1, m_2, m_3 \in M$ (or even just for $m_1 = m_2 = m_3 = 0$) gives the second condition on c.

To prove the converse, one must work backwards, but the computation is exactly the same. □

normalized 2-cocycle

Definition 7.9 We say that a map $c\colon G \times G \to M$, where G is a group and M is a $\mathbb{Z}[G]$-module, is a **normalized 2-cocycle** when it satisfies condition (2) of the proposition. In this chapter, we will also say "cocycle" for short.

When c is a cocycle, we will write $M \times_c G$ for the group Γ defined as in the proposition. ∎

The group $M \times_c G$ is an extension of G by M. Conversely, the work before the proposition shows that any group Γ which is an extension of G by M is isomorphic to a group $M \times_c G$ for some cocycle c. We shall make this relationship more precise, but after some examples.

Example 7.10 We have already encountered the group D_8, which we define to be the group of 3×3-matrices with coefficients in $\mathbb{F}_2$, which are upper-triangular with 1s on the diagonal. Let us define maps $\pi_1, \pi_2, t \colon D_8 \to \mathbb{F}_2$ by writing an element $g \in D_8$ as

$$g = \begin{pmatrix} 1 & \pi_1(g) & t(g) \\ 0 & 1 & \pi_2(g) \\ 0 & 0 & 1 \end{pmatrix}.$$

Note that π_1, π_2 are group homomorphisms, but t is not. We define $\pi = (\pi_1, \pi_2)\colon D_8 \to \mathbb{F}_2 \times \mathbb{F}_2$. Finally, define $\iota \colon \mathbb{F}_2 \to D_8$ by

$$x \mapsto \begin{pmatrix} 1 & 0 & x \\ 0 & 1 & 0 \\ 0 & 0 & 1 \end{pmatrix}.$$

We caution that, in the example we are building, we have $\Gamma := D_8$ in multiplicative notation, but $M := \mathbb{F}_2$ and $G := \mathbb{F}_2 \times \mathbb{F}_2$ both naturally show up in additive notation! Anyway, there is an exact sequence

$$0 \longrightarrow \mathbb{F}_2 \xrightarrow{\ \iota\ } D_8 \xrightarrow{\ \pi\ } \mathbb{F}_2 \times \mathbb{F}_2 \longrightarrow 1.$$

Here $\iota(\mathbb{F}_2)$ is precisely the center of D_8, so we view $\mathbb{F}_2$ as a trivial G-module. We define a section $s \colon G \to D_8$ by

$$s(x_1, x_2) = \begin{pmatrix} 1 & x_1 & 0 \\ 0 & 1 & x_2 \\ 0 & 0 & 1 \end{pmatrix},$$

yielding the bijection $\Phi \colon \mathbb{F}_2 \times \mathbb{F}_2 \times \mathbb{F}_2 \to D_8$ given by

$$\varphi(z, x, y) = \begin{pmatrix} 1 & x & z \\ 0 & 1 & y \\ 0 & 0 & 1 \end{pmatrix},$$

with inverse $\Phi^{-1}(g) = (t(g), \pi_1(g), \pi_2(g))$.

To find out the cocycle c controlling the multiplication on $\mathbb{F}_2 \times \mathbb{F}_2 \times \mathbb{F}_2$, we multiply two matrices and find

$$(z_1, x_1, y_1)(z_2, x_2, y_2) = (z_1 + z_2 + x_1 y_2, x_1 + x_1, y_1 + y_2).$$

Thus $c((x_1, y_1), (x_2, y_2)) = x_1 y_2$.

There is in fact a general phenomenon here. Suppose $p_i\colon G \to R$ is a group homomorphism, for $i = 1,2$ where R is a trivial G-module but also a ring. Define $c(\sigma,\tau) = p_1(\sigma)p_2(\tau)$, using the product in R. Then c is always a cocycle. It is called the *cup-product* of p_1 and p_2, usually written $c = p_1 \smile p_2$.

Example 7.11 Let us try to think of a cocycle directly, and use Proposition 7.8 to get a group extension. Well, the easiest possible example is to take c to be identically 0. In this case $M \times_c G$ becomes a group with

$$(m_1,\sigma)(m_2,\tau) = (m_1 + \sigma \cdot m_2, \sigma\tau)$$

and also $(m,\sigma)^{-1} = (-\sigma^{-1} \cdot m, \sigma^{-1})$. This is called the *semidirect product* of M and G, written $M \rtimes G$, and we also speak of the *trivial extension*. When the action of G on M is trivial to boot, then we recognize the direct product $M \times G$ of the two groups.

The reader may be aware that semidirect products can also be defined when M is not abelian, but is an arbitrary group equipped with a map $\varphi\colon G \to \mathrm{Aut}(M)$. The notation $M \rtimes_\varphi G$ is sometimes employed.

In order to state the relationship between cocycles and extensions properly, we arrange both in equivalence classes.

Definition 7.12 Two extensions of the group G by the $\mathbb{Z}[G]$-module M are called **equivalent extensions** when there is a commutative diagram of the form

equivalent extensions

$$\begin{array}{ccccccccc} 0 & \longrightarrow & M & \longrightarrow & \Gamma & \longrightarrow & G & \longrightarrow & 1 \\ & & {\scriptstyle m\mapsto m}\big\downarrow & & \big\downarrow{\scriptstyle \varphi} & & \big\downarrow{\scriptstyle \sigma\mapsto\sigma} & & \\ 0 & \longrightarrow & M & \longrightarrow & \Gamma' & \longrightarrow & G & \longrightarrow & 1 \end{array}$$

for some isomorphism φ. The set of all equivalence classes of extensions of G by M is denoted by $\mathrm{EXT}(G,M)$.

Recall that a diagram is called commutative when, for any two groups taken from it, all the sequences of homomorphisms leading from one to the other in the diagram produce the same homomorphism when composed. In the example here, commutativity of the whole diagram amounts to the commutativity of the two inner squares.

Next we define some cocycles which we consider trivial.

Lemma 7.13 *Let $f\colon G \to M$ be any map with $f(1) = 0$, where M is a G-module. Define*

$$c(\sigma,\tau) = f(\sigma) + \sigma \cdot f(\tau) - f(\sigma\tau)$$

for $\sigma,\tau \in G$. Then c is a 2-cocycle.

Proof. Direct computation. □

coboundaries

Definition 7.14 Cocycles obtained as in the previous lemma (for at least one f which in general is not unique) are called **coboundaries**.

We write $\mathscr{Z}^2(G,M)$ for the group of cocycles, under pointwise addition, and $\mathscr{B}^2(G,M)$ for the subgroup of coboundaries. Finally we put

$$\mathrm{H}^2(G,M) = \mathscr{Z}^2(G,M)/\mathscr{B}^2(G,M)$$

second cohomology group

and call it the **second cohomology group**. ■

Theorem 7.15 *The association $c \mapsto M \times_c G$ gives rise to a bijection*

$$\mathrm{H}^2(G,M) \cong \mathrm{EXT}(G,M).$$

Proof. By the discussion above, we have a surjective map

$$\mathscr{Z}^2(G,M) \longrightarrow \mathrm{EXT}(G,M).$$

It remains to prove that two cocycles define equivalent extensions if and only if they differ by a coboundary.

So assume that we have a commutative diagram

$$\begin{array}{ccccccccc} 0 & \longrightarrow & M & \longrightarrow & M \times_c G & \longrightarrow & G & \longrightarrow & 1 \\ & & \Big\downarrow{\scriptstyle m \mapsto m} & & \Big\downarrow{\scriptstyle \varphi} & & \Big\downarrow{\scriptstyle \sigma \mapsto \sigma} & & \\ 0 & \longrightarrow & M & \longrightarrow & M \times_{c'} G & \longrightarrow & G & \longrightarrow & 1 \end{array}$$

for some isomorphism φ. The commutativity implies that $\varphi(m,\sigma) = (\theta(m,\sigma),\sigma)$ for some map θ and that $\theta(m,1) = m$. Writing that $(m,1)(0,\sigma) = (m,\sigma)$ in $M \times_c G$, applying φ, and expressing that it is a homomorphism, we draw $\theta(m,\sigma) = m + \theta(0,\sigma)$.

Finally, put $f(\sigma) = \theta(0,\sigma)$, so that $f(1) = 0$. Write that $(0,\sigma)(0,\tau) = (c(\sigma,\tau),\sigma\tau)$ in $M \times_c G$, apply the homomorphism φ, and draw that

$$f(\sigma) + \sigma \cdot f(\tau) + c'(\sigma,\tau) = c(\sigma,\tau) + f(\sigma\tau).$$

In other words, c and c' differ by the coboundary defined by f. For the converse, one has to work backwards. □

A peculiarity of this bijection is that $\mathrm{H}^2(G,M)$ is naturally an abelian group, while there is no "obvious" composition law on $\mathrm{EXT}(G,M)$. Here is a description of the induced group structure (by forcing the bijection of the theorem to be an isomorphism). Let Γ, Γ' define extensions of G by M by means of maps ι, π, ι', π'. Form the "pull-back" diagram

$$\begin{array}{ccc} \Gamma \times_G \Gamma' & \xrightarrow{p'} & \Gamma' \\ \Big\downarrow{\scriptstyle p} & & \Big\downarrow{\scriptstyle \pi'} \\ \Gamma & \xrightarrow{\pi} & G \end{array}$$

where $\Gamma \times_G \Gamma' = \{(\gamma,\gamma') \in \Gamma \times \Gamma' : \pi(\gamma) = \pi'(\gamma')\}$. The maps p and p' are the left and right projections; the kernel of each is isomorphic to M. The composition $\pi \circ p = \pi' \circ p'$ has a kernel isomorphic to $M \times M$. Let $\Delta = \{(m,-m) : m \in M\} = \{(m,0) - (0,m) : m \in M\}$, a subgroup of this kernel, and define

$$\Gamma'' = \left(\Gamma \times_G \Gamma'\right)/\Delta.$$

Lemma 7.16 *The sum of the extensions involving Γ and Γ' respectively is, up to equivalence, the extension*

$$0 \longrightarrow M \xrightarrow{m\mapsto(m,0)=(0,m)} \Gamma'' \xrightarrow{\pi\circ p=p'\circ\pi'} G \longrightarrow 1.$$

Proof. We may assume that $\Gamma = M \times_c G$ and $\Gamma' = M \times_{c'} G$ for some cocycles c and c'. We must show that the extension suggested above corresponds to the sum $c + c'$.

First consider the extension

$$0 \longrightarrow M \times M \longrightarrow \Gamma \times_G \Gamma' = (M \times M) \times_{c''} G \longrightarrow G \longrightarrow 1$$

defining the cocycle $c'' \in \mathscr{Z}^2(G, M \times M)$. A direct computation shows that

$$c''(\sigma,\tau) = (c(\sigma,\tau), c'(\sigma,\tau)).$$

The map $M \times M \to M$ mapping (m_1,m_2) to $m_1 + m_2$ has Δ as its kernel, and is obviously surjective, so $(M \times M)/\Delta \cong M$. Under this identification, the extension obtained by factoring out Δ, that is

$$0 \longrightarrow (M \times M)/\Delta \longrightarrow (\Gamma \times_G \Gamma)/\Delta \longrightarrow G \longrightarrow 1$$

is just

$$0 \longrightarrow M \longrightarrow M \times_{c+c'} G \longrightarrow G \longrightarrow 1.$$

More properly, the extension proposed in the statement of the lemma is equivalent to the last one displayed. The latter by definition is the "sum", for the group law in $\mathrm{EXT}(G,M)$, of the two given extensions. □

The neutral element for the composition law is then the semidirect product $M \rtimes G$.

First applications: cyclic groups

We will give an easy description of $\mathrm{H}^2(\mathbb{Z}/n\mathbb{Z}, M)$, for any $n \geq 1$ and any $\mathbb{Z}/n\mathbb{Z}$-module M. Things start with the following observation.

Example 7.17 Let $f : \mathbb{Z} \to \mathbb{Z}$ be defined by $f(x) = [x/n]$, the integral part of x/n, for some fixed $n \geq 1$; in other words, if x is written in a long division as $x = nq + r$ with $0 \leq r < n$, then $f(x) = q$. The expression

$$c(x,y) = f(x) + f(y) - f(x+y) = [x/n] + [y/n] - [(x+y)/n] \qquad (*)$$

is then a cocycle for the group $\mathbb{Z}$, with values in the trivial module $\mathbb{Z}$ (Lemma 7.13). Of course this cocycle is a coboundary, by construction.

However, a useful observation is that the right-hand side of (*) only depends on x modulo n and y modulo n, as a quick inspection reveals. Therefore, we may use (*) to define $c \in \mathscr{Z}^2(\mathbb{Z}/n\mathbb{Z},\mathbb{Z})$. We shall see that this cocycle is no longer a coboundary. What is more, if now M is any $\mathbb{Z}/n\mathbb{Z}$-module, and if $m \in M$ is fixed by the action, then we have an equivariant map $\mathbb{Z} \to M$ mapping 1 to m. Using the induced map $\mathscr{Z}^2(\mathbb{Z}/n\mathbb{Z},\mathbb{Z}) \to \mathscr{Z}^2(\mathbb{Z}/n\mathbb{Z},M)$, we can consider the image of c, which is a cocycle given by

$$x,y \mapsto ([x/n]+[y/n]-[(x+y)/n])\cdot m\,.$$

■

Proposition 7.18 *Let G be a finite cyclic group, let M be any G-module, and let M^G denote the submodule of elements fixed by the action of G. Finally, for $m \in M$, put $\mathrm{N}(m) = \sum_{g\in G} g\cdot m$, and write $\mathrm{N}(M)$ for the image of the map $\mathrm{N}\colon M \to M$.*

Then there is an isomorphism

$$\mathrm{H}^2(G,M) \cong M^G/\mathrm{N}(M)\,.$$

More precisely, for each generator ρ of G, one such isomorphism is induced by $\varphi_\rho\colon \mathscr{Z}^2(G,M) \longrightarrow M^G$ defined by

$$\varphi_\rho(c) = \sum_{\tau\in G} c(\tau,\rho)\,.$$

Proof. In this proof G is written multiplicatively. Let c be a cocycle. Let us add up all the relations

$$\sigma\cdot c(\tau,\rho)+c(\sigma,\tau\rho) = c(\sigma\tau,\rho)+c(\sigma,\tau)$$

for all $\tau \in G$, with $\sigma,\rho \in G$ fixed. Noting that

$$\sum_\tau c(\sigma,\tau\rho) = \sum_\tau c(\sigma,\tau)\,,$$

we are left with

$$\sigma\cdot\left(\sum_\tau c(\tau,\rho)\right) = \sum_\tau \sigma\cdot c(\tau,\rho) = \sum_\tau c(\sigma\tau,\rho) = \sum_\tau c(\tau,\rho)\,,$$

or in other words the element $\sum_\tau c(\tau,\rho)$ belongs to M^G. Choose ρ to be a generator for G, and define $\varphi = \varphi_\rho\colon \mathscr{Z}^2(G,M) \to M^G$ by $\varphi(c) = \sum_\tau c(\tau,\rho)$.

To show that φ is surjective, let $m \in M^G$, and define c by

$$c(\rho^x,\rho^y) = ([x/n]+[y/n]-[(x+y)/n])\cdot m\,,$$

where n is the order of G. This is a cocycle, as shown in the example preceding the proposition (rewritten in multiplicative notation here). We have

$$\varphi(c) = \sum_{x\in\mathbb{Z}/n\mathbb{Z}} c(\rho^x,\rho) = \left[\sum_{x\in\mathbb{Z}/n\mathbb{Z}} ([x/n]+[1/n]-[(x+1)/n]) \right] \cdot m.$$

Leaving the case $n = 1$ to the reader, we see for $n \geq 2$ that $[1/n] = 0$, and that, after cancellations happen, what remains of the sum between brackets is $[0/n] - [n/n] = -1$. So $\varphi(x) = -m$, and it follows readily that the homomorphism φ is surjective.

Next, we need to show that c is a cocycle if and only if $f(c)$ belongs to the subgroup $\mathrm{N}(M)$. First, suppose that $c \in \mathscr{B}^2(G,M)$, and write

$$c(\sigma,\tau) = f(\sigma) + \sigma \cdot f(\tau) - f(\sigma\tau).$$

Then

$$\varphi(c) = \sum_{\tau} \tau \cdot f(\rho) = \mathrm{N}(f(\rho)).$$

For the converse, suppose $\varphi(c) = \mathrm{N}(m)$ for some $m \in M$. We define a map $f\colon G \to M$ by setting $f(1) = 0, f(\rho) = m$, and requiring

$$f(\rho^x) = f\left(\rho^{x-1}\right) + \rho^{x-1} \cdot m - c\left(\rho^{x-1},\rho\right), \qquad (*)$$

or equivalently we set

$$f(\rho^x) = \sum_{h=0}^{x-1} \rho^h \cdot m - \sum_{h=0}^{x-1} c\left(\rho^h,\rho\right),$$

an expression which depends only on x modulo n; for example $f(\rho^n) = \mathrm{N}(m) - \varphi(c) = 0 = f(1)$. Let c' be the associated cocycle, that is

$$c'(\sigma,\tau) = f(\sigma) + \sigma \cdot f(\tau) - f(\sigma\tau),$$

and let us prove that $c' = c$. For this we note that it is sufficient to prove that $c(\tau,\rho) = c'(\tau,\rho)$ for all τ, for then the very definition of "cocycle" will allow us to show inductively that $c(\tau,\rho^h) = c'(\tau,\rho^h)$ for all h.

However, for $\tau = \rho^{x-1}$, the relation (*) says precisely that $c(\rho^{x-1},\rho) = c'(\rho^{x-1},\rho)$. This completes the proof. □

Corollary 7.19 *Let E/F be a finite Galois extension such that* $\mathrm{Gal}(E/F)$ *is cyclic. Then*

$$\mathrm{H}^2\left(\mathrm{Gal}(E/F),E^\times\right) \cong F^\times/\mathrm{N}_{E/F}\left(E^\times\right).$$

We give a series of applications. All the results we shall obtain are classic, easy exercises in group theory, but they will nonetheless give the reader a sense for how much information is contained in the explicit description of a cohomology group.

Let us start with examples in which M is a trivial module, that is, the elements of $G = \mathbb{Z}/n\mathbb{Z}$ act as the identity. Then one has

$$\mathrm{H}^2(G,M) = M/nM,$$

by the proposition. Specializing further, if we assume that M is finite, with order prime to n, then $\mathrm{H}^2(G,M) = \{0\}$. The only element must correspond, in $\mathrm{EXT}(G,M)$, to the direct product $M \times G$, which always exists. Thus, if we have an extension

$$0 \longrightarrow \mathbb{Z}/m\mathbb{Z} \longrightarrow \Gamma \longrightarrow \mathbb{Z}/n\mathbb{Z} \longrightarrow 0$$

where m is prime to n, and where $\mathbb{Z}/m\mathbb{Z}$ is central in Γ, we deduce that $\Gamma \cong \mathbb{Z}/m\mathbb{Z} \times \mathbb{Z}/n\mathbb{Z}$ (by the Chinese Remainder lemma, this is also $\mathbb{Z}/nm\mathbb{Z}$ of course).

We stick to trivial actions, but now we pick $n = p^r$ for some prime number p, and $M = \mathbb{Z}/p\mathbb{Z}$. Here, $\mathrm{H}^2(G,M) = \mathrm{H}^2(\mathbb{Z}/p^r\mathbb{Z}, \mathbb{Z}/p\mathbb{Z}) \cong \mathbb{Z}/p\mathbb{Z}$. Thus there are precisely p nonequivalent extensions of the form

$$0 \longrightarrow \mathbb{Z}/p\mathbb{Z} \longrightarrow \Gamma \longrightarrow \mathbb{Z}/p^r\mathbb{Z} \longrightarrow 0,$$

where $\mathbb{Z}/p\mathbb{Z}$ is central in Γ. Let us try to identify them. One possibility is the trivial extension, with $\Gamma \cong \mathbb{Z}/p\mathbb{Z} \times \mathbb{Z}/p^r\mathbb{Z}$. Another one, which is certainly not equivalent, is the extension

$$0 \longrightarrow \mathbb{Z}/p\mathbb{Z} \xrightarrow{x \mapsto \lambda p^r x} \mathbb{Z}/p^{r+1}\mathbb{Z} \xrightarrow{\pi} \mathbb{Z}/p^r\mathbb{Z} \longrightarrow 0.$$

where $1 \leq \lambda \leq p-1$, and π is the obvious projection. In fact, as λ varies, these extensions are not equivalent to one another, as the reader will check as an exercise. We have found all p extensions. All in all, there are just two possibilities for Γ: either $\mathbb{Z}/p\mathbb{Z} \times \mathbb{Z}/p^r\mathbb{Z}$ or $\mathbb{Z}/p^{r+1}\mathbb{Z}$.

Remark 7.20 The reader will also guess how one could relax the notion of "equivalence" of short exact sequences, allowing three arbitrary vertical isomorphisms, and if we did that, the last $p-1$ sequences just defined would be "isomorphic". However, it just happens that this "relaxed" notion does not lead to anything as well-behaved as the group $\mathrm{H}^2(G,M)$. ■

Let us try some nontrivial actions. Say we pick $M = \mathbb{Z}/q\mathbb{Z}$ and $G = \mathbb{Z}/p\mathbb{Z}$, where p and q are both prime. For there to be a nontrivial action at all, we need a map $\alpha\colon \mathbb{Z}/p\mathbb{Z} \to \mathrm{Aut}(\mathbb{Z}/q\mathbb{Z}) \cong \mathbb{Z}/(q-1)\mathbb{Z}$, and for this map not to be constant, we need p to divide $q-1$. Having found such a map, we have $M^G = \{0\}$ since the order of M is prime. It follows that $\mathrm{H}^2(\mathbb{Z}/p\mathbb{Z}, M) = \{0\}$, and so if we have an exact sequence

$$0 \longrightarrow \mathbb{Z}/q\mathbb{Z} \longrightarrow \Gamma \longrightarrow \mathbb{Z}/p\mathbb{Z} \longrightarrow 0$$

where the induced action on $\mathbb{Z}/q\mathbb{Z}$ is nontrivial, then Γ is the semidirect product $\mathbb{Z}/q\mathbb{Z} \rtimes_\alpha \mathbb{Z}/p\mathbb{Z}$. The index α is here to signal the dependence on the module structure, as provided by the map α, but the reader will prove that, as it turns out, the groups obtained by varying α are all isomorphic to one another. (This is not provided by the computation of the cohomology group, and must be verified separately.)

Crossed product algebras

We reverse-engineer the considerations of the introduction. We let E/F be a finite Galois extension of fields, and we pick a cocycle $c \in \mathscr{Z}^2(G,E^\times)$ where we have put $G = \mathrm{Gal}(E/F)$. From these data, we build an algebra A_c, called the crossed product algebra associated to c.

First, we want A_c to be an E-vector space with a basis in bijection with the elements of G, so that as a vector space, we have $A_c \cong E[G]$. To really bridge the notation with that of the introduction, we let a_σ be a formal symbol associated to $\sigma \in G$, and define A_c to be the free vector space on the set of all a_σ, or equivalently the vector space of formal sums

$$\sum_{\sigma \in G} e_\sigma a_\sigma ,$$

with $e_\sigma \in E$. Next we define a multiplication on A_c by

$$\left(\sum_\sigma e_\sigma a_\sigma\right) \cdot \left(\sum_\tau e'_\tau a_\tau\right) = \sum_{\sigma,\tau} e_\sigma \sigma(e'_\tau) c(\sigma,\tau) a_{\sigma\tau} .$$

(These formulae are of course obtained by mimicking Proposition 7.1.) Let Γ_c be the subset of A_c comprised of the sums with exactly one nonzero term, that is

$$\Gamma_c = \left\{ e a_\sigma : e \in E^\times, \sigma \in G \right\} \subset A_c .$$

Then Γ_c is $E^\times \times G$ as a set, and moreover, the proposed multiplication on A_c preserves Γ_c, and indeed, precisely turns it into what we have written $E^\times \times_c G$ above.

Theorem 7.21 *The crossed product algebra A_c is a central, simple F-algebra. The field E is a splitting field for A_c, so $[A_c] \in \mathrm{Br}(E/F)$. The association $c \mapsto [A_c]$ induces a bijection*

$$\mathrm{H}^2(\mathrm{Gal}(E/F), E^\times) \cong \mathrm{Br}(E/F) .$$

Proof. The proposed multiplication $A_c \times A_c \longrightarrow A_c$ is visibly bilinear. To prove that it is associative, it is therefore enough to check the condition on elements of Γ_c; however, we have already established that $\Gamma_c = E^\times \times_c G$ is a group when c is a cocycle. Thus A_c is a ring. Its unit is $1a_1$, to be written 1 from now on. We see E, and so also F, as a subring of A_c.

To find out about the center of A_c, we compute, for $\tau \in G, x \in E$:

$$\left(\sum_\sigma e_\sigma a_\sigma\right) x a_\tau = \sum_\sigma e_\sigma \sigma(x) c(\sigma,\tau) a_{\sigma\tau} = \sum_\sigma e_{\sigma\tau^{-1}} \left(\sigma\tau^{-1}\right)(x) c\left(\sigma\tau^{-1},\tau\right) a_\sigma$$

and

$$x a_\tau \left(\sum_\sigma e_\sigma a_\sigma \right) = \sum_\sigma x\tau(e_\sigma) c(\tau,\sigma) a_{\tau\sigma} = \sum_\sigma x\tau\left(e_{\tau^{-1}\sigma}\right) c\left(\tau, \tau^{-1}\sigma\right) a_\sigma .$$

Thus the element $\sum_\sigma e_\sigma a_\sigma$ is central if and only if

$$e_{\sigma\tau^{-1}} \left(\sigma\tau^{-1}\right)(x) c\left(\sigma\tau^{-1}, \tau\right) = x\tau(e_{\tau^{-1}\sigma}) c\left(\tau, \tau^{-1}\sigma\right)$$

for all σ, all τ, all x. For $\tau = 1$ we draw $e_\sigma \sigma(x) = x e_\sigma$, so that, if $\sigma \neq 1$, we can find an x with $x \neq \sigma(x)$ and we conclude $e_\sigma = 0$. As for e_1, we pick $\sigma = \tau$ and $x = 1$, and see that the condition is now $e_1 = \tau(e_1)$ for all τ, implying $e_1 \in F$. Clearly, the elements of F are central, conversely. We have $\mathcal{Z}(A_c) = F$, and A_c is a central F-algebra.

Let I be a two-sided ideal of A_c. If $I \neq \{0\}$, let $\sum_\sigma e_\sigma a_\sigma$ be a nonzero element of I. Arguing as in the proof of Proposition 7.1, we find that I contains an element of the form $e_\sigma a_\sigma$ for some $\sigma \in G$ and $e_\sigma \in E^\times$, that is, an element of Γ_c. These elements are units, so $I = A_c$. The algebra A_c is simple.

If the order of G is n, we have $[E : F] = n$ since E/F is Galois, and $[A_c : F] = n^2$. By Corollary 6.27, we deduce that E is a maximal commutative subalgebra of A_c, and by Proposition 6.35, we know that E is a splitting field for A_c. In the end, we have defined an element $[A_c]$ of $\mathrm{Br}(E/F)$.

We prove that $[A_c] = [A_{c'}]$ if and only if c and c' differ by a coboundary. First, suppose that A_c and $A_{c'}$ are Brauer equivalent, so that, for reasons of dimension, there is an isomorphism $\varphi\colon A_c \to A_{c'}$. The image $\varphi(E)$ is a simple subalgebra of $A_{c'}$ which is isomorphic to E as an F-algebra, so the Skolem–Noether theorem allows us to compose φ with an automorphism α of $A_{c'}$ so that $\alpha \circ \varphi(E) = E$, and indeed so that the restriction of $\alpha \circ \varphi$ to E is the identity. Write $\psi = \alpha \circ \varphi$. Characterizing Γ_c and $\Gamma_{c'}$ as in Lemma 7.2, we see that $\psi(\Gamma_c) = \Gamma_{c'}$, and indeed that ψ can be inserted into a commutative diagram providing an equivalence between the two extensions of G by $E^\times$ defined by Γ_c and $\Gamma_{c'}$. By Theorem 7.15, the cocycles c and c' differ by a coboundary.

Conversely, suppose $c - c' \in \mathscr{Z}^2(G, E^\times)$, so that Theorem 7.15 gives the existence of an isomorphism $\psi\colon \Gamma_c \to \Gamma_{c'}$ fitting in an appropriate commutative diagram; in particular ψ is the identity on $E^\times$. We define a map $A_c \to A_{c'}$ by

$$\sum_\sigma e_\sigma a_\sigma \longrightarrow \sum_\sigma e_\sigma \psi(a_\sigma) .$$

This is easily seen to be an isomorphism of F-algebras, as the reader will check.

We have established that $\mathrm{H}^2(G, E^\times)$ injects into $\mathrm{Br}(E/F)$. On the other hand, we have begun this chapter with arguments establishing that any element of $\mathrm{Br}(E/F)$ is Brauer equivalent to some A_c. □

It is remarkable that the next corollary does not involve cohomology in its statement.

Corollary 7.22 *Suppose E/F is a finite Galois extension of fields, such that $\mathrm{Gal}(E/F)$ is cyclic. Then*

$$\mathrm{Br}(E/F) \cong F^\times / \mathrm{N}_{E/F}(E^\times).$$

Proof. Combine the theorem with Corollary 7.19. □

In both the theorem and the corollary, we have bijections of sets. As the reader may suspect, we shall improve these to isomorphisms of groups.

Example 7.23 Let us describe $\mathrm{Br}(\mathbb{R}) = \mathrm{Br}(\mathbb{C}/\mathbb{R})$ (since $\mathbb{C}$ is algebraically closed). We have $\mathrm{N}_{\mathbb{C}/\mathbb{R}}(\mathbb{C}^\times) = \{z\bar{z} : z \in \mathbb{C}^\times\} = \mathbb{R}^{>0}$, so $\mathrm{Br}(\mathbb{R}) \cong \{\pm 1\}$ has order 2. Thus, there exist precisely two skewfields, finite-dimensional over $\mathbb{R}$, whose center is $\mathbb{R}$. We know them already, of course, these are $\mathbb{R}$ and $\mathbb{H}$, but now we know that there are no others. ■

Example 7.24 Here is another proof of Wedderburn's theorem. Let K be a finite skewfield, and let $F = \mathcal{Z}(K)$. Of course F is a finite field. We claim that all the groups $\mathrm{Br}(E/F)$ are trivial, so that $\mathrm{Br}(F)$ is trivial, and thus there is only one finite skewfield whose center is F, namely F itself. As a result $K = F$ is commutative.

To prove the claim, we use of course that $\mathrm{Br}(E/F) \cong F^\times / \mathrm{N}_{E/F}(E^\times)$ and rely on a simple counting argument (together with basic facts about finite fields). Let $|F| = q$ and $|E| = q^n$. The Galois group $\mathrm{Gal}(E/F)$ is generated by $x \mapsto x^q$, so $\mathrm{N}_{E/F}(x) = xx^qx^{q^2} \cdots x^{q^{n-1}} = x^{\frac{q^n-1}{q-1}}$ for $x \in E^\times$. However, the group $E^\times$ is cyclic of order $q^n - 1$, so the elements of $\frac{q^n-1}{q-1}$-torsion comprise the unique subgroup of index $q - 1$. It follows that the norm map is surjective. ■

Compatibilities

We keep the notation as above. We want to prove that the bijection of Theorem 7.21 is really an isomorphism of groups. We also intend to understand how the various cohomology groups can be glued together to form $\mathrm{Br}(F)$, which is itself the union of the subgroups $\mathrm{Br}(E/F)$.

We start with a lemma, helping us build modules over crossed product algebras. As an analogy, a module over a group algebra $F[G]$ can be alternatively described as an F-vector space V endowed with an action of the group G by F-linear maps, that is, a map $G \to \mathrm{Aut}_F(V)$. Likewise, modules over the crossed product algebra A_c, which bears a resemblance to both $F[\Gamma_c]$ and $E[G]$, can be characterized as vector spaces with special group actions:

Lemma 7.25 *Let V be an abelian group. Then a structure of A_c-module on V is specified precisely by the structures (1) and (2) below, satisfying condition (3).*

1. *A structure of E-vector space on V. In particular, V is an F-vector space, and one has a group homomorphism $\rho_1 : E^\times \to \mathrm{Aut}_F(V)$.*

2. *An action of Γ_c on V by means of F-linear maps, that is, a group homomorphism $\rho_2\colon \Gamma_c \to \mathrm{Aut}_F(V)$.*
3. *The map ρ_1 coincides with the restriction of ρ_2 to $E^\times$.*

The obvious proof is left to the reader.

Proposition 7.26 *The bijection* $\mathrm{H}^2(\mathrm{Gal}(E/F), E^\times) \cong \mathrm{Br}(E/F)$ *induced by* $c \mapsto A_c$ *is a group isomorphism.*

Proof. Let $A = A_c$, $B = A_{c'}$, and $C = A_{cc'}$. (Note: since $E^\times$ is written multiplicatively, we also write cocycles multiplicatively, since they are maps $G \times G \to E^\times$, where we put $G = \mathrm{Gal}(E/F)$.) We want to show that C is Brauer equivalent to $A \otimes_F B$.

Our strategy is this. Put $n = |G| = [E:F]$, so $[A:F] = [B:F] = [C:F] = n^2$ and $[A \otimes B : F] = n^4$. We seek to define a vector space V of dimension $\leq n^3$ over F, which is simultaneously a C-module on the left, and an $A \otimes B$-module on the right, in such a way that the actions commute (in other words, a C–$(A \otimes B)^{op}$-bimodule, or a $C \otimes (A \otimes B)^{op}$-module!). Suppose we had achieved this. Then the simple algebras C and $(A \otimes B)^{op}$ are both identified with subalgebras of $\mathrm{End}_F(V)$, and each is included in the commutant of the other. However,

$$[C:F][(A \otimes B)^{op}:F] = n^2 \cdot n^4 = n^6 \geq [\mathrm{End}_F(V):F],$$

so by the commutant theorem, we see that C and $(A \otimes B)^{op}$ are each other's commutants, and in passing that V has dimension n^3. By Lemma 6.37 we have $[C] = [(A \otimes B)^{op}]^{-1} = [A \otimes B]$.

We turn to the construction of V. Essentially we take $V = A \otimes_E B$, but to conduct the proof it is easier to reconstruct this as a quotient of $A \otimes_F B$. (In the rest of this proof $\otimes = \otimes_F$.) We write the elements of A as $\sum_\sigma x_\sigma a_\sigma$ with $x_\sigma \in E$, and those of B as $\sum_\sigma y_\sigma b_\sigma$ with $y_\sigma \in E$; both units are written as 1, and we see E as a subalgebra of both A and B. Now let U denote the *right* ideal of $A \otimes B$ generated by all the elements $e \otimes 1 - 1 \otimes e$ for $e \in E$, and $V = (A \otimes B)/U$. It is immediate that V is a right $A \otimes B$-module.

To turn V into a C-module (on the left), we use Lemma 7.25. First we note that U is stable by left multiplication by elements of the form $x \otimes 1$ with $x \in E$, for

$$(x \otimes 1)(e \otimes 1 - 1 \otimes e) = xe \otimes 1 - x \otimes e = (e \otimes 1 - 1 \otimes e)(x \otimes 1).$$

Likewise for elements of the form $1 \otimes x$. Thus, it makes sense to multiply elements of V by expressions of the form $x \otimes 1$ or $1 \otimes x$ with $x \in E$, and since $x \otimes 1 - 1 \otimes x \in U$, we have in fact

$$(x \otimes 1) \cdot v = (1 \otimes x) \cdot v$$

for $v \in V$. So V is an E-vector space in a natural way, using either expression. We may now consider the map $A \times B \to V$ mapping (a,b) to $a \otimes b \bmod U$. This is E-bilinear (we consider A and B as E-vector spaces using multiplication on the left), so we have a linear map $A \otimes_E B \to V$ which is surjective. This shows that $\dim_E(V) \leq n^2$, so $\dim_F(V) \leq n^3$. As the rest of the proof will show that $\dim_F(V) = n^3$, as already pointed out, we have in fact $V \cong A \otimes_E B$; yet we keep working with the definition of V as $(A \otimes_F B)/U$.

Next, we observe the slightly more sophisticated fact that U is stable by multiplication on the left by elements of the form $a_\sigma \otimes b_\sigma$, for some $\sigma \in G$. Indeed,

$$(a_\sigma \otimes b_\sigma)(e \otimes 1 - 1 \otimes e) = a_\sigma e \otimes b_\sigma - a_\sigma \otimes b_\sigma e = (\sigma(e) \otimes 1 - 1 \otimes \sigma(e))(a_\sigma \otimes b_\sigma).$$

Let us bring some groups into the picture. Consider the subgroup of $(A \otimes B)^\times$ generated by $E^\times \otimes 1$, $1 \otimes E^\times$, and all elements $a_\sigma \otimes b_\tau$ for $\sigma, \tau \in G$: It is clearly isomorphic to $\Gamma_c \times \Gamma_{c'}$. The subgroup generated by $E^\times \otimes 1$, $1 \otimes E^\times$, and all elements $a_\sigma \otimes b_\sigma$ for $\sigma \in G$ is the pullback group $\Gamma_c \times_G \Gamma_{c'}$ as in the discussion before Lemma 7.16.

Combining the above observations, we see that $\Gamma_c \times_G \Gamma_{c'}$ acts on V by linear maps, and also that

$$\left(e \otimes e^{-1}\right) \cdot v = (e \otimes 1)\left(1 \otimes e^{-1}\right) \cdot v = (e \otimes 1)\left(e^{-1} \otimes 1\right) \cdot v = (1 \otimes 1) \cdot v = v.$$

So the action factors through the quotient $(\Gamma_c \times_G \Gamma_{c'})/\Delta$ as in Lemma 7.16. The same lemma shows that this group is $\Gamma_{cc'}$. The conditions of Lemma 7.25 are visibly satisfied, so V is a C-module. The two module actions commute by construction. □

Here is a little application: An easy statement about cohomology groups giving a much less obvious property of Brauer groups.

Lemma 7.27 *Let G be a group of order n, and let M be a $\mathbb{Z}[G]$-module. Then any cohomology class in $\mathrm{H}^2(G,M)$ is n-torsion.*

As a result, if $x \in \mathrm{Br}(E/F)$ with $[E : F] = n$, then $x^n = 1$. What is more, any element in $\mathrm{Br}(F)$ has finite order.

Proof. Let $c \in \mathscr{Z}^2(G,M)$ be a cocycle, and let us prove that nc is a coboundary. Write the defining condition

$$\sigma \cdot c(\tau,\rho) + c(\sigma,\tau\rho) = c(\sigma\tau,\rho) + c(\sigma,\tau).$$

Take the sum over all ρ, and after a little rearrangement we get

$$\sigma \cdot \left(\sum_\rho c(\tau,\rho)\right) + \left(\sum_\rho c(\sigma,\rho)\right) = \left(\sum_\rho c(\sigma\tau,\rho)\right) + nc(\sigma,\tau).$$

So if we put $f(\sigma) = \sum_\rho c(\sigma,\rho)$, then $nc(\sigma,\tau) = f(\sigma) + \sigma \cdot f(\tau) - f(\sigma\tau)$, and nc is indeed a coboundary.

The application to Brauer groups is clear: $\mathrm{Br}(E/F)$ is a cohomology group for $\mathrm{Gal}(E/F)$, whose order is $[E : F]$, and $\mathrm{Br}(F)$ is the union of all the subgroups $\mathrm{Br}(E/F)$. □

Having described $\mathrm{Br}(E/F)$ as a cohomology group, we would like to turn to $\mathrm{Br}(F)$, which is the union of them all. We need to investigate how $\mathrm{H}^2(G,M)$ varies when we change either G or M.

When there is a map $M \to M'$ of $\mathbb{Z}[G]$-modules, we also have a map $\mathrm{H}^2(G,M) \to \mathrm{H}^2(G,M')$ defined by composing cocycles (we have used it already in Example 7.17). On the other hand, when there is a group homomorphism $G' \to G$, there is a map $\mathrm{H}^2(G,M) \to \mathrm{H}^2(G',M)$, obtained by precomposing cocycles. The two can be combined, when we are in the presence of a $\mathbb{Z}[G]$-module M, together with a map $G' \to G$ allowing us to see M as a $\mathbb{Z}[G']$-module, and another $\mathbb{Z}[G']$-module M' with a G'-equivariant map $M \to M'$. Then we have an induced homomorphism $\mathrm{H}^2(G,M) \to \mathrm{H}^2(G',M')$, which is really the composition of $\mathrm{H}^2(G,M) \to \mathrm{H}^2(G',M)$ followed by $\mathrm{H}^2(G',M) \to \mathrm{H}^2(G',M')$.

We do not introduce any special notation for these maps (for now), since they are usually the only homomorphisms which make any sense. However, they are collectively, and somewhat loosely, called *inflation maps* (especially when $G' \to G$ is surjective).

The typical situation for us arises when $F \subset E_0 \subset E$ with E/F and E_0/F both finite and Galois. Then $\mathrm{Gal}(E_0/F)$ is a quotient of $\mathrm{Gal}(E/F)$, and $E_0^\times$ is a sub-$\mathrm{Gal}(E/F)$-module of $E^\times$. Thus we have an inflation map of the form $\mathrm{H}^2(\mathrm{Gal}(E_0/F),E_0^\times) \to \mathrm{H}^2(\mathrm{Gal}(E/F),E^\times)$.

Example 7.28 To get acquainted with these induced maps, without the context of Galois theory, we propose the following exercise for the reader. (The result will not be used in this chapter, but it will be at the end of the next one.) Suppose G and G' are both cyclic, with chosen generators ρ and ρ' respectively, and that we have a (surjective) map $G' \to G$ taking ρ' to ρ. Then the following diagram is commutative:

$$\begin{array}{ccc} \mathrm{H}^2(G,M) & \xrightarrow{\varphi_\rho} & M^G/\mathrm{N}_G(M) \\ {\scriptstyle\mathrm{inf}}\big\downarrow & & \big\downarrow{\scriptstyle m\mapsto \frac{|G'|}{|G|}m} \\ \mathrm{H}^2(G',M) & \xrightarrow{\varphi_{\rho'}} & M^G/\mathrm{N}_{G'}(M). \end{array}$$

The notation is as in Proposition 7.18; we have written N_G and $\mathrm{N}_{G'}$ rather than just N, for clarity. Note that it takes a little verification to make sure that the proposed map, that is multiplication by the index $\frac{|G'|}{|G|}$, takes $\mathrm{N}_G(M)$ into $\mathrm{N}_{G'}(M)$. On the other hand $M^G = M^{G'}$ (no typo!).

Here is a second one. We assume that G is cyclic of order dn, generated by ρ, and consider the cyclic subgroup G' of order n generated by ρ^d. This time we have a commutative diagram as follows:

$$\begin{array}{ccc} \mathrm{H}^2(G,M) & \xrightarrow{\varphi_\rho} & M^G/\mathrm{N}_G(M) \\ \big\downarrow & & \big\downarrow \\ \mathrm{H}^2(G',M) & \xrightarrow{\varphi_{\rho^d}} & M^{G'}/\mathrm{N}_{G'}(M). \end{array}$$

The vertical maps are the "obvious" ones (people usually call the leftmost one the "restriction" map, for obvious reasons). It is a little trickier to prove the commutativity, so here is a hint. Starting with $c_0 \in \mathscr{Z}^2(G,M)$, let $m = \varphi_\rho(c_0)$. Then the proof of Proposition 7.18 shows that the formula

$$c\left(\rho^x, \rho^y\right) = ([(x+y)/n] - [x/n] - [y/n]) \cdot m$$

defines a new cocycle c with $\varphi_\rho(c) = m$; so c and c_0 differ by a coboundary, and we discard c_0 altogether to work with c. Having the cocycle in this "canonical form", it is now straightforward to restrict it to G', apply φ_{ρ^d}, and see that we again obtain m. ∎

Returning to the context of Brauer groups, our goal is to prove:

Proposition 7.29 *The following diagram is commutative:*

$$\begin{array}{ccc} \mathrm{H}^2\left(\mathrm{Gal}(E_0/F), E_0^\times\right) & \xrightarrow{\text{inf}} & \mathrm{H}^2\left(\mathrm{Gal}(E/F), E^\times\right) \\ \cong\downarrow & & \downarrow\cong \\ \mathrm{Br}(E_0/F) & \xrightarrow{\subset} & \mathrm{Br}(E/F) \end{array}$$

Here the vertical maps are the isomorphisms considered in this chapter, the top horizontal map is the inflation, and the bottom horizontal map is the inclusion.

Proof. Let us write $G = \mathrm{Gal}(E/F)$ and $\overline{G} = \mathrm{Gal}(E_0/F)$, so $\overline{G}$ is a quotient of G. Let $c \in \mathscr{Z}^2(\overline{G}, E_0^\times)$ be a cocycle, and let $c' \in \mathscr{Z}^2(G, E^\times)$ be the cocycle obtained by the process described above, representing the inflation (so c' is c precomposed with $G \to \overline{G}$, and postcomposed with $E_0^\times \to E^\times$). We need to prove that the algebras A_c and $A_{c'}$ are Brauer equivalent.

The strategy is the same as that employed in the proof of Proposition 7.26, so we look for an F-vector space V which is an $A_{c'}$-module on the left, and an A_c-module on the right. Recall that the elements of A_c are formal sums $\sum_{\bar\sigma \in \overline{G}} x_{\bar\sigma} a_{\bar\sigma}$ with $x_{\bar\sigma} \in E_0$, and those of $A_{c'}$ are formal sums $\sum_{\sigma \in G} y_\sigma b_\sigma$ with $y_\sigma \in E$. We define V to be halfway between these, by taking it to be the E-vector space of formal sums $\sum_{\bar\sigma \in \overline{G}} y_{\bar\sigma} v_{\bar\sigma}$ where we do allow $y_{\bar\sigma} \in E$. Alternatively, $V = A_{c'}/U$ where U is the (left) E-vector space spanned in $A_{c'}$ by the elements $b_\sigma - b_\tau$ where σ and τ map to the same element in $\overline{G}$.

This U is stable by left multiplication, since

$$b_\rho(b_\sigma - b_\tau) = c'(\rho,\sigma)b_{\rho\sigma} - c'(\rho,\tau)b_{\rho\tau} = c'(\rho,\sigma)(b_{\rho\sigma} - b_{\rho\tau}),$$

noting that $c'(\rho,\sigma) = c'(\rho,\tau)$, given the definition of c'. Thus $A_{c'}$ acts on V on the left. As for the action of A_c on the right, we let $x_{\bar\sigma} a_{\bar\sigma}$ act on $e_{\bar\tau} v_{\bar\tau}$ via

$$e_{\bar\tau} v_{\bar\tau} \cdot x_{\bar\sigma} a_{\bar\sigma} = e_{\bar\tau} \bar\tau(x_{\bar\sigma}) c(\bar\tau, \bar\sigma) v_{\bar\tau\bar\sigma}\,.$$

This mimics the multiplication on A_c, and therefore associativity is clear. We have our bimodule.

We let the reader perform the simple dimension count, and finish the argument based on the commutant theorem. □

The cohomological Brauer group

We want to express that $\mathrm{Br}(F)$ can be recovered from all the cohomology groups $\mathrm{H}^2(\mathrm{Gal}(E/F), E^\times)$, where E/F ranges through the finite Galois extensions of F. For this, we need a little algebraic gadget which some readers may not know, namely, the language of colimits. We have met limits in Chapter 3 when discussing profinite groups, and what we will do here is the exact same thing *reversing all the arrows*. We cut corners by restricting to colimits of *abelian* groups, though.

Suppose I is a poset, suppose that we have an abelian group G_i for each $i \in I$, and finally suppose that whenever $i \leq j$ we have a homomorphism $\varphi_{ij} : G_i \to G_j$ (with φ_{ii} the identity of G_i). Consider the direct sum

$$\bigoplus_{i \in I} G_i,$$

which is typically " large" as I can very well be a complicated set, and let R be the subgroup generated by all the elements of the form $x - \varphi_{ij}(x)$, for $x \in G_i$. We put

$$\mathrm{colim}_{i \in I}\, G_i := \left(\bigoplus_{i \in I} G_i \right) / R.$$

This is called the *colimit*, or *direct limit*, or *inductive limit*, of the groups G_i. Note that we have maps $\alpha_j : G_j \to \mathrm{colim}_i\, G_i$, for each $j \in I$, induced by the inclusion of G_j in the direct sum.

Recall that I is called a directed set when, given $i,j \in I$, there always exists $k \in I$ with $i \leq k$ and $j \leq k$. Much as with limits, we have a better grip over colimits when the indexing set is directed.

Lemma 7.30 *Let H be an abelian group, let $\theta_i : G_i \to H$ be a homomorphism for each $i \in I$, and suppose that $\theta_j \circ \varphi_{ij} = \theta_i$ for all i,j. Then there is a unique homomorphism $\theta : \mathrm{colim}_i\, G_i \to H$ such that $\theta \circ \alpha_i = \theta_i$.*

Moreover, if I is directed, and if each θ_i is injective, then θ is also injective.

Proof. The first statement, which is usually called the "universal property of colimits", is obvious by construction. We turn to the second.

Let $x = x_{i_1} + \cdots + x_{i_s} \in \ker(\theta)$, where $x_{i_j} = \alpha_{i_j}(z_{i_j})$ for some $z_{i_j} \in G_{i_j}$. We can find $k \in I$ such that $i_j \leq k$ for all j, and we put $y_j = \varphi_{i_j,k}(z_{i_j}) \in G_k$. Finally, set $y = y_1 + \cdots + y_s \in G_k$.

We have then $x = \alpha_k(y)$, so that $0 = \theta(x) = \theta_k(y)$, implying that $y = 0$ since θ_k is assumed to be injective. As a result, $x = 0$. □

Example 7.31 Let F be a fixed abelian group, let $I = \mathbb{N}$, and for each n let $V_n = F^n$. Whenever $n \leq m$, we define $\varphi_{nm} : F^n \to F^m$ by $\varphi_{nm}(x_1, \ldots, x_n) = (x_1, \ldots, x_n, 0, \ldots, 0)$. One can form $V^\infty = \mathrm{colim}_n\, V_n$, and from the lemma, the reader will prove easily that it is isomorphic to the group of all sequences $(x_n)_{n\geq 1}$ of elements of F which are 0 for all but finitely many indices. ■

Remark 7.32 When each G_i is a vector space over some fixed field F, and each φ_{ij} is F-linear, we see that R is a subspace of the direct sum, and thus the colimit is naturally an F-vector space, too. In the preceding example, one can take F to be a field, and the isomorphism with the space of sequences is in fact linear. (In the sequel, this will not matter much.) ■

Example 7.33 Let G be an abelian group, with a collection of subgroups $(G_i)_{i\in I}$ such that G is the union of the various G_i, and such that G_i and G_j are always both contained in some G_k. Then from the lemma one proves immediately that $G \cong \operatorname{colim}_i G_i$, using inclusion maps as the φ_{ij}. ■

This last example shows that $\mathrm{Br}(F) \cong \operatorname{colim}_E \mathrm{Br}(E/F)$, where the indexing set is that of all the fields E with $F \subset E \subset \overline{F}$ and E/F finite and Galois (cf. Theorem 6.42). (We restrict to $E \subset \overline{F}$ just to be sure that the indexing set is indeed a set. If you don't see the issue, ignore it.)

Now, we have identified $\mathrm{Br}(E/F)$ with a cohomology group, and the inclusion map $\mathrm{Br}(E_0/F) \to \mathrm{Br}(E/F)$ with an inflation map (Propositions 7.26 and 7.29). This identifies the colimit just considered with a colimit of cohomology groups, or in symbols:

$$\mathrm{Br}(F) \cong \operatorname{colim}_E \mathrm{H}^2\left(\mathrm{Gal}(E/F), E^\times\right).$$

We are going to state this as a theorem, but we introduce some notation first.

Definition 7.34 Let G be a profinite group, and let M be an abelian group with a map $G \to \mathrm{Aut}(M)$ defining an action. For each open, normal subgroup U, let $M^U = \{m \in M : x \cdot m = m \text{ for all } x \in U\}$. When $U \subset V$ we have inflation maps

$$\mathrm{H}^2\left(G/V, M^V\right) \longrightarrow \mathrm{H}^2\left(G/U, M^U\right).$$

topological cohomology group

The **topological cohomology group** is defined to be

$$\mathrm{H}^2_{top}(G,M) := \operatorname{colim}_U \mathrm{H}^2\left(G/U, M^U\right).$$

■

A few comments are in order. First, recall that G/U is finite when U is open in the profinite group G; in fact, we have shown that $G = \lim_U G/U$ (in the proof of Proposition 3.13). Second, in later chapters we will need to impose some conditions on the map $G \to \mathrm{Aut}(M)$ (for example, if there is a topology on M, we may wish for this map to be continuous); to leave the door open for this, we have not merely stated that M is a $\mathbb{Z}[G]$-module. Finally, we note that in practice, one often drops the "top" part of the notation, when it is clear that G is profinite and we do not mean to consider the cohomology of the "abstract" group.

For now, we will merely look at this in the example of $G = \mathrm{Gal}(\overline{F}/F)$ and $M = \overline{F}^\times$ with its usual action. Each open, normal subgroup U is of the form $U = \mathrm{Gal}(\overline{F}/E)$, and $G/U = \mathrm{Gal}(E/F)$, where E/F is finite and Galois, by the fundamental theorem of Galois theory. Moreover, we have $M^U = E^\times$. So here we have

$$\mathrm{H}^2_{top}(\mathrm{Gal}(\overline{F}/F), \overline{F}^\times) \cong \mathrm{colim}_E\, \mathrm{H}^2(\mathrm{Gal}(E/F), E^\times) \cong \mathrm{Br}(F).$$

The topological cohomology group on the left-hand side is usually written as $\mathrm{H}^2(F, \overline{F}^\times)$ (or sometimes $\mathrm{H}^2(*/F, \overline{F}^\times)$, or sometimes $\mathrm{H}^2(F, \mathbb{G}_m)$, where $\mathbb{G}_m$ means "the multiplicative group"). In words, we call it the *cohomological Brauer group*. Thus we have:

Theorem 7.35 *The cohomological Brauer group of F is isomorphic to the Brauer group. That is, there is an isomorphism*

$$\mathrm{Br}(F) \cong \mathrm{H}^2\left(F, \overline{F}^\times\right).$$

Also, each map $\mathrm{H}^2(\mathrm{Gal}(E/F), E^\times) \to \mathrm{H}^2(F, \overline{F}^\times)$ *is injective.*

Naturality

We conclude this chapter with the following question: How does $\mathrm{Br}(F)$, as identified with a colimit of cohomology groups, vary with F? For simplicity, we restrict attention to a finite extension F_1/F_0, and wish to describe the homomorphism $\mathrm{Br}(F_0) \to \mathrm{Br}(F_1)$.

A little general fact about colimits will help. Suppose we want to compare $\mathrm{colim}_{i \in I} G_i$ and $\mathrm{colim}_{i \in J} G_i$ where $J \subset I$. There is a simple case when the two colimits will be the same (or rather, naturally isomorphic), as the reader will check: When any $i \in I$ satisfies $i \leq j$ for some $j \in J$, which one summarizes by saying that J is "cofinal" in I.

In the colimit defining the cohomological Brauer group of F_0, the indexing set is comprised of all the fields $E \subset \overline{F}_0$ with E/F_0 finite and Galois. We build a cofinal set as follows. Fix a field $F_1 \subset \overline{F}_0$ which is a finite extension of F_0, so that $\overline{F}_1 = \overline{F}_0$. Consider then the set S of fields E with $F_1 \subset E \subset \overline{F}_0$ such that E/F_0 is finite and Galois (and thus E/F_1 is finite and Galois, too). Then S is cofinal in *both* the indexing set for $\mathrm{Br}(F_0)$ and that for $\mathrm{Br}(F_1)$, so that

$$\mathrm{Br}(F_0) \cong \mathrm{colim}_{E \in S}\, \mathrm{H}^2(\mathrm{Gal}(E/F_0), E^\times), \quad \mathrm{Br}(F_1) \cong \mathrm{colim}_{E \in S}\, \mathrm{H}^2(\mathrm{Gal}(E/F_1), E^\times).$$

We wish to compare the two, and clearly the map

$$\mathrm{H}^2\left(\mathrm{Gal}(E/F_0), E^\times\right) \longrightarrow \mathrm{H}^2\left(\mathrm{Gal}(E/F_1), E^\times\right)$$

will be instrumental. It is obtained by restricting cocycles from the group $G = \mathrm{Gal}(E/F_0)$ to the subgroup $G' = \mathrm{Gal}(E/F_1)$, and thus is often called the "restriction map".

With this setup, we have the following proposition. Recall that the map $\mathrm{Br}(F_0) \to \mathrm{Br}(F_1)$ is induced by the tensor product operation, as in the previous chapter, and that it maps $\mathrm{Br}(E/F_0)$ into $\mathrm{Br}(E/F_1)$, for $E \in S$.

Proposition 7.36 *For any $E \in S$ there is a commutative diagram*

$$\begin{array}{ccc} \mathrm{Br}(E/F_0) & \longrightarrow & \mathrm{Br}(E/F_1) \\ \cong\downarrow & & \downarrow\cong \\ \mathrm{H}^2\left(\mathrm{Gal}(E/F_0), E^\times\right) & \longrightarrow & \mathrm{H}^2\left(\mathrm{Gal}(E/F_1), E^\times\right), \end{array}$$

where all the maps are those introduced in this chapter. In particular, $\mathrm{Br}(F_1/F_0)$ *can be identified with the kernel of the "restriction map" in cohomology.*

It follows that there is a commutative diagram

$$\begin{array}{ccc} \mathrm{Br}(F_0) & \longrightarrow & \mathrm{Br}(F_1) \\ \cong\downarrow & & \downarrow\cong \\ \mathrm{H}^2\left(F_0, \overline{F}_0^\times\right) & \longrightarrow & \mathrm{H}^2\left(F_1, \overline{F}_1^\times\right). \end{array}$$

Proof. We start by explaining how the first statement implies the others. Indeed, consider the second diagram, which we want to show is commutative. What it claims is that two maps defined on $\mathrm{Br}(F_0)$ agree. However, since $\mathrm{Br}(F_0)$ is the union of the subgroups $\mathrm{Br}(E/F_0)$ for $E \in S$, it is enough to check that the restrictions of our two maps agree on each $\mathrm{Br}(E/F_0)$; and of course, this is implied by the commutativity of the first diagram.

Also, the group $\mathrm{Br}(F_1/F_0)$ is contained in $\mathrm{Br}(E/F_0)$, and it is the kernel of $\mathrm{Br}(E/F_0) \to \mathrm{Br}(E/F_1)$, as we see by examining the definitions. So the "in particular" statement is clear.

The key point is then the commutativity of the first diagram, and we turn to this. After a rapid translation, what we must prove is this: Let $c \in \mathscr{Z}^2(\mathrm{Gal}(E/F_0), E^\times)$ be a cocycle, and let c' be its restriction to the subgroup $\mathrm{Gal}(E/F_1)$. We have the crossed product algebra A_c, over F_0, and the crossed product algebra $A_{c'}$ over F_1, and we must establish that $A_c \otimes_{F_0} F_1$ is Brauer equivalent to $A_{c'}$.

Unsurprisingly (by now), this is done by exhibiting an appropriate bimodule V. Here, we take V to be the F_1-vector space underlying A_c. Note that, in the current situation, the algebra $A_{c'}$ is a subalgebra of A_c, so it acts on it on the right by multiplication, and we see V as a right $A_{c'}$-module.

Further, A_c acts on itself on the left; the field F_1 is viewed as a subalgebra of A_c, and we make it act *on the right* by multiplication, so that the operation commutes with the previous one, and V is now viewed as an $A_c \otimes_{F_0} F_1^{op}$-module on the left. Of course $F_1^{op} = F_1$, so we have the desired bimodule. We let the reader complete the argument (dimension count, and commutant theorem, for the third time in this chapter). □

Problems

7.1. Check the claim made in Example 7.10, that cup-products define cocycles.

7.2. Prove the claims made in Example 7.28.

7.3. Finish the argument in the proof of Proposition 7.29.

7.4. Finish the argument in the proof of Proposition 7.36.

7.5. Let

$$0 \longrightarrow M \longrightarrow \Gamma \overset{\pi}{\longrightarrow} G \longrightarrow 1$$

be an extension of the group G by the $\mathbb{Z}[G]$-module M, and let $\alpha \in \mathrm{H}^2(G,M)$ be its cohomology class. Let $f\colon G' \to G$ be a homomorphism, where G' is another group, and let us write $f^*\colon \mathrm{H}^2(G,M) \to \mathrm{H}^2(G',M)$ for the induced "inflation" map. Prove the equivalence of the two statements below:

1. $f^*(\alpha) = 0$;
2. there exists $\overline{f}\colon G' \to \Gamma$ which is a lift of f, that is, $\pi \circ \overline{f} = f$.

Deduce in particular that $\pi^*(\alpha) = 0$.

7.6. Let G be a cyclic group of order p^k, where p is a prime number. Define an $\mathbb{F}_p[G]$-module M by letting a generator of G act on $\mathbb{F}_p^n$ via a Jordan block of size n, with eigenvalue 1, where $1 \le n \le p^k$ (you should check that this makes sense). For which values of n do we have $\mathrm{H}^2(G,M) = 0$?

7.7. Let F be a field of characteristic not 2, and let $a \in F^\times$. Assume that a is not a square, so that $G := \mathrm{Gal}(F[\sqrt{a}]/F)$ has order 2, and write σ for its nontrivial element. Finally, let $b \in F^\times$.

1. Define $c\colon G \times G \to F[\sqrt{a}]^\times$ by $c(x,y) = 1$ unless $x = y = \sigma$, in which case $c(\sigma,\sigma) = b$. Check that c is a cocycle (which takes its values in $F^\times$), and that it is a coboundary if and only if b is a norm from $F[\sqrt{a}]$.
2. Use c to define the crossed product algebra A_c, and check that it is isomorphic to the quaternion algebra $(a,b)_F$. Revisit Problem 5.3.

8 The Brauer group of a local number field

In this short chapter, we compute $\mathrm{Br}(F)$ when F is a local number field, and more precisely we show that it is isomorphic to $\mathbb{Q}/\mathbb{Z}$, one of the most important results in this book. The work in all the preceding chapters was arranged so that this would appear relatively easy, but this could be deceiving. The reader should bear in mind that a lot of the material introduced up to this point comes into play. Later, when we study local class field theory, this becomes a major ingredient.

Preliminaries

So let p be a prime number, let F be a p-adic field, and let v denote its normalized valuation. As usual, $\mathcal{O}_F$ will be the ring of integers of F. Assume that K is a *skewfield* containing F, with $[K : F] < \infty$, and let us start with elementary remarks. A crucial tool is provided by Theorem 2.25, which gives the existence of a (unique) valuation w on K extending v. Here we must point out that the definition of a "valuation" on a skewfield is no different than the one we have for fields. This puts a topology on K, using the norm $\|x\| = p^{-w(x)}$, for example (or you may want to change the constant so that the restriction to F is the absolute value). The F-vector space K is then complete by Theorem 3.28.

Trying to recreate the usual apparatus of local fields, we put

$$\mathcal{O}_K = \{x \in K : w(x) \geq 0\}, \qquad \mathfrak{P} = \{x \in K : w(x) > 0\}.$$

After a little thinking, we realize that $\mathcal{O}_K$ is still a ring, and $\mathfrak{P}$ is a two-sided ideal of this ring, even in this non-commutative setting. If $\Pi \in \mathcal{O}_K$ is an element such that $w(\Pi)$ is minimal and >0, then it is also true that $\mathfrak{P}$ is generated by Π, and indeed that any $x \in K$ can be written $x = \Pi^n u$ with $n \in \mathbb{Z}$ and $w(u) = 0$. In particular K is generated by $\mathcal{O}_K$ and Π^{-1}.

Lemma 8.1 *Suppose that $F \subset \mathcal{Z}(K)$, and that K is not commutative. Then there exists a field E with $F \subset E \subset K$ such that E/F is unramified, and $E \neq F$.*

Proof. Assume the lemma fails. Then, for any $x \in K$, the field $E = F(x)$ is totally ramified over F (since $E_0 = F$). This means that E and F have the same residue field, and so if $x \in \mathcal{O}_K$, then we can find $a_0 \in \mathcal{O}_F$ with $x - a_0 \in \mathfrak{p}$.

Apply this again to the element $(x - a_0)\Pi^{-1} \in \mathcal{O}_K$. We find $a_1 \in \mathcal{O}_F$ such that $(x - a_0)\Pi^{-1} \equiv a_1 \bmod \mathfrak{p}$, or

$$x \equiv a_0 + a_1 \Pi \bmod \mathfrak{p}^2 .$$

Continuing, for each $n \geq 1$ we find $a_n \in \mathcal{O}_F$ such that

$$x \equiv a_0 + a_1 \Pi + a_2 \Pi^2 + \cdots + a_n \Pi^n \bmod \mathfrak{p}^{n+1} .$$

This implies that the field $F(\Pi)$ is *dense* in K. As a result, the skewfield K must be commutative, which we assumed it was not. This contradiction shows that the lemma holds. □

Corollary 8.2 *Assume $\mathcal{Z}(K) = F$. There exists a field E with $F \subset E \subset K$ which is maximal, and such that E/F is unramified. It follows that $[K] \in \mathrm{Br}(E/F)$.*

As a result, the group $\mathrm{Br}(F)$ is the union, or the colimit, of the groups $\mathrm{Br}(E/F)$ where E ranges through the fields $F \subset E \subset \overline{F}$, with E/F unramified. The cohomological Brauer group is the colimit of the groups $\mathrm{H}^2(\mathrm{Gal}(E/F), E^\times)$ for the same Es.

Proof. Copy the proof of Proposition 6.41 word for word, replacing "separable" by "unramified", and also replacing the references to Lemma 6.40 by references to the last lemma. The rest is clear. □

This is of course excellent news. From Corollary 2.42 we have a complete control over the unramified extensions of F contained in $\overline{F}$, since they paint the same picture as the (finite) extensions of a finite field do. Namely, for each integer $n \geq 1$, there is a field $E_n \subset \overline{F}$ with E_n/F unramified with $[E_n : F] = n$, and $E_n \subset E_m$ if and only if n divides m; there are no other unramified extensions of F in $\overline{F}$. An unramified extension of F is thus determined up to isomorphism by its dimension over F, and as a result, if $[K : F] = n^2$, then the field E in the corollary is F-isomorphic to E_n (its dimension over F being n, from the intensively used Corollary 6.27).

Thus the indexing set, in the colimit defining the cohomological Brauer group, is rather simple (it is the poset of positive integers under the relation of divisibility). The groups appearing are pretty simple, too, as we proceed to show. We need a lemma first.

Lemma 8.3 *Let E/F be an unramified extension of local number fields, and let $\mathcal{O}_E$ and $\mathcal{O}_F$ denote the respective rings of integers. Let $\mathrm{Tr}_{E/F}$ denote the trace map. Then $\mathrm{Tr}_{E/F}(\mathcal{O}_E) = \mathcal{O}_F$.*

Proof. The Galois action preserves valuations by Corollary 2.26, so that $\mathrm{Tr}_{E/F}(\mathcal{O}_E) \subset \mathcal{O}_F$. The image of this $\mathcal{O}_F$-linear map is an $\mathcal{O}_F$-submodule, that is, an ideal of $\mathcal{O}_F$. Suppose this ideal is not all of $\mathcal{O}_F$, but is, on the contrary, contained in the maximal ideal $\mathfrak{p}$. Then we pass to the residue fields $\mathbb{E}$ and $\mathbb{F}$, and realize that the trace map $\mathrm{Tr}_{\mathbb{E}/\mathbb{F}}$ is the zero map (this uses the isomorphism $\mathrm{Gal}(E/F) \cong \mathrm{Gal}(\mathbb{E}/\mathbb{F})$, since E/F is unramified). However, this contradicts Dedekind's lemma on the "independence of characters". □

Proposition 8.4 *Let E_n be the unramified extension of F of degree n, as above. The subgroup $\mathrm{N}_{E_n/F}(E_n^\times)$ of $F^\times$ is precisely*

$$\mathrm{N}_{E_n/F}\left(E_n^\times\right) = \{f \in F^\times : n \text{ divides } v(f)\}.$$

Proof. ([Bla72], Théorème V-4.) In this proof we write $\mathrm{N} = \mathrm{N}_{E_n/F}$, and also $\mathrm{Tr} = \mathrm{Tr}_{E_n/F}$.

Since the Galois action again preserves valuations, we see that $v(\mathrm{N}(x)) = nw(x)$ where w is the valuation of E_n extending v. We have $v(F^\times) = \mathbb{Z}$ as we assumed v to be normalized in this chapter, and E_n/F is unramified, so $w(E_n^\times) = v(F^\times) = \mathbb{Z}$; we conclude that $v(\mathrm{N}(E_n^\times)) = n\mathbb{Z}$. To prove the proposition, it suffices now to prove that, whenever $f \in \mathcal{O}_F^\times$ (that is, $v(f) = 0$), then f is a norm from E_n.

For this, we shall construct a sequence $(z_k)_{k\geq 1}$ of elements of E_n with

$$\mathrm{N}(z_k) \equiv f \bmod \mathfrak{p}^k \qquad \text{and} \qquad z_k \equiv z_{k-1} \bmod \mathfrak{P}^{k-1}$$

where $\mathfrak{p}$ and $\mathfrak{P}$ are the maximal ideals of $\mathcal{O}_F$ and $\mathcal{O}_{E_n}$ respectively. The second condition guarantees that (z_k) is a Cauchy sequence, so it converges to some $z \in E_n$, while the first condition shows $\mathrm{N}(z) = f$.

We proceed by induction. Suppose we have $f\mathrm{N}(z_{k-1})^{-1} = 1 + x\pi^{k-1}$, where $\mathfrak{p} = (\pi)$ and $x \in \mathcal{O}_F$ (it makes sense to write the induction hypothesis in this way, since f and so also $\mathrm{N}(z_{k-1})$ are units). We look for a z_k of the form $z_k = z_{k-1}(1 + y\pi^{k-1})$ with $y \in \mathcal{O}_{E_n}$ (note that π is also a generator for $\mathfrak{P}$, since E_n/F is unramified). We must have

$$\mathrm{N}\left(1 + y\pi^{k-1}\right) \equiv 1 + x\pi^{k-1} \bmod \mathfrak{p}^k;$$

when $k \geq 2$, this amounts clearly to $\mathrm{Tr}(y) = x$, after expanding. By the last lemma, this can always be achieved.

It remains to treat the case $k = 1$, which starts the induction. We need to find $z_1 \in E_n$ such that $\mathrm{N}(z_1) \equiv f \bmod \mathfrak{p}$. This is equivalent, when we pass to the residue fields $\mathbb{E}$ and $\mathbb{F}$, to finding an element $z \in \mathbb{E}^\times$ such that $\mathrm{N}_{\mathbb{E}/\mathbb{F}}(z) = \overline{f}$. We can certainly manage that, however, since the norm map between finite fields is always surjective, as we discovered with Example 7.24. □

Corollary 8.5 *The group* $\mathrm{Br}(E_n/F)$ *is cyclic of order n.*

Proof. The map $v\colon F^\times \to \mathbb{Z}$ induces an isomorphism

$$F^\times / \mathrm{N}_{E_n/F}\left(E_n^\times\right) \cong \mathbb{Z}/n\mathbb{Z},$$

by the proposition. Now apply Corollary 7.22. □

The basic idea is starting to emerge: $\mathrm{Br}(F)$ is the union of the cyclic groups $\mathrm{Br}(E_n/F)$, and when n divides m, the inclusion $\mathrm{Br}(E_n/F) \subset \mathrm{Br}(E_m/F)$ is that of the unique subgroup of order n, by the theory of cyclic groups. In particular, $\mathrm{Br}(E_n/F)$ is the unique subgroup of order n in $\mathrm{Br}(F)$, for any $n \geq 1$. It is intuitive that $\mathrm{Br}(F)$ should probably be isomorphic to $\mathbb{Q}/\mathbb{Z}$, with the unique subgroup of order n corresponding to $\frac{1}{n}\mathbb{Z}/\mathbb{Z}$. It would be awkward, though, to prove it using just the material developed so far, and we shall rather directly define a map $\mathrm{Br}(F) \to \mathbb{Q}/\mathbb{Z}$, in an elementary way.

Before we do this, however, we can collect some interesting facts about skewfields over F. Recall that, when K is a skewfield with $\mathcal{Z}(K) = F$ and $[K : F] = n^2$, we call n the *degree* of K.

Lemma 8.6 *Let F be a local number field.*

1. *Let K be a skewfield with $\mathcal{Z}(K) = F$. Then the degree of K is the smallest integer n such that $K \in \mathrm{Br}(E_n/F)$. If $K \in \mathrm{Br}(E_m/F)$, then n divides m.*
2. *The number of (isomorphism classes of) skewfields K with $\mathcal{Z}(K) = F$, of degree dividing n, is n.*
3. *The number of (isomorphism classes of) skewfields K with $\mathcal{Z}(K) = F$, of degree precisely n, is $\varphi(n)$, the order of $(\mathbb{Z}/n\mathbb{Z})^\times$.*

Of course the notation φ is standard, and this function is usually called the Euler function.

Proof. If n is the degree of K, as in (1), then we have seen above that E_n is a splitting field. The rest of (1) follows from Corollary 6.39.

This observation allows us to think, conversely, of the elements of $\mathrm{Br}(E_n/F)$ as the (isomorphism classes of) skewfields whose center is F and whose degree divides n. Since $\mathrm{Br}(E_n/F)$ has order n, we have (2).

The skewfields described in (3) correspond to elements of $\mathrm{Br}(E_n/F)$ which do not belong to any proper subgroup, that is, to generators of this group. □

Example 8.7 Since $\varphi(2) = 1$, there is (up to isomorphism) just one skewfield over F of dimension 4, whose center is F (or equivalently, which is not a field). In Problem 8.1, we ask you to prove (at least for p odd) that it is a quaternion algebra. A more difficult result asserts that, if $a \in F^\times$ is not a square in the local number field F, then there always exists $b \in F^\times$ such that $(a,b)_F$ is a skewfield; this is called *Tate duality*, and you will have to wait until Theorem 13.21 for this.

The Hasse invariant of a skewfield

We now associate an element of $\mathbb{Q}/\mathbb{Z}$ to any skewfield K with $\mathcal{Z}(K) = F$. The construction will appear to depend on choices, and we will have to prove that it does not.

Choose a field E with $F \subset E \subset K$ which is maximal, and with E/F unramified. The corresponding extension $\mathbb{E}/\mathbb{F}$ of residue fields has a cyclic Galois group, generated by a canonical generator, namely the Frobenius map $x \mapsto x^q$ where $\mathbb{F}$ has q elements. The group $\mathrm{Gal}(E/F) \cong \mathrm{Gal}(\mathbb{E}/\mathbb{F})$ thus also has a Frobenius element, call it σ. (We considered the Frobenius element in Corollary 2.41.) By Skolem–Noether, there exists $a \in K$ such that $axa^{-1} = \sigma(x)$ for $x \in E$. Finally, recall that there is a unique extension w to K of the valuation v defined on F. The number $w(a)$ is *a priori* not an integer, but an element of $\mathbb{Q}$. We let $\mathrm{Inv}(K,E,a)$ denote the class of $w(a)$ in $\mathbb{Q}/\mathbb{Z}$.

Lemma 8.8 *The element* $\mathrm{Inv}(K,E,a)$ *does not depend on E or a, and only depends on K up to isomorphism of F-algebras. It yields a map*

$$\mathrm{Inv}\colon \mathrm{Br}(F) \longrightarrow \mathbb{Q}/\mathbb{Z}.$$

The element $\mathrm{Inv}(K)$ is called the *Hasse invariant* of K (or sometimes simply its invariant).

Proof. First we show that, when E has been chosen, the choice of a is not important. Indeed, another a' inducing the same automorphism of E would be of the form $a' = ea$ with $e \in E^\times$, by (1) of Proposition 7.1. However, $w(E^\times) = v(F^\times) = \mathbb{Z}$ since E/F is unramified and v is normalized, so $w(ea) = w(e) + w(a) \equiv w(a) \bmod \mathbb{Z}$. We can now write $\mathrm{Inv}(K,E)$ instead of $\mathrm{Inv}(K,E,a)$.

We make an observation. Let $\theta\colon K \to L$ be an isomorphism of F-algebras. We claim that $\mathrm{Inv}(K,E,a) = \mathrm{Inv}(L,\theta(E),\theta(a))$ (so $\mathrm{Inv}(K,E) = \mathrm{Inv}(L,\theta(E))$). For this, we write that

$$axa^{-1} \equiv x^q \bmod \mathfrak{P},$$

where $\mathfrak{P}$ is the maximal ideal of $\mathcal{O}_E$, for all $x \in \mathcal{O}_E$. Applying θ we draw

$$\theta(a)\theta(x)\theta(a)^{-1} \equiv \theta(x)^q \bmod \theta(\mathfrak{P}).$$

Of course θ gives an isomorphism $\mathcal{O}_E \to \mathcal{O}_{\theta(E)}$, and $\theta(\mathfrak{P})$ is the maximal ideal of $\mathcal{O}_{\theta(E)}$. The last congruence then characterizes the Frobenius automorphism of $\theta(E)$, so we have at least proved that it makes sense to talk about $\mathrm{Inv}(L,\theta(E),\theta(a))$. However, the rest is easy, as θ must preserve the valuations on K and L, by uniqueness of extensions, so $w_K(a) = w_L(\theta(a))$ in obvious notation. It does follow that $\mathrm{Inv}(K,E,a) = \mathrm{Inv}(L,\theta(E),\theta(a))$.

Now if $E' \subset K$ is another candidate for the role of E, then by Skolem–Noether there is an automorphism θ of K with $\theta(E) = E'$. By the observation, we have $\mathrm{Inv}(K,E) = \mathrm{Inv}(K,E')$, so we may use the notation $\mathrm{Inv}(K)$. The same observation implies that $\mathrm{Inv}(K) = \mathrm{Inv}(L)$ when K and L are isomorphic. □

Theorem 8.9 *The invariant*

$$\mathrm{Inv}\colon \mathrm{Br}(F) \longrightarrow \mathbb{Q}/\mathbb{Z}$$

is an isomorphism of groups. It induces an isomorphism between $\mathrm{Br}(E_n/F)$ *and* $\frac{1}{n}\mathbb{Z}/\mathbb{Z}$.

Proof. Let $n \geq 1$, let K be a skewfield with $[K] \in \mathrm{Br}(E_n/F)$, and suppose that $[K]$ does not belong to a subgroup $\mathrm{Br}(E_d/F)$ for any proper divisor d of n. This implies, as in Lemma 8.6, that $[K : F] = n^2$; we may choose E such that $F \subset E \subset K$, with E/F unramified, and $[E : F] = n$.

The skewfield K is then isomorphic to a crossed product algebra A_c, for some $c \in Z^2(\mathrm{Gal}(E/F), E^\times)$, as established in the introduction to the previous chapter. (This is really "isomorphic" and not "Brauer equivalent".) Of course E is isomorphic to E_n, and since the invariant of K only depends on its isomorphism class, we may as well consider that $E = E_n$ in what follows, and that $K = A_c$.

When $a \in K$, recall from the statement of Theorem 2.25 that its valuation $w(a)$ is computed by

$$w(a) = \frac{1}{n^2} v(\det(m_a)),$$

where m_a is multiplication by a (on the left or on the right; the two are conjugate). Let $e_1, \ldots, e_n$ be a basis of E_n over F, let the elements of $\mathrm{Gal}(E_n/F)$ be $1, \rho, \ldots, \rho^{n-1}$ where ρ is the Frobenius, and let us use $e_i a_{\rho^j}$ as the basis for K over F. We are after $\mathrm{Inv}(K)$, which is the class of $w(a_\rho)$ in $\mathbb{Q}/\mathbb{Z}$.

Multiplication by a_ρ on the right is given in this basis by

$$e_i a_{\rho^j} a_\rho = c(\rho^j, \rho) e_i a_{\rho^{j+1}}.$$

Thus the matrix of m_{a_ρ} in this basis has only one nonzero coefficient in each row and each column, and this coefficient is a value of c. We only care about $\det(m_{a_\rho})$ up to sign, since we are after $v(\det(a_\rho))$, so we may freely permute the rows and columns to get a diagonal matrix, whose determinant is

$$\left(\prod_{0 \leq i < n} c(\rho^i, \rho) \right)^n. \tag{*}$$

We have encountered a similar expression with Proposition 7.18. Recall that this computes $\mathrm{H}^2(G, M)$ when G is cyclic, and when $G = \mathrm{Gal}(E_n/F)$ and $M = E_n^\times$, the answer is $\mathrm{H}^2(\mathrm{Gal}(E_n/F), E_n^\times) \cong F^\times / \mathrm{N}_{E_n/F}(E_n^\times)$. Moreover, a specific isomorphism, still by the same proposition, is that induced by the homomorphism

$$\varphi_\rho \colon Z^2\left(\mathrm{Gal}(E_n/F), E_n^\times\right) \longrightarrow F^\times.$$

Looking up the definition of φ_ρ, we recognize that (*) is simply $\varphi_\rho(c)^n$ (pay attention to the fact that $E_n^\times$ is in multiplicative notation, unlike the "general" module M). Therefore,

$$w\left(a_\rho\right) = \frac{1}{n} v\left(\varphi_\rho(c)\right).$$

Proposition 8.4 shows that v induces an isomorphism $F^\times/\mathrm{N}_{E_n/F}(E_n^\times) \cong \mathbb{Z}/n\mathbb{Z}$, and so $\frac{1}{n}v$ induces an isomorphism $F^\times/\mathrm{N}_{E_n/F}(E_n^\times) \cong \frac{1}{n}\mathbb{Z}/\mathbb{Z}$. The composition

$$I'_n\colon \mathrm{H}^2\left(\mathrm{Gal}(E_n/F), E_n^\times\right) \xrightarrow{\varphi_\rho} F^\times/\mathrm{N}_{E_n/F}\left(E_n^\times\right) \xrightarrow{\frac{1}{n}v} \tfrac{1}{n}\mathbb{Z}/\mathbb{Z}$$

is an isomorphism of groups. If we call

$$\gamma_n\colon \mathrm{Br}(E_n/F) \longrightarrow \mathrm{H}^2\left(\mathrm{Gal}(E_n/F), E_n^\times\right)$$

the inverse of our usual isomorphism, then $I_n := I'_n \circ \gamma_n$ is an isomorphism between $\mathrm{Br}(E_n/F)$ and $\frac{1}{n}\mathbb{Z}/\mathbb{Z}$. The fact that $\mathrm{Inv}(K)$ is the image of $w(a_\rho)$ in $\mathbb{Q}/\mathbb{Z}$, with K corresponding to the cocycle c, means that $\mathrm{Inv}(K) = I_n(K)$. In other words, the map

$$\mathrm{Inv}\colon \mathrm{Br}(E_n/F) \longrightarrow \mathbb{Q}/\mathbb{Z}$$

takes its values in $\frac{1}{n}\mathbb{Z}/\mathbb{Z}$, and agrees with the isomorphism I_n, when you restrict it to those elements K generating $\mathrm{Br}(E_n/F)$.

To prove the theorem, we must pay attention to the following diagram, where d divides n.

$$\begin{array}{ccccccc}
\mathrm{Br}(E_d/F) & \xrightarrow{\gamma_d} & \mathrm{H}^2\left(\mathrm{Gal}(E_d/F), E_d^\times\right) & \xrightarrow{\varphi_\rho} & F^\times/\mathrm{N}_{E_d/F}\left(E_d^\times\right) & \xrightarrow{\frac{1}{d}v} & \frac{1}{d}\mathbb{Z}/\mathbb{Z} \\
\subset\downarrow & & \mathrm{inf}\downarrow & & f\mapsto f^{\frac{n}{d}}\downarrow & & \subset\downarrow \\
\mathrm{Br}(E_n/F) & \xrightarrow{\gamma_n} & \mathrm{H}^2\left(\mathrm{Gal}(E_n/F), E_n^\times\right) & \xrightarrow{\varphi_\rho} & F^\times/\mathrm{N}_{E_n/F}\left(E_n^\times\right) & \xrightarrow{\frac{1}{n}v} & \frac{1}{n}\mathbb{Z}/\mathbb{Z}
\end{array}$$

The letter ρ is used twice here, to denote the Frobenius element of the group $\mathrm{Gal}(E_n/F)$, and also the Frobenius element of $\mathrm{Gal}(E_d/F)$ (one could call then ρ_n and ρ_d). The natural quotient map $\mathrm{Gal}(E_n/F) \to \mathrm{Gal}(E_d/F)$ sends one to the other.

The arguments above, applied with d replacing n, establish the following: When $[K] \in \mathrm{Br}(E_n/F)$ has order d, or equivalently generates $\mathrm{Br}(E_d/F)$, then in order to compute $\mathrm{Inv}(K)$ we may use the composition of the isomorphisms on the first line of the diagram, which is I_d. If we could only prove that the diagram commutes (for all d), then we would know that the restriction of I_n to $\mathrm{Br}(E_d/F)$ is none other than I_d, so that in the end the map Inv coincides with I_n on all of $\mathrm{Br}(E_n/F)$, and so it is an isomorphism.

Thus we are reduced to proving that the three squares in the diagram commute. For the leftmost square, this is Proposition 7.29. For the rightmost square, this is trivial. For the inner square, this is part of Example 7.28. We remark that, in the case of this inner square, it is crucial that the generators taken for the cyclic groups be compatible. The fact that *all* the Galois groups in sight have a canonical generator, namely the Frobenius, is thus essential. □

Naturality

Theorem 8.10 *Let F'/F be a finite extension of local number fields. Then there is a commutative diagram*

$$\begin{array}{ccc} \mathrm{Br}(F) & \xrightarrow{\mathrm{Inv}} & \mathbb{Q}/\mathbb{Z} \\ \downarrow & & \downarrow{\scriptstyle x\mapsto [F':F]x} \\ \mathrm{Br}(F') & \xrightarrow{\mathrm{Inv}} & \mathbb{Q}/\mathbb{Z}. \end{array}$$

Since there exists an intermediate field $F \subset F'_0 \subset F'$ with F'_0/F unramified, and F'/F'_0 totally ramified, and since $[F' : F] = [F' : F'_0][F'_0 : F]$, it is enough to consider separately the cases of unramified extensions and totally ramified extensions.

Proof when F'/F is unramified. We assume that $F' \subset \overline{F}$, so $\overline{F'} = \overline{F}$. Let $E'_n \subset \overline{F}$ be the unramified extension of F' of degree n. Then E'_n/F is unramified, of degree dn where $d = [F' : F]$. So $E'_n = E_{dn}$, keeping the same notation for the unramified extensions of F. The group $\mathrm{Gal}(E'_n/F')$ is the unique subgroup of order n in $\mathrm{Gal}(E'_n/F)$, which is itself cyclic of order dn.

The commutativity of the diagram asserts that two maps coincide, and it is enough to prove that they coincide on all the subgroups $\mathrm{Br}(E'_n/F) = \mathrm{Br}(E_{dn}/F)$, since the union of these is all of $\mathrm{Br}(F)$. We are reduced to proving the commutativity of

$$\begin{array}{ccccc} \mathrm{Br}\left(E'_n/F\right) & \xrightarrow{\gamma_{dn}} & \mathrm{H}^2\left(\mathrm{Gal}\left(E'_n,F\right),\left(E'_n\right)^\times\right) & \xrightarrow{I'_{dn}} & \mathbb{Q}/\mathbb{Z} \\ \downarrow & & \downarrow & & \downarrow{\scriptstyle x\mapsto dx} \\ \mathrm{Br}\left(E'_n/F'\right) & \longrightarrow & \mathrm{H}^2\left(\mathrm{Gal}\left(E'_n,F'\right),\left(E'_n\right)^\times\right) & \longrightarrow & \mathbb{Q}/\mathbb{Z}, \end{array}$$

since the previous proof has established that the composition of the maps on the first line is Inv, and likewise for the bottom line. The left square commutes by Proposition 7.36, and the right square can be expanded as follows, using the definition of I'_n and its analog for F'.

$$\begin{array}{ccccc} \mathrm{H}^2\left(\mathrm{Gal}\left(E'_n,F\right),\left(E'_n\right)^\times\right) & \xrightarrow{\varphi_\rho} & F^\times/\mathrm{N}_{E'_n/F}\left(\left(E'_n\right)^\times\right) & \xrightarrow{\frac{1}{dn}v} & \frac{1}{dn}\mathbb{Z}/\mathbb{Z} \\ \downarrow & & \downarrow & & \downarrow{\scriptstyle x\mapsto dx} \\ \mathrm{H}^2\left(\mathrm{Gal}\left(E'_n,F'\right),\left(E'_n\right)^\times\right) & \xrightarrow{\varphi_{\rho^d}} & (F')^\times/\mathrm{N}_{E'_n/F'}\left(\left(E'_n\right)^\times\right) & \xrightarrow{\frac{1}{n}v'} & \frac{1}{n}\mathbb{Z}/\mathbb{Z} \end{array}$$

Here, ρ is the Frobenius for E'_n/F, so ρ^d is the Frobenius for E'_n/F'. The commutativity of the left square is the second part of Example 7.28. In the right square, we are compelled to use v', the *normalized* valuation of F'. Crucially, though, since F'/F is unramified, this v' is simply the extension of v to F'. Having made this point, the commutativity of the right square is obvious. □

Proof when F'/F is totally ramified. Again assume that $F' \subset \overline{F}$, let E_n/F be unramified of degree n with $E_n \subset \overline{F}$ as usual, and put $E'_n = E_nF'$. The situation looks as follows.

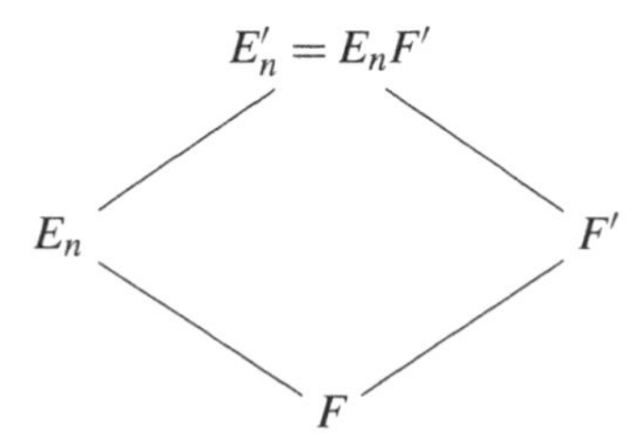

We have $n = [E_n : F]$, and let $e = [F' : F]$. The field $E_n \cap F'$ is both unramified and totally ramified over F, as follows from the definitions, so we must have $E_n \cap F' = F$. Further, since E_n/F is Galois, Proposition 1.15 shows that E'_n/F' is Galois, and that the restriction gives an isomorphism $\mathrm{Gal}(E'_n/F') \cong \mathrm{Gal}(E_n/F)$. For a start, this gives $[E'_n : F'] = [E_n : F] = n$, so $[E'_n : F] = en$.

Now, the ramification index $e(E'_n/F)$ must be a multiple of e, and the inertia degree $f(E'_n/F)$ must be a multiple of n. From $en = [E'_n : F] = e(E'_n/F)f(E'_n/F)$, we deduce that $e(E'_n/F) = e$ and $f(E'_n/F) = n$. We see that E_n/F is the largest unramified extension contained in E'_n/F, and also that E'_n/F' is unramified of degree n.

To prove the theorem, it suffices to establish the commutativity of

$$\begin{array}{ccc} \mathrm{Br}(E_n/F) & \xrightarrow{\mathrm{Inv}} & \mathbb{Q}/\mathbb{Z} \\ \downarrow & & \downarrow{\scriptstyle x\mapsto ex} \\ \mathrm{Br}\left(E'_n/F'\right) & \xrightarrow{\mathrm{Inv}} & \mathbb{Q}/\mathbb{Z}, \end{array}$$

the left vertical map being really the composition $\mathrm{Br}(E_n/F) \to \mathrm{Br}(E'_n/F) \to \mathrm{Br}(E'_n/F')$. Using both Proposition 7.29 and Proposition 7.36, we can replace the Brauer groups by cohomology groups. Here we have the map

$$\mathrm{H}^2\left(\mathrm{Gal}(E_n/F), E_n^\times\right) \longrightarrow \mathrm{H}^2\left(\mathrm{Gal}\left(E'_n/F'\right), \left(E'_n\right)^\times\right)$$

induced by the isomorphism $\mathrm{Gal}(E'_n/F') \cong \mathrm{Gal}(E_n/F)$ considered above, together with the inclusion $E_n^\times \subset (E'_n)^\times$. Things boil down to the commutativity of:

$$\begin{array}{ccccc} \mathrm{H}^2\left(\mathrm{Gal}(E_n/F), E_n^\times\right) & \xrightarrow{\varphi_\rho} & F^\times/\mathrm{N}_{E_n/F}\left(E_n^\times\right) & \xrightarrow{\frac{1}{n}v} & \frac{1}{n}\mathbb{Z}/\mathbb{Z} \\ \downarrow & & \downarrow & & \downarrow{\scriptstyle x\mapsto ex} \\ \mathrm{H}^2\left(\mathrm{Gal}\left(E'_n/F'\right), \left(E'_n\right)^\times\right) & \xrightarrow{\varphi_{\rho'}} & (F')^\times/\mathrm{N}_{E'_n/F'}\left(\left(E'_n\right)^\times\right) & \xrightarrow{\frac{1}{n}v'} & \frac{1}{n}\mathbb{Z}/\mathbb{Z}. \end{array}$$

Here, ρ resp. ρ' is the Frobenius for E_n/F resp. E'_n/F'. It is routine to check that they correspond to each other via our isomorphism of Galois groups, and that the left square commutes (this is not one of the cases covered by Example 7.28; it is easier).

For the rightmost, and very last, square, we have to recall that v' is the *normalized* valuation of F'. If w is the extension of v to F', then $v' = ew$, and we are done. □

The theorem allows us to generalize something we knew about unramified extensions only:

Corollary 8.11 *Let F'/F be any finite extension of local number fields. Then $\mathrm{Br}(F'/F)$ is cyclic of order $[F' : F]$.*

Proof. Let $n = [F' : F]$. By definition $\mathrm{Br}(F'/F)$ is the kernel of $\mathrm{Br}(F) \to \mathrm{Br}(F')$, and the theorem allows us to identify this homomorphism with multiplication by n on $\mathbb{Q}/\mathbb{Z}$. The kernel of this is the cyclic group $\frac{1}{n}\mathbb{Z}/\mathbb{Z}$. □

Problem

8.1. Let F be a p-adic field, where p is odd, and let v be its normalized valuation. Using a counting argument, prove that there exists an element $a \in F^{\times}$ such that $v(a)$ is even, but $a \notin F^{\times 2}$. Letting π be a uniformizer, deduce that the equation

$$ax^2 + \pi y^2 = 1$$

does not have any solutions $x, y \in F$ (Hensel's lemma is relevant). Conclude that $(a, \pi)_F$ is a skewfield, and return to Example 8.7.

Part III

Galois cohomology

In this part of the book, we define and study some abelian groups denoted, for $n \geq 0$, by

$$\mathrm{H}^n(G,M),$$

and attached to a finite (or profinite) group G and a $\mathbb{Z}[G]$-module M. These are called the *cohomology groups* of G with coefficients in M, and they generalize our definition of $\mathrm{H}^2(G,M)$ (see Definition 7.14). Special attention will be paid to the case when $G = \mathrm{Gal}(K/F)$ is a Galois group.

The power of these is understood when we realize that we have many different ways of computing them, all giving the same answer. Thus, whenever we make a general statement about cohomology groups, it can be translated in various ways, gaining depth instantaneously. The best example, without question, is *Hilbert's Theorem 90*, which says that

$$\mathrm{H}^1(\mathrm{Gal}(K/F),K^\times) = 0$$

whenever K/F is a finite Galois extension. The list of applications appears endless, as a great number of seemingly unrelated problems all turn out to amount to the computation of this cohomology group, and Hilbert predicts that they will all lead to 0.

Historically, group cohomology developed as part of algebraic topology, and much (but not all) of the research conducted in this area is done by topologists. The focus is sometimes different from what it is when the application in view is Galois theory: In topology, one typically seeks complete computations of groups of the form $\mathrm{H}^n(G,\mathbb{F}_p)$, where p is a prime, for all n and as many different groups G as possible. In *Galois cohomology*, which is the study of groups of the form $\mathrm{H}^n(\mathrm{Gal}(\overline{F},F),M)$, it is sometimes more important to establish general properties of the groups rather than actually compute them. Of course, the distinction is not as clean-cut as that.

The sort of algebra developed in this part is called *homological algebra*. A notable feature of the subject is that, while some statements can be striking, the proofs tend to be easy (and repetitive). In the first and second editions of Serge Lang's book *Algebra* (but not in the third!), one chapter ends with a single problem: "Take any book on homological algebra, and prove all the theorems without looking at the proofs given in that book". This may be inherent to the subject itself, but is certainly also a manifestation of Grothendieck's work in the area, which was greatly influential, and always written in this style – clever statements, easy proofs.

The bottom line is that we will leave some proofs to the reader, contrary to our habit in the rest of this book.

Many readers will already know some homological algebra, and will want to skip this part of the book, or some of it (say, the coming chapter). For this reason, we have been careful to avoid idiosyncrasies, especially in the beginning.

Those readers who know the convenient language of categories and functors (and derived functors) may be surprised at first to see that we avoid these entirely. This is merely because the extra machinery would not bring anything of direct use to us – and other readers, for sure, will want to keep the "abstract nonsense" to a minimum. (Indeed, while the author expects some criticism for this lack of sophistication, he has already received suggestions to remove all references to $\mathrm{H}^n(G,M)$ when $n > 2$. A line had to be drawn, somewhere.) Still, the next chapters constitute an introduction to the more general phenomena of homological algebra.

9 Ext and Tor

The Ext groups arise when one studies the operation $M \rightsquigarrow \mathrm{Hom}_R(M,N)$, where M and N are R-modules, while the Tor groups arise when we scrutinize $M \rightsquigarrow N \otimes_R M$. In either case, the operation fails to preserve *exact sequences*, unless R, or N, is special, and the groups in question measure the extent of that failure.

The cohomology groups of a finite group, at the beginning of the next chapter, are defined as Ext groups, and we will also define the homology groups as Tor groups.

Preliminaries

In this chapter, the letter R denotes a ring, which is not necessarily commutative. The typical examples will be $R = \mathbb{Z}$ and $R = \mathbb{Z}[G]$, the group algebra of the finite group G. (Algebras over a field, while certainly included in the discussion, are not our prime object of interest, so we drop the letter A and the connotation it has gained in the previous part.) We will study R-modules (on the left), including those which are not finitely generated. The constructions to follow would be mostly trivial, were R assumed to be semisimple; in a sense, they give a measure of the failure of semi-simplicity.

Definition 9.1 A sequence of R-modules and homomorphisms as below

$$\cdots \longrightarrow M_{n+1} \xrightarrow{f_{n+1}} M_n \xrightarrow{f_n} M_{n-1} \longrightarrow \cdots$$

exact sequence

is said to be **exact sequence** when $\mathrm{Im}(f_{n+1}) = \ker f_n$ for all $n \in \mathbb{Z}$. A similar definition applies when there are only finitely many modules, or for sequences indexed by $\mathbb{N} = \mathbb{Z}_{\geq 0}$ or by $\mathbb{Z}_{\leq 0}$. In particular, an exact sequence of the form

$$0 \longrightarrow A \xrightarrow{\iota} B \xrightarrow{\pi} C \longrightarrow 0$$

is called a **short exact sequence**. We sometimes speak of a **long exact sequence** when the modules are indexed by $\mathbb{Z}$.

We have already encountered short exact sequences when studying group extensions (the objects were groups rather than modules).

Lemma 9.2 *Given a short exact sequence*

$$0 \longrightarrow A \xrightarrow{\iota} B \xrightarrow{\pi} C \longrightarrow 0,$$

the following conditions are equivalent:

1. *There exists a homomorphism $s\colon C \to B$ such that $\pi \circ s$ is the identity.*
2. *There exists a homomorphism $r\colon B \to A$ such that $r \circ \iota$ is the identity.*
3. *There exists a commutative diagram*

$$\begin{array}{ccccccccc} 0 & \longrightarrow & A & \xrightarrow{\iota} & B & \xrightarrow{\pi} & C & \longrightarrow & 0 \\ & & \downarrow{=} & & \downarrow{\cong} & & \downarrow{=} & & \\ 0 & \longrightarrow & A & \xrightarrow{a \mapsto (a,0)} & A \oplus C & \xrightarrow{(a,c) \mapsto c} & C & \longrightarrow & 0. \end{array}$$

Proof. Suppose s exists as in (1). Then $p = s \circ \pi$ satisfies $p \circ p = s \circ (\pi \circ s) \circ \pi = p$, so it is a projector. As a result, we have $B = \ker(p) \oplus \mathrm{Im}(p)$, and clearly $\ker(p) = \ker(\pi) = \iota(A)$. The map $1 - p$ maps B onto $\ker(p)$, and if we compose it with ι^{-1}, we obtain the map r as in (2). The proof of (2) $\Longrightarrow$ (1) is similar.

Keeping the notation, we have $\ker(p) \cong A$ and $\mathrm{Im}(p) \cong C$, so (3) follows easily. That (3) implies both (1) and (2) is trivial. □

Definition 9.3 A short exact sequence satisfying the equivalent properties of the lemma is said to be **split exact sequence**. ■

split exact sequence

Example 9.4 The short exact sequence

$$0 \longrightarrow \mathbb{Z}/2\mathbb{Z} \xrightarrow{x \mapsto 2x} \mathbb{Z}/4\mathbb{Z} \xrightarrow{\text{mod } 2} \mathbb{Z}/2\mathbb{Z} \longrightarrow 0$$

is not split. ■

As in Chapter 5, we shall write R^1 for R itself, viewed as an R-module using left multiplication. Likewise, R^n is the module of n-tuples of elements of R, or in other words, the direct sum of n copies of R^1. More generally, a module is called *free* when it is isomorphic to $\oplus_{i \in I} R^1$, that is, a direct sum of copies of R^1, over an indexing set I which may well be infinite. Equivalently, M is free if there exists a basis, that is, a collection of elements $(e_i)_{i \in I}$ of M such that any $m \in M$ can be written uniquely $m = \sum_i \lambda_i e_i$ with $\lambda_i \in R$, and with only finitely many of the λ_i nonzero.

Lemma 9.5 *Let P be an R-module. The following conditions are equivalent.*

1. *There exists a module M such that $P \oplus M$ is free.*
2. *For any surjective map $\pi : B \to C$ between modules, and any given homomorphism $f\colon P \to C$, there exists a lift $\tilde{f}\colon P \to B$, that is, a map satisfying $\pi \circ \tilde{f} = f$.*
3. *Any exact sequence*

$$0 \longrightarrow A \longrightarrow B \longrightarrow P \longrightarrow 0$$

is split.

Proof. Suppose (1), so that $F = P \oplus M$ is free with basis $(e_i)_{i\in I}$. Let the situation of (2) present itself, and consider the map

$$g\colon F \longrightarrow P \xrightarrow{f} C$$

given by the projection, followed by f. If b_i is any element of B such that $\pi(b_i) = g(e_i)$, then we may define $\widetilde{g}\colon F \to B$ by requiring $\widetilde{g}(e_i) = b_i$. Since F is free, this makes sense. Clearly $\pi \circ \widetilde{g} = g$. The composition

$$\widetilde{f}\colon P \longrightarrow F \xrightarrow{\widetilde{g}} B$$

given by the inclusion, followed by $\widetilde{g}$, is as requested.

Now suppose (2), and assume we are given an exact sequence as in (3). We call π the given map $B \to P$. Using (2) with $C = P$ and $f =$ the identity, we obtain $\widetilde{f}$, which is the map called s in the definition of "split". So (3) holds.

Finally, assume (3). Let $(e_i)_{i\in I}$ be a family of generators of P (this certainly exists, if only because we can take $I = P$ itself). The map

$$\pi : B = \bigoplus_{i\in I} R^1 \longrightarrow P$$

with $\pi((\lambda_i)_{i\in I}) = \sum_i \lambda_i e_i$ is surjective. We let $A = \ker\pi$, obtaining a short exact sequence, which by (3) is split. This means that $B \cong A \oplus P$, so (1) holds because B is free. □

projective

Definition 9.6 A module satisfying the equivalent conditions of the lemma is said to be **projective**.

Example 9.7 Free modules are of course projective. Now, take $R = \mathbb{Z}$, and let P be finitely generated, for simplicity. If P is projective, then as in the proof of (3) $\implies$ (1) in the lemma, we see that P is a direct sum in $\mathbb{Z}^n$ for some integer n. From the theory of modules over $\mathbb{Z}$, that is, abelian groups, we deduce that P is itself isomorphic to $\mathbb{Z}^m$ for some $m \leq n$. *Over $\mathbb{Z}$, projective modules are just free modules.*

Note that the module $\mathbb{Z}^1$ is thus projective, but it is not semisimple, since there is no complement for the submodule $2\mathbb{Z}^1$ (contrary to Proposition 5.31); the module $\mathbb{Z}/2\mathbb{Z}$ is simple by inspection, so also semisimple, but it is not projective (= not free).

These examples show clearly that projective modules and semisimple modules are different things. However, consider the following.

Lemma 9.8 *Suppose R is a finite-dimensional algebra over a field F. The following conditions are then equivalent.*

1. *All R-modules are semisimple (that is, R is semisimple).*
2. *All R-modules are projective.*
3. *All short exact sequences of R-modules are split.*

The condition on R is here to simplify the proof, but is not essential. Since the result is purely motivational, this will suffice.

Proof. Suppose (1), and let

$$0 \longrightarrow A \xrightarrow{\iota} B \xrightarrow{\pi} C \longrightarrow 0$$

be a short exact sequence. Proposition 5.31 shows, since B is semisimple, that there exists W such that $\iota(A) \oplus W = B$. Clearly, this shows that the sequence is split, so (3) holds. Obviously this gives (2).

Now assume that all modules are projective, and let us prove (1). For this, it suffices to prove that R^1 is semisimple. Let S be a simple submodule of R^1, and consider the exact sequence

$$0 \longrightarrow S \longrightarrow R^1 \longrightarrow R^1/S \longrightarrow 0.$$

This must be split, since R^1/S is projective, and so $R^1 \cong S \oplus R^1/S$. We finally use the hypothesis on R (which was useless up to now), in order to conclude comfortably by induction on the dimension of R^1 (as an F-vector space) that it can be written as a sum of simple modules. (The reader who knows about Noetherian rings will extend the result to these at once.) □

For example, all $F[G]$-modules, where F is a field of characteristic not dividing the order of the finite group G, are projective, since all $F[G]$-modules are semisimple, by the arguments of Example 5.25.

If we are to measure the extent to which a ring R fails to be semisimple, then it is reasonable to use the existence of nonprojective modules. In fact, we will be able to comment on an individual module, describing "how far" it is from being projective. This is the object of the next section.

Resolutions

When a module is not projective, we seek to approximate it by projective modules, as in the next definition.

projective resolution

Definition 9.9 A **projective resolution** of the R-module M is an exact sequence of the form

$$\cdots \longrightarrow P_n \longrightarrow P_{n-1} \longrightarrow \cdots \longrightarrow P_1 \longrightarrow P_0 \longrightarrow M \longrightarrow 0,$$

where each P_n is projective, for $n \geq 0$. ■

Example 9.10 If P is itself projective, it has a resolution of the form

$$0 \longrightarrow P \longrightarrow P \longrightarrow 0,$$

that is, with $P_0 = P$ and $P_n = 0$ for $n > 0$.

In general, we will have to consider projective resolutions with more terms, depending on "how far" the module M is from being projective. In this section we study projective resolutions naively, before giving some intuition from the realm of topology in the next.

Lemma 9.11 *Every module M possesses a projective resolution.*

Proof. For any set X at all, let us write $R[X]$ for the free module "on X", that is the direct sum $\oplus_{x \in X} R^1$, or the module of formal sums $\sum_x \lambda_x x$ with $\lambda_x \in R$ and only finitely many coefficients nonzero. (The construction of $R[G]$, the group algebra of G already considered, is an example of this, if we forget the multiplication.)

If we take $X = M$, we obtain the free module $P_0 = R[M]$, and there is a map of R-modules $R[M] \to M$ mapping $m \in M$ to m.

Let $M_0 \subset P_0$ be the kernel of this map $P_0 \to M$. Then we put $P_1 = R[M_0]$, which is free and comes equipped with a surjective map $P_1 \to M_0$. The kernel of this map is called M_1, we put $P_2 = R[M_1]$, and continue in this fashion. Note that we are in fact constructing a resolution by *free* modules. □

Example 9.12 Of course this is not always the most economical way of constructing the resolution. For example when $R = \mathbb{Z}$, we can take $P_0 = R[M]$ as prescribed, but then $M_0 \subset P_0$ is automatically free, as a submodule of a free $\mathbb{Z}$-module. So it is already projective, and we have a resolution of the form

$$0 \longrightarrow M_0 \longrightarrow P_0 \longrightarrow M \longrightarrow 0.$$

Thus over $\mathbb{Z}$, each module has a two-step resolution. This reflects the "good" properties of $\mathbb{Z}$, or any principal ideal domain, in the sense that the theory of $\mathbb{Z}$-modules is not too far from plain linear algebra.

Example 9.13 This example is important for the sequel, as well as very simple. Put $R = \mathbb{Z}[C_r]$, where $C_r = \{1, T, T^2, \ldots, T^{r-1}\}$ is a cyclic group of order r, with generator T, in multiplicative notation. (Any ring could replace $\mathbb{Z}$ as the base here.) We have $\mathbb{Z}[C_r] = \mathbb{Z}[T]/(T^r - 1)$.

A $\mathbb{Z}[C_r]$-module is nothing but an abelian group with an action of C_r, and we shall study in particular the module $\mathbb{Z}$ with trivial action (that is, T acts as the identity). This is often called the "trivial $\mathbb{Z}[G]$-module", but do not be misled by that name, for a lot of nontrivial information will be discovered by studying it. Let us start here with a projective resolution.

For all $n \geq 0$, let $P_n = \mathbb{Z}[C_r]^1$, the free module of rank 1. The projective resolution will be of the form

$$\cdots \xrightarrow{1-T} \mathbb{Z}[C_r]^1 \xrightarrow{N} \mathbb{Z}[C_r]^1 \xrightarrow{1-T} \mathbb{Z}[C_r]^1 \xrightarrow{\varepsilon} \mathbb{Z} \longrightarrow 0.$$

Let us explain the notation. The map ε is the so-called *augmentation map*, for which $\varepsilon(T^k) = 1$. In other words, ε is "evaluation at $T = 1$", and its kernel is visibly the ideal generated by $(1 - T)$. Of course, "ideal" is the same as "submodule".

The map $P_{2n+1} \to P_{2n}$ is taken to be multiplication by $1 - T$ (for example $P_1 \to P_0$ is pictured above). And the map $P_{2n} \to P_{2n-1}$ is taken to be multiplication by $N = 1 + T + \cdots + T^{r-1}$, the "norm element". Thus the whole resolution is periodic of period 2.

It is of course essential that

$$(1 + T + \cdots + T^{r-1})(1 - T) = 1 - T^r = 0,$$

so the composition of two consecutive maps is zero. To prove, more precisely, that we have an exact sequence, we first take an element $m \in P_{2n-1}$ of the form $m = \sum_k a_k T^k$, and write that

$$(1 - T)m = (1 - T)\left(\sum_{k=0}^{r-1} a_k T^k\right) = \sum_{k=0}^{r-1} (a_k - a_{k-1})T^k,$$

with $a_{-1} := a_r$. So if $(1 - T)m = 0$ then $a_k = a_{k-1}$ for all k, and $m = a_0 N$, so that m is in the image of the "multiplication by N" map. We have exactness "at P_{2n-1}".

To prove exactness at P_{2n}, we take the same generic m and assume that $Nm = 0$. Here we find that $Nm = \varepsilon(m)N$, where $\varepsilon(m) = a_0 + \cdots + a_{n-1}$ as above. If $Nm = 0$ then $\varepsilon(m) = 0$, so m is in the ideal generated by $1 - T$, and we have exactness at P_{2n} as well (this includes P_0). ∎

Later we shall prove that this resolution is actually minimal, in some sense; in particular, the module $\mathbb{Z}$ does not admit finite resolutions. This shows that $\mathbb{Z}[C_n]$ is considerably more complicated than $\mathbb{Z}$.

It is natural to ask about the uniqueness of projective resolutions. Here we prove the technical statement behind the answer. In the next section we shall rephrase things in a more intuitive manner.

Proposition 9.14 *Let M be an R-module with a projective resolution by modules $(P_n)_{n\geq 0}$ and maps $\partial_n \colon P_n \to P_{n-1}$, and likewise let N be a module with a resolution by modules $(Q_n)_{n\geq 0}$ and maps $\delta_n \colon Q_n \to Q_{n-1}$. Also, let $f \colon M \to N$ be a homomorphism.*

Then there exist homomorphisms $f_n \colon P_n \to Q_n$ such that the following diagram commutes:

$$\begin{array}{ccccccccccccc}
\cdots & \longrightarrow & P_n & \xrightarrow{\partial_n} & P_{n-1} & \longrightarrow & \cdots & \longrightarrow & P_0 & \longrightarrow & M & \longrightarrow & 0 \\
 & & \downarrow{\scriptstyle f_n} & & \downarrow{\scriptstyle f_{n-1}} & & & & \downarrow{\scriptstyle f_0} & & \downarrow{\scriptstyle f} & & \\
\cdots & \longrightarrow & Q_n & \xrightarrow{\delta_n} & Q_{n-1} & \longrightarrow & \cdots & \longrightarrow & Q_0 & \longrightarrow & N & \longrightarrow & 0
\end{array}$$

Moreover, suppose $(f'_n)_{n\geq 0}$ is another family of homomorphisms making the diagram commute. Then there exist maps $s_n\colon P_n \to Q_{n+1}$ such that

$$f_n - f'_n = s_{n-1} \circ \partial_n + \delta_{n+1} \circ s_n$$

for $n \geq 1$, and $f_0 - f'_0 = \delta_1 \circ s_0$.

Proof. We prove the first statement by induction on n, the existence of f_0 being guaranteed since P_0 is projective. Suppose $f_0,\ldots,f_{n-1}$ are defined, making the diagram commute. The map $f_{n-1} \circ \partial_n\colon P_n \to Q_{n-1}$ takes its values in $\mathrm{Im}(\delta_n) = \ker(\delta_{n-1})$. To see this, write $\delta_{n-1} f_{n-1}\partial_n = (f_{n-2}\partial_{n-1})\partial_n$ by commutativity, and this is the zero map as $\partial\partial = 0$. (We have started to omit the symbol $\circ$, for simplicity, and the indices when they are irrelevant.) As P_n is projective, we see that f_n exists. This proves the first part.

The maps s_n are also constructed by induction, and for s_0 we simply use the fact that $f_0 - f'_0$ takes its values in $\ker(\delta_0)$, by definition. So assume s_{n-1} is defined, and write

$$\begin{aligned}\delta_n(f_n - f'_n) &= (f_{n-1} - f'_{n-1})\partial_n \\ &= (s_{n-2}\partial_{n-1} + \delta_n s_{n-1})\partial_n \\ &= \delta_n s_{n-1}\partial_n\,.\end{aligned}$$

Thus the map $h_n = f_n - f'_n - s_{n-1}\partial_n$ satisfies $\delta_n h_n = 0$. We find a lift of $h_n\colon P_n \to \ker(\delta_n)$ using the fact that P_n is projective, and we call it s_n. □

Complexes

We proceed to introduce a lot of vocabulary, which will help us understand the situation. In particular, we shall see why the last proposition is indeed a uniqueness statement, in a certain sense.

chain complex

Definition 9.15 A **chain complex** is a collection of modules and maps

$$\cdots \longrightarrow C_{n+1} \xrightarrow{\partial_{n+1}} C_n \xrightarrow{\partial_n} C_{n-1} \longrightarrow \cdots$$

such that $\partial\partial = 0$, that is, $\mathrm{Im}(\partial_{n+1}) \subset \ker(\partial_n)$ for all $n \in \mathbb{Z}$. The elements of $\ker(\partial_n)$ are called ***n*-cycles**, and the elements of $\mathrm{Im}(\partial_{n+1})$ are called ***n*-boundaries**. The maps ∂_n are collectively called **boundary operators**.

n-cycles

n-boundaries

Remark 9.16 We pause to introduce the use of the symbol $*$, which occupies a special place in homological algebra. On the one hand, we will use it to denote an entire chain complex simply by C_* or (C_*, ∂_*); notice how this is not adorned by "$* \in \mathbb{Z}$" or any other quantifier, so $*$ is different from letters of the alphabet, say. However, it will be acceptable to write something like "$C_* = 0$ when $* < 0$" (a property often verified for us), and thus treat $*$ like other "variables". One must be flexible, and in practice this causes no confusion and is very helpful. Consider the following definition.

homology groups

Definition 9.17 Let (C_*, ∂_*) be a chain complex. Its **homology groups** are defined to be

$$\mathrm{H}_n(C_*, \partial_*) := \ker(\partial_n)/\operatorname{Im}(\partial_{n+1})$$

for $n \in \mathbb{Z}$.

Example 9.18 A long exact sequence is exactly the same as a chain complex whose homology groups are all zero.

Example 9.19 This is the archetypal chain complex, and this example will explain the terminology. Let $\mathcal{G}$ be a *graph*, that is, a set V of elements called *vertices*, and a set E of pairs $\{v, w\}$ with $v, w \in V$, $v \neq w$, called *edges*. We stick to this combinatorial definition, but of course graphs can be defined by drawing pictures.

Let $R = \mathbb{F}_2$. We define $C_0 := \mathbb{F}_2[V]$, the free vector space on the vertices, and $C_1 := \mathbb{F}_2[E]$, the free vector space on the edges. Also, we define the map $\partial_1 : \mathbb{F}_2[E] \to \mathbb{F}_2[V]$ by $\partial_1(\{v, w\}) = v + w$, extending by linearity. We have a simple chain complex

$$0 \longrightarrow \mathbb{F}_2[E] \xrightarrow{\partial_1} \mathbb{F}_2[V] \longrightarrow 0.$$

(So $\partial_n = 0$ for $n \neq 1$.) Here of course the "boundary" of an edge consists precisely of the two vertices at its ends. What are the homology groups? We write them $\mathrm{H}_1(\mathcal{G}) = \ker(\partial_1)$ and $\mathrm{H}_0(\mathcal{G}) = \mathbb{F}_2[V]/\operatorname{Im}(\partial_1)$ (the others being 0).

As an amusing exercise, the reader will check that an element $c = \sum_i e_i \in \mathbb{F}_2[E]$ is a "1-cycle" if and only if the edges e_i showing up in c can be arranged into closed paths around the graph. It is customary to call these "cycles" in graph theory. The number $\beta_1 = \dim_{\mathbb{F}_2} \mathrm{H}_1(\mathcal{G})$ is called the *first Betti number* of the graph; it is 0 for a tree, 1 for a "circle", 2 for the "figure 8" graph, and so on.

Likewise, the reader will prove that two vertices in the graph, seen as elements of $\mathbb{F}_2[V]$, differ by a "0-boundary" if and only if they can be joined by a path. The number $\beta_0 = \dim_{\mathbb{F}_2} \mathrm{H}_0(\mathcal{G})$ is simply the number of connected components of the graph.

Higher dimensional analogs are easy to envisage. Suppose we have a set T of *triangles*, that is, sets of the form $\{u, v, w\}$, with u, v, w distinct elements of V, and such that all three edges $\{u, v\}$, $\{v, w\}$, and $\{u, w\}$ are part of our graph. Such data can be instantaneously conveyed with a picture – for example, imagine a tetrahedron.

One can then consider $\mathbb{F}_2[T]$ and the map $\partial_2 : \mathbb{F}_2[T] \to \mathbb{F}_2[E]$ defined by

$$\partial_2(\{u, v, w\}) = \{u, v\} + \{v, w\} + \{u, w\}.$$

(The "boundary" of a triangle consisting of the sum of the three edges on the intuitive boundary.) One has a complex

$$0 \longrightarrow \mathbb{F}_2[T] \xrightarrow{\partial_2} \mathbb{F}_2[E] \xrightarrow{\partial_1} \mathbb{F}_2[V] \longrightarrow 0,$$

with homology groups H_2, H_1, and H_0. The triangles actually do not alter homology in degree 0, but the other two groups require a computation. We urge the reader to perform it in the case of the tetrahedron, where one finds $H_2 \cong \mathbb{F}_2$, $H_1 = 0$, $H_0 = \mathbb{F}_2$ (the latter by connectedness). This reflects the intuitive sense that, on the surface of a tetrahedron, no nontrivial loops can be drawn, but there is "a two-dimensional hole".

Generally speaking, the intuition with these geometric examples is that the homology groups are obstructions to the contractibility of the space under consideration. ■

chain morphism

Definition 9.20 Let (C_*, ∂_*) and (C'_*, ∂'_*) be chain complexes. A **chain morphism** between them, written $f_* : C_* \to C'_*$, is a collection of homomorphisms $f_n : C_n \to C'_n$ for $n \in \mathbb{Z}$ such that the following diagram commutes:

$$\begin{array}{ccccccc} \cdots & \longrightarrow & C_n & \xrightarrow{\partial_n} & C_{n-1} & \longrightarrow & \cdots \\ & & \downarrow{\scriptstyle f_n} & & \downarrow{\scriptstyle f_{n-1}} & & \\ \cdots & \longrightarrow & C'_n & \xrightarrow{\partial'_n} & C'_{n-1} & \longrightarrow & \cdots \end{array}$$

When all the homomorphisms f_n are isomorphisms, we say that f_* is an **chain equivalence**. ■

chain equivalence

The following is then obvious, given the definitions.

Lemma 9.21 *A morphism $f_* : C_* \to C'_*$ induces canonically a family of homomorphisms $H_n(f_*) : H_n(C_*) \to H_n(C'_*)$ for $n \in \mathbb{Z}$. One has $H_n(f_* \circ g_*) = H_n(f_*) \circ H_n(g_*)$, and the identity of C_* induces the identity of $H_n(C_*)$.* □

Definition 9.22 When $f_* : C_* \to C'_*$ is a morphism such that for each $n \in \mathbb{Z}$, the induced homomorphism $H_n(f_*) : H_n(C_*) \to H_n(C'_*)$ is an isomorphism, we say that f_* is a **weak equivalence**. ■

weak equivalence

The word "quasi-isomorphism" is sometimes used instead of weak equivalence, but such mixtures of Latin and Greek roots should be avoided (it is not by accident that the word "isomorphism" was accepted by the mathematical community, rather than "equimorphism" or "isovalence").

We can at last return to projective resolutions, and recast the material of the previous section. Indeed, instead of presenting a resolution of M as

$$\cdots \longrightarrow P_2 \xrightarrow{\partial_2} P_1 \xrightarrow{\partial_1} P_0 \xrightarrow{\varepsilon} M \longrightarrow 0,$$

we organize things as follows:

$$\begin{array}{ccccccccccc} \cdots & \longrightarrow & P_2 & \xrightarrow{\partial_2} & P_1 & \xrightarrow{\partial_1} & P_0 & \longrightarrow & 0 & \longrightarrow & \\ & & \downarrow & & \downarrow & & \downarrow{\scriptstyle \varepsilon} & & \downarrow & & \\ \cdots & \longrightarrow & 0 & \longrightarrow & 0 & \longrightarrow & M & \longrightarrow & 0 & \longrightarrow & \end{array}$$

The first line of this diagram defines a chain complex P_*, which is 0 for $* < 0$, and the bottom line defines another chain complex, written M_*, with $M_0 = M$ while $M_n = 0$

for $n \neq 0$. The vertical maps, which are all 0 except ε in degree 0, form a morphism. Stating that the P_n, the ∂_n, and ε together form a projective resolution of M is equivalent to stating:

1. each P_n is projective;
2. P_* is indeed a chain complex (so $\partial\partial = 0$);
3. the morphism above is a weak equivalence between P_* and M_*.

Indeed, that $\mathrm{H}_n(P_*) = 0$ for $n > 1$ is equivalent to exactness at P_n, and the fact that ε induces an isomorphism $\mathrm{H}_0(P_0) \to \mathrm{H}_0(M_0) = M$ is equivalent to exactness at M in the original sequence. Gradually we will see that this is a better point of view. For the rest of the section, when we speak of a projective resolution of M, we mean the chain complex P_* as above (concentrated in nonnegative degrees and not involving M), which is assumed to be equipped with a weak equivalence $P_* \to M_*$. We may write $(P_*, \partial_*, \varepsilon)$ for the resolution, emphasizing the special role of ε.

Definition 9.23 Two morphisms $f_*, f'_*\colon C_* \to C'_*$ are said to be **(chain) homotopic** when there exists a collection of maps $s_n\colon C_n \to C'_{n+1}$ such that

homotopic

$$f_n - f'_n = s_{n-1} \circ \partial_n + \delta_{n+1} \circ s_n$$

for all $n \in \mathbb{Z}$. The family $(s_n)_{n\in\mathbb{Z}}$ itself is called a **(chain) homotopy** between f_* and f'_*.

homotopy

A morphism $f_*\colon C_* \to C'_*$ is said to be a **homotopy equivalence** when there exists a morphism $g_*\colon C'_* \to C_*$ such that $g_* \circ f_*$ is homotopic to the identity of C_*, and $f_* \circ g_*$ is homotopic to the identity of C'_*. The morphism g_* is called a homotopy inverse of f_*.

homotopy equivalence

In the expression "homotopy equivalence", the term "homotopy" is thus a diminutive (not-quite-an-equivalence). Of course the same can be said of "weak" equivalences. An actual equivalence is both a weak equivalence and a homotopy equivalence (given our definitions, there is a little something to check here). In general, we have:

Lemma 9.24 1. *When f_* and f'_* are homotopic, the induced maps are the same, that is $\mathrm{H}_n(f_*) = \mathrm{H}_n(f'_*)$ for all $n \in \mathbb{Z}$.*
2. *A homotopy equivalence is a weak equivalence.*

Proof. (1) Let $x \in \ker(\partial_n) \subset C_n$ be a cycle in degree n. We write

$$f_n(x) - f'_n(x) = s_{n-1}(\partial_n x) + \delta_{n+1}(s_n x) = \delta_{n+1}(s_n x),$$

so $f_n(x)$ and $f'_n(x)$, which differ by a boundary, define the same element of $\mathrm{H}_n(C'_*)$.

(2) Let g_* be a homotopy inverse of f_*. We have

$$\mathrm{H}_n(g_*) \circ \mathrm{H}_n(f_*) = \mathrm{H}_n(g_* \circ f_*) = \mathrm{H}_n(id) = \mathrm{id},$$

by the first point. Likewise, we see that $\mathrm{H}_n(f_*) \circ \mathrm{H}_n(g_*)$ is the identity, and in the end $\mathrm{H}_n(f_*)$ and $\mathrm{H}_n(g_*)$ are inverses of each other, so $\mathrm{H}_n(f_*)$ is an isomorphism. By definition, f_* is a weak equivalence. □

With this new language, Proposition 9.14 can be restated as:

Proposition 9.25 *Suppose P_* is a projective resolution of M, and that Q_* is a projective resolution of N. Suppose also that $f: M \to N$ is given. Then there exists a morphism $f_*: P_* \to Q_*$ such that $\mathrm{H}_0(f_*) = f$, under the given identifications $\mathrm{H}_0(P_*) \cong M$ and $\mathrm{H}_0(Q_*) \cong N$. Moreover, the morphism f_* is unique up to homotopy.*

At long last, we can now add:

Corollary 9.26 *Let P_* and Q_* be two projective resolutions of M. Then there exists a homotopy equivalence $f_*: P_* \to Q_*$. Moreover, the morphism f_* can be chosen so that $\mathrm{H}_0(f_*)$ is the identity of M under the given identifications $\mathrm{H}_0(P_*) \cong M$ and $\mathrm{H}_0(Q_*) \cong M$, and this characterizes f_* up to homotopy.*

And so with the introduction of the word "homotopy" here and there, we see that projective resolutions are unique up to unique homotopy equivalence (sic).

Proof of the corollary. Apply the proposition to the identity of M. This gives $f_*: P_* \to Q_*$, inducing the identity on H_0, and the proposition already asserts that f_* is then unique up to homotopy. Now, apply the proposition again, this time reversing the roles of Q_* and P_*, to obtain $g_*: Q_* \to P_*$ with $\mathrm{H}_0(g_*) = \mathrm{id}$. The composition $g_* \circ f_*: P_* \to P_*$ is a morphism with $\mathrm{H}_0(g_* \circ f_*) = \mathrm{H}_0(g_*) \circ \mathrm{H}_0(f_*) = \mathrm{id} \circ \mathrm{id} = \mathrm{id}$.

By the uniqueness statement in the proposition, this implies that $g_* \circ f_*$ is homotopic to the identity of P_* (which also induces the identity on H_0, of course). Likewise, we see that $f_* \circ g_*$ is homotopic to the identity of Q_*. Thus f_* and g_* are homotopy inverses of one another. □

The definitions in this section are parallel to those of algebraic topology. It is easy to get confused with the analogy, so here is how the topological metaphor works. Let the ring R be fixed. The realm of chain complexes over R is an analog of that of topological spaces (we could just as well restrict attention to those chain complexes C_* with $C_* = 0$ for $* < 0$). When we consider *all* chain complexes, we must at the same time think of *all* topological spaces, including degenerate offsprings of the Mandelbrot set and the graph of $x \mapsto \sin(\frac{1}{x})$.

Projective complexes, that is, chain complexes P_* for which each module P_n is projective, correspond in our analogy to "nice" topological spaces, meaning spaces with some combinatorial structure, such as those considered in Example 9.19. If you know about such things, think of CW-complexes; if not, think of polyhedra. (A CW-complex is to a polyhedron what a general graph, allowing multiple edges between a pair of vertices, and loops at a single vertex, is to a "simple" graph as we have defined them.)

Morphisms are continuous maps, the metaphor extends. A homotopy between morphisms is the analog of a continuous family $X \times [0,1] \to Y$ of maps between the spaces X and Y. A homotopy equivalence between topological spaces, formally defined just like a homotopy equivalence between chain complexes, is intuitively a deformation of a space X until it assumes the shape of Y.

If you know about these, the *homotopy groups* $\pi_n(X)$ are the analogs of the homology groups $\mathrm{H}_n(C_*)$ of complexes. A weak equivalence in topology is one which induces an isomorphism on all homotopy groups. If you do not know what homotopy groups are, just think of weak equivalences, on the topological side, as maps which are close to being homotopy equivalences, without quite being so.

Given a topological space X, there are various theorems providing us with a "nice" (in the above sense) space X' and a weak equivalence $X' \to X$. Analogously, we have shown that a module M, after it has been turned into a chain complex M_* concentrated in degree 0, is weakly equivalent to a projective complex.

Here is an example, in topology, when such a "replacement" can be useful. When X and Y are topological spaces, one may wish to consider the set $\mathrm{Hom}(X,Y)$ of all continuous maps $X \to Y$. If we want to put a topology on $\mathrm{Hom}(X,Y)$, usually we employ the *compact-open topology* (see [Bre97], chapter VII, section 2). However, this is well-behaved only when X and Y are reasonable; for example the evaluation map $\mathrm{Hom}(X,Y) \times X \to Y$ is only continuous when X is locally compact and Hausdorff. If X is pathological, and if one is not bothered by replacing it by a weakly equivalent space X', then it is a good idea to study $\mathrm{Hom}(X',Y)$ instead.

This is *precisely* what we are going to do in the rest of this chapter. We will study the abelian group $\mathrm{Hom}_R(M,N)$ of all R-linear maps between M and N, and we will discover some of its weaknesses, which would not be visible if M were assumed projective. Then we will decide to pick a projective resolution P_* of M, and to study the various groups $\mathrm{Hom}_R(P_n,N)$ instead. The work we have just done will serve to prove that what we get is independent of the choice of resolution.

There is one last thing we need to introduce before we move on to that:

cochain complex

Definition 9.27 A **cochain complex** is a collection of modules and maps

$$\cdots \longrightarrow C^{n-1} \xrightarrow{d^{n-1}} C^n \xrightarrow{d^n} C^{n+1} \longrightarrow \cdots$$

such that $dd = 0$, that is, $\mathrm{Im}(d^{n-1}) \subset \ker(d^n)$ for all $n \in \mathbb{Z}$. The elements of $\ker(d^n)$ are called ***n*-cocycles**, and the elements of $\mathrm{Im}(d^{n-1})$ are called ***n*-coboundaries**. The maps d^n are collectively called the **coboundary operators**.

cohomology groups

The **cohomology groups** of the complex are

$$\mathrm{H}^n(C^*,d^*) = \ker(d^n)/\,\mathrm{Im}(d^{n-1}),$$

for all $n \in \mathbb{Z}$. ∎

Something must be pointed out at once. Let C^* be a cochain complex, and put $C_n := C^{-n}$, for all $n \in \mathbb{Z}$, as well as $\partial_n := d^{-n} \colon C^{-n} = C_n \to C^{-n+1} = C_{n-1}$.

Then (C_*, ∂_*) is a chain complex, and its homology groups are the cohomology groups of (C^*, d^*). Thus *cochain complexes are chain complexes in a different notation.*

The reason why the terminology is useful at all is that many chain complexes – essentially all complexes in this book – are either 0 in negative degrees, or 0 in positive degrees. The latter is less attractive, since many signs appear, and so it has become customary to renumber everything and add "co-" everywhere. And to be fair, there are examples which naturally appear as cochain complexes (like the de Rham complex in differential geometry, if you know what that is).

However, all notions and results about cochain complexes can be deduced from those between chain complexes. Thus we will not spell out the definitions of morphisms, homotopies, and homotopy equivalences between cochain complexes.

The Ext groups

We have encountered $\mathrm{Hom}_R(M,N)$, when M and N are R-modules, with Definition 5.14. In the absence of any extra hypothesis on R, we simply view it as an abelian group.

Definition 9.28 Let $f\colon N_1 \to N_2$ be a homomorphism of R-modules. Then we write

$$f_0^M\colon \mathrm{Hom}_R(M,N_1) \longrightarrow \mathrm{Hom}_R(M,N_2)$$

for the homomorphism of abelian groups defined by $f_0^M(\varphi) = f \circ \varphi$.

For $f\colon M_1 \to M_2$, we write

$$f_N^0\colon \mathrm{Hom}_R(M_2,N) \longrightarrow \mathrm{Hom}_R(M_1,N)$$

for the homomorphism defined by $f_N^0(\varphi) = \varphi \circ f$.

induced homomorphism

We will say that f_0^M and f_N^0 are the homomorphisms **induced homomorphism** by f. ■

We will immediately get into the habit of writing just f_0 and f^0 whenever the extra information is obvious from the context. The next lemma is a direct consequence of the definitions.

Lemma 9.29 *We have $(f \circ g)_0 = f_0 \circ g_0$, and $(f \circ g)^0 = g^0 \circ f^0$. Also, if f is an identity, so are all induced maps f_0 and f^0.*

Further, we have $(f+g)_0 = f_0 + g_0$ and $(f+g)^0 = f^0 + g^0$ (here f and g have the same source and target). Finally, if f is the zero homomorphism between two modules, then all induced maps f_0 and f^0 are also zero. □

Corollary 9.30 *Suppose $M = M_1 \oplus M_2$. Then*

$$\mathrm{Hom}_R(M,N) \cong \mathrm{Hom}_R(M_1,N) \oplus \mathrm{Hom}(M_2,N),$$

If $N = N_1 \oplus N_2$, then

$$\mathrm{Hom}_R(M,N) \cong \mathrm{Hom}_R(M,N_1) \oplus \mathrm{Hom}_R(M,N_2)\,.$$

Proof. We make a point of deducing this from the properties given by the lemma, ignoring entirely the definition of the group $\mathrm{Hom}_R(M,N)$. (Later, this will allow us to claim in all honesty that other proofs can be copied from this one.) Let $\iota_1 : M_1 \to M$ be the obvious inclusion, let $\pi_1 : M \to M_1$ be the projection with kernel M_2, and define ι_2, π_2 symmetrically.

One has $\pi_1 \circ \iota_1 = \mathrm{id}$, so $(\pi_1 \circ \iota_1)^0 = \iota_1^0 \circ \pi_1^0 = id^0 = \mathrm{id}$ (where we write uniformly id for all identity homomorphisms). So π_1^0 is injective; likewise, we see that π_2^0 is injective. Let us show that $H = \mathrm{Hom}_R(M,N)$ is the direct sum of its two subgroups $H_1 = \pi_1^0(\mathrm{Hom}_R(M_1,N))$ and $H_2 = \pi_2^0(\mathrm{Hom}_R(M_2,N))$.

For this, write that $\iota_1 \circ \pi_1 + \iota_2 \circ \pi_2 = \mathrm{id}$, as is obvious. So

$$\pi_1^0 \circ \iota_1^0 + \pi_2^0 \circ \iota_2^0 = \mathrm{id}\,. \qquad (*)$$

Thus $H = H_1 + H_2$, as any $x \in H$ can be written $x = \pi_1^0(x_1) + \pi_2^0(x_2)$ with $x_i = \iota_i^0(x) \in \mathrm{Hom}_R(M_i,N)$, for $i = 1,2$.

To see that the sum is direct, write finally that $\pi_2 \circ \iota_1 = 0$, so $\iota_1^0 \circ \pi_2^0 = 0$, and similarly $\iota_2^0 \circ \pi_1^0 = 0$. If $x \in H$ belongs to both H_1 and H_2, we see that $\iota_1^0(x) = 0$ and $\iota_2^0(x) = 0$. From (*), we deduce $x = 0$, and the sum is indeed direct.

The second case is similar. □

The phenomenon described in the next lemma is fundamental.

Lemma 9.31 *Suppose*

$$0 \longrightarrow A \xrightarrow{\ \iota\ } B \xrightarrow{\ \pi\ } C \longrightarrow 0$$

is an exact sequence of R-modules, and suppose M,N are R-modules. Then there is an exact sequence

$$0 \longrightarrow \mathrm{Hom}_R(M,A) \xrightarrow{\ \iota_0\ } \mathrm{Hom}_R(M,B) \xrightarrow{\ \pi_0\ } \mathrm{Hom}_R(M,C)\,,$$

and an exact sequence

$$0 \longrightarrow \mathrm{Hom}_R(C,N) \xrightarrow{\ \pi^0\ } \mathrm{Hom}_R(B,N) \xrightarrow{\ \iota^0\ } \mathrm{Hom}_R(A,N)\,.$$

It is essential that you provide the easy proof yourself – no amount of staring at a written account will bring the same wisdom. Perhaps the most important thing is to try, and fail, to prove that π_0 and ι^0 are surjective; that is, you should briefly try, and fail, to prove that the next lemma holds without any hypothesis.

Lemma 9.32 *Resume the notation of the previous lemma. If C is projective, we have an exact sequence*

$$0 \longrightarrow \mathrm{Hom}_R(M,A) \xrightarrow{\ \iota_0\ } \mathrm{Hom}_R(M,B) \xrightarrow{\ \pi_0\ } \mathrm{Hom}_R(M,C) \longrightarrow 0\,,$$

and an exact sequence

$$0 \longrightarrow \mathrm{Hom}_R(C,N) \xrightarrow{\pi^0} \mathrm{Hom}_R(B,N) \xrightarrow{\iota^0} \mathrm{Hom}_R(A,N) \longrightarrow 0.$$

The first sequence is also exact under the assumption that M is projective (and C is any module).

Proof. When C is projective, the exact sequence

$$0 \longrightarrow A \xrightarrow{\iota} B \xrightarrow{\pi} C \longrightarrow 0$$

is split, by Lemma 9.5. Lemma 9.2 gives us the existence of $s\colon C \to B$ such that $\pi \circ s = \mathrm{id}$. Thus $\pi_0 \circ s_0 = \mathrm{id}$, and π_0 is surjective, as we wanted.

Similarly, Lemma 9.2 shows that there exists $r\colon B \to A$ such that $r \circ \iota = \mathrm{id}$, so $\iota^0 \circ r^0 = \mathrm{id}$, and ι^0 is surjective.

Now suppose M is projective, rather than C. The fact that any map $f\colon M \to C$ is of the form $\pi^0(\widetilde{f})$ for some $\widetilde{f} \in \mathrm{Hom}_R(M,B)$ is exactly the definition of "projective". □

In this proof, we see that the last argument actually relies on the interplay between the definition of "projective", and the definition of the group $\mathrm{Hom}_R(M,B)$. In a sense, it is more anecdotal than the first part of the proof, which chiefly uses the properties described in Lemma 9.29. The same argument will work in great generality.

Having realized that the operation $\mathrm{Hom}_R(-,N)$, for fixed N, works "better when applied to projective modules", in the sense that the last lemma can then be invoked and gives finer information than the previous one, we can deploy the strategy described in the previous section. Namely, instead of studying $\mathrm{Hom}_R(M,N)$, we will pick a projective resolution $(P_*, \partial_*, \varepsilon)$ of M and study collectively all the abelian groups $\mathrm{Hom}_R(P_n,N)$ for all $n \geq 0$.

More precisely, we have maps

$$d^n := \partial^0_{n+1}\colon \mathrm{Hom}_R(P_n,N) \longrightarrow \mathrm{Hom}_R(P_{n+1},N),$$

and $d \circ d = 0$ follows from $\partial \circ \partial = 0$. In other words, $\mathrm{Hom}_R(P_*,N)$ becomes a *cochain complex.*

Definition 9.33 The cohomology groups of $(\mathrm{Hom}_R(P_*,N),d^*)$ are denoted by

$$\mathrm{Ext}^n_R(M,N) = \mathrm{H}^n(\mathrm{Hom}_R(P_*,N),d^*),$$

for $n \geq 0$. ■

We must at once take care of:

Lemma 9.34 *If Q_* is another projective resolution of M, then the groups*

$$\mathrm{H}^n(\mathrm{Hom}_R(P_*,N),d^*) \quad \text{and} \quad \mathrm{H}^n(\mathrm{Hom}_R(Q_*,N),d^*)$$

are canonically isomorphic, for all $n \geq 0$, and so the notation $\mathrm{Ext}^n_R(M,N)$ makes sense.

What "canonically isomorphic" means is explained in the proof.

Proof. Using Corollary 9.26, we pick a homotopy equivalence $f_*: P_* \to Q_*$, with homotopy inverse $g_*: Q_* \to P_*$. Then $f_*^0: \mathrm{Hom}_R(Q_*,N) \to \mathrm{Hom}_R(P_*,N)$ is a morphism of cochain complexes. It is in fact a homotopy equivalence, with homotopy inverse g_*^0. These facts are immediate from the definitions.

This proves that the groups proposed are indeed isomorphic (by (2) of Lemma 9.24). If we agree to pick f_* inducing the identity of M, as in Corollary 9.26 again, then it is well-defined up to chain homotopy; the same can be said of f_*^0. By (1) of Lemma 9.24, the induced isomorphisms are well-defined. So there is a canonical choice, as claimed. □

We will examine a number of examples to illustrate the definition. It may be good to remind the reader that a projective resolution $(P_*, \partial_*, \varepsilon)$ of M, as above, is a chain complex with $P_* = 0$ for $* < 0$, and it does *not* involve M; rather, it is assumed that there is a weak equivalence $\varepsilon: P_* \to M_*$, where M_* is M in degree 0, and 0 otherwise.

Lemma 9.35 *In degree* 0 *we have (canonically)*

$$\mathrm{Ext}_R^0(M,N) \cong \mathrm{Hom}_R(M,N).$$

Proof. It is assumed that we have an exact sequence

$$P_1 \xrightarrow{\partial_1} P_0 \xrightarrow{\varepsilon} M \longrightarrow 0. \qquad (*)$$

We apply $\mathrm{Hom}_R(-,N)$ and obtain

$$0 \longrightarrow \mathrm{Hom}_R(M,N) \xrightarrow{\varepsilon^0} \mathrm{Hom}_R(P_0,N) \xrightarrow{\partial_1^0} \mathrm{Hom}_R(P_1,N) \qquad (**)$$

which is a complex (the composition of any two maps is zero). By definition, we are asked to study the cohomology group $\mathrm{H}^0(\mathrm{Hom}_R(P_*,N))$, so this is $\ker(\partial_1^0)$ (careful: It is not $\ker(\partial_1^0)/\mathrm{Im}(\varepsilon^0)$, cf. the remark made just before the proof).

Now, we claim that (**) is exact. This is a common trick: we replace (*) by

$$0 \longrightarrow \mathrm{Im}(\partial_1) \longrightarrow P_0 \xrightarrow{\varepsilon} M \longrightarrow 0,$$

then we use Lemma 9.31 to obtain the exact sequence

$$0 \longrightarrow \mathrm{Hom}_R(M,N) \xrightarrow{\varepsilon^0} \mathrm{Hom}_R(P_0,N) \longrightarrow \mathrm{Hom}_R(\mathrm{Im}(\partial_1),N).$$

Finally, we note that

$$\partial_1^0: \mathrm{Hom}_R(\mathrm{Im}(\partial_1),N) \longrightarrow \mathrm{Hom}_R(P_1,N)$$

is injective by another application of Lemma 9.31. The claim follows. In the end, we see that $\ker(\partial_1^0) = \varepsilon^0(\mathrm{Hom}_R(M,N)) \cong \mathrm{Hom}_R(M,N)$. □

Example 9.36 Suppose M is projective. Then $(M_*, 0, id)$ is a projective resolution of M (as we observed during Example 9.10). The cochain complex $\mathrm{Hom}_R(M_*,N)$ is "concentrated" in degree 0 (meaning it is 0 in other degrees), so $\mathrm{Ext}_R^n(M,N) = 0$ for $n > 0$, and for any module N. We see now how the Ext groups are, for a general M, a measure of the extent to which M fails to be projective. The existence of nonzero Ext groups at all is an indication of the extent to which the ring R fails to be semisimple. ∎

Example 9.37 Recall Example 9.12: Every $\mathbb{Z}$-module has a projective resolution P_* for which only P_0 and P_1 are nonzero. As a result, we have $\mathrm{Ext}^n_{\mathbb{Z}}(M,N) = 0$ when $n > 1$, regardless of the modules M and N. Traditionally, one writes $\mathrm{Ext}(M,N) = \mathrm{Ext}^1_{\mathbb{Z}}(M,N)$.

The previous example shows that $\mathrm{Ext}(\mathbb{Z},N) = 0$. Let us now describe the group $\mathrm{Ext}(\mathbb{Z}/m\mathbb{Z},N)$. We use the projective resolution

$$0 \longrightarrow \mathbb{Z} \xrightarrow{x \mapsto mx} \mathbb{Z} \xrightarrow{\varepsilon} \mathbb{Z}/m\mathbb{Z} \longrightarrow 0,$$

and we must consider the cohomology groups of the complex

$$0 \longrightarrow \mathrm{Hom}(\mathbb{Z},N) \xrightarrow{\varphi \mapsto m\varphi} \mathrm{Hom}(\mathbb{Z},N) \longrightarrow 0.$$

Of course we may identify $\mathrm{Hom}(\mathbb{Z},N)$ with N itself: the isomorphism maps φ to $\varphi(1)$. The complex becomes

$$0 \longrightarrow N \xrightarrow{x \mapsto mx} N \longrightarrow 0.$$

The zeroth cohomology group is $\{x \in N : mx = 0\}$ which, as predicted by the last lemma, can be identified with $\mathrm{Hom}(\mathbb{Z}/m\mathbb{Z},N)$. The first cohomology group is $\mathrm{Ext}(\mathbb{Z}/m\mathbb{Z},N) \cong N/mN$. ∎

We have in fact done enough to describe entirely $\mathrm{Ext}(M,N)$ when M is finitely generated (so it is a direct sum of copies of $\mathbb{Z}$ and of cyclic groups). Indeed, consider the following construction, and the accompanying lemma.

Definition 9.38 Let $f\colon M_1 \to M_2$ be a homomorphism of R-modules. Let P_* be a resolution of M_1, and Q_* be a resolution of M_2. Pick a morphism $h_*\colon P_* \to Q_*$ such that $H_0(h_*) = f$, as in Proposition 9.25; up to homotopy, there is only one such h_*. The induced collection of homomorphisms $h^0_n\colon \mathrm{Hom}_R(Q_n,N) \to \mathrm{Hom}_R(P_n,N)$ is a morphism of cochain complexes for any module N. The maps induced on the level of cohomology are simply written

$$f^n = f^n_N\colon \mathrm{Ext}^n_R(M_2,N) \longrightarrow \mathrm{Ext}^n_R(M_1,N),$$

for $n \geq 0$. ∎

The next lemma is left as an exercise.

Lemma 9.39 *The homomorphisms f^n do not depend on the choices of resolutions P_* and Q_*. In degree 0, this new definition of f^0 coincides with the previous one, when $\mathrm{Ext}^n_R(-,N)$ is identified with $\mathrm{Hom}_R(-,N)$.*

In each degree, this construction has the properties stated in Lemma 9.29 for the degree 0. It follows that $\mathrm{Ext}^n_R(M_1 \oplus M_2,N) \cong \mathrm{Ext}^n_R(M_1,N) \oplus \mathrm{Ext}^n_R(M_2,N)$, for all $n \geq 0$. □

The reader will construct (it is easier) a collection of homomorphisms

$$f_n\colon \operatorname{Ext}^n_R(M,N_1) \longrightarrow \operatorname{Ext}^n_R(M,N_2),$$

when $f\colon N_1 \to N_2$. The formal properties are the same again, and it follows that $\operatorname{Ext}^n_R(M,N_1 \oplus N_2) \cong \operatorname{Ext}^n_R(M,N_1) \oplus \operatorname{Ext}^n_R(M,N_2)$.

Long exact sequences

We present some classics of homological algebra.

exact sequence of chain complexes

Definition 9.40 A **exact sequence of chain complexes** is a sequence of morphisms

$$0 \longrightarrow C'_* \xrightarrow{f_*} C_* \xrightarrow{g_*} C''_* \longrightarrow 0,$$

where C_*, C'_* and C''_* are chain complexes, such that for each degree $n \in \mathbb{Z}$ the sequence

$$0 \longrightarrow C'_n \xrightarrow{f_n} C_n \xrightarrow{g_n} C''_n \longrightarrow 0,$$

is exact. ■

In other words, the situation is that of an infinite commutative diagram of the form

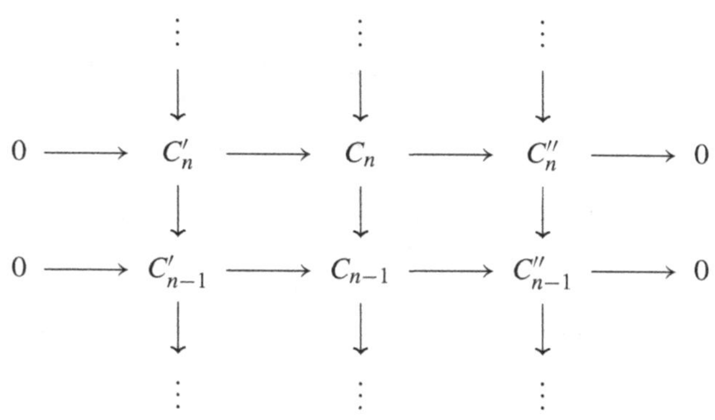

in which the columns are chain complexes, and the rows are short exact sequences.

Theorem 9.41 (zig-zag lemma, snake lemma) *If*

$$0 \longrightarrow C'_* \xrightarrow{f_*} C_* \xrightarrow{g_*} C''_* \longrightarrow 0,$$

is a short exact sequence of chain complexes, then there exists a long exact sequence

$$\cdots \longrightarrow \mathrm{H}_n(C'_*) \xrightarrow{\mathrm{H}_n(f_*)} \mathrm{H}_n(C_*) \xrightarrow{\mathrm{H}_n(g_*)} \mathrm{H}_n(C''_*) \xrightarrow{\delta_n} \mathrm{H}_{n-1}(C'_*)$$

$$\xrightarrow{\mathrm{H}_{n-1}(f_*)} \mathrm{H}_{n-1}(C_*) \xrightarrow{\mathrm{H}_{n-1}(g_*)} \mathrm{H}_{n-1}(C''_*) \xrightarrow{\delta_{n-1}} \mathrm{H}_{n-2}(C'_*) \longrightarrow \cdots$$

The morphisms δ_n, for $n \in \mathbb{Z}$, are called the connecting homomorphisms.

Sketch. This is an excellent exercise. Here are a few hints, in case you get stuck (refer to the last commutative diagram). Let $x \in C''_n$ be a cycle; lift it to $y \in C_n$; push y to $z \in C_{n-1}$; this z maps to 0 in C''_{n-1}, so it is the image of $t \in C'_{n-1}$; finally, this t is seen to be a cycle. The association $x \mapsto t$ induces δ_n.

You have to check that δ_n is independent of all choices made. When this is done, exactness must be checked everywhere. □

Proofs such as the one we have just sketched are called proofs by *diagram chasing*. (If you do not understand the phrase, then you have probably not tried hard enough to prove the theorem.) We point out that, in the sequel, we shall occasionally have to refer to the specific construction of the "connecting homomorphisms" δ_* sketched in this proof. Still, many consequences can be drawn from the mere existence of the long exact sequence.

Note that, when all three complexes involved are zero in negative degrees, the long exact sequence ends with

$$\cdots \longrightarrow \mathrm{H}_0(C_*) \longrightarrow \mathrm{H}_0(C''_*) \longrightarrow 0.$$

On the other hand, we can apply the theorem in the case of cochain complexes, which are particular chain complexes in a different notation. We obtain:

Corollary 9.42 (zig-zag lemma for cochain complexes) *If*

$$0 \longrightarrow C_1^* \xrightarrow{f^*} C_2^* \xrightarrow{g^*} C_3^* \longrightarrow 0,$$

is a short exact sequence of cochain complexes, then there exists a long exact sequence

$$\cdots \longrightarrow \mathrm{H}^n(C_1^*) \xrightarrow{\mathrm{H}^n(f^*)} \mathrm{H}^n(C_2^*) \xrightarrow{\mathrm{H}^n(g^*)} \mathrm{H}^n(C_3^*) \xrightarrow{\delta^n} \mathrm{H}^{n+1}(C_1^*)$$

$$\xrightarrow{\mathrm{H}^{n+1}(f^*)} \mathrm{H}^{n+1}(C_2^*) \xrightarrow{\mathrm{H}^{n+1}(g^*)} \mathrm{H}^{n+1}(C_3^*) \xrightarrow{\delta^{n+1}} \mathrm{H}^{n+2}(C_1^*) \longrightarrow \cdots$$

When we have $C_1^* = C_2^* = C_3^* = 0$ for $* < 0$, this sequence starts with

$$0 \longrightarrow \mathrm{H}^0(C_1^*) \longrightarrow \mathrm{H}^0(C_2^*) \longrightarrow \cdots$$

□

Here is the first application.

Theorem 9.43 *Let*

$$0 \longrightarrow A \xrightarrow{\iota} B \xrightarrow{\pi} C \longrightarrow 0$$

be a short exact sequence of R-modules, and let M be another module. Then there is a long exact sequence

$$0 \longrightarrow \mathrm{Hom}_R(M,A) \xrightarrow{\iota_0} \mathrm{Hom}_R(M,B) \xrightarrow{\pi_0} \mathrm{Hom}_R(M,C)$$

$$\xrightarrow{\delta_0} \mathrm{Ext}^1_R(M,A) \xrightarrow{\iota_1} \mathrm{Ext}^1_R(M,B) \xrightarrow{\pi_1} \mathrm{Ext}^1_R(M,C)$$

$$\xrightarrow{\delta_1} \mathrm{Ext}^2_R(M,A) \xrightarrow{\iota_2} \mathrm{Ext}^2_R(M,B) \xrightarrow{\pi_2} \mathrm{Ext}^2_R(M,C) \longrightarrow \cdots$$

Proof. Let $(P_*, \partial_*, \varepsilon)$ be a projective resolution of M. Each P_n being projective, we obtain short exact sequences

$$0 \longrightarrow \mathrm{Hom}_R(P_n,A) \xrightarrow{\iota_0} \mathrm{Hom}_R(P_n,B) \xrightarrow{\pi_0} \mathrm{Hom}_R(P_n,C) \longrightarrow 0,$$

by Lemma 9.32. We have in fact a short exact sequence of cochain complexes

$$0 \longrightarrow \mathrm{Hom}_R(P_*,A) \longrightarrow \mathrm{Hom}_R(P_*,B) \longrightarrow \mathrm{Hom}_R(P_*,C) \longrightarrow 0,$$

so that the long exact sequence is produced by the zig-zag lemma. The reader is simply invited to check that the maps involved in the sequence are indeed those proposed (also note that Lemma 9.35 is used to identify the cohomology in degree 0). □

Thus, the *Ext* groups are just what we need to complement Lemma 9.31, or at least one-half it. We will speak of "the sequence obtained by applying $\mathrm{Hom}_R(M,-)$" to the original short exact sequence.

One may also "apply $\mathrm{Hom}_R(-,N)$" and obtain a similar long exact sequence, but we need the following.

Proposition 9.44 (the horseshoe lemma) *Suppose*

$$0 \longrightarrow M' \xrightarrow{\iota} M \xrightarrow{\pi} M'' \longrightarrow 0$$

is a short exact sequence. Let $(P'_*, \partial', \varepsilon')$, resp. $(P''_*, \partial'', \varepsilon'')$ be a projective resolution of M' resp. M''. Then there exists a projective resolution $(P_*, \partial, \varepsilon)$ of M, and a short exact sequence of chain complexes

$$0 \longrightarrow P'_* \longrightarrow P_* \longrightarrow P''_* \longrightarrow 0$$

which is compatible with the original sequence, in the sense that we have a commutative diagram

$$\begin{array}{ccccccccc} 0 & \longrightarrow & P'_0 & \longrightarrow & P_0 & \longrightarrow & P''_0 & \longrightarrow & 0 \\ & & \downarrow{\scriptstyle \varepsilon'} & & \downarrow{\scriptstyle \varepsilon} & & \downarrow{\scriptstyle \varepsilon''} & & \\ 0 & \longrightarrow & M' & \xrightarrow{\iota} & M & \xrightarrow{\pi} & M'' & \longrightarrow & 0 \end{array}$$

Sketch. We let the reader fill the details of the following argument. We let $P_n = P'_n \oplus P''_n$ (we have to, since the short exact sequences which we are to construct will be automatically split, the module P''_n being projective for all n). We construct ε using the definition of "projective" directly (we see that we will not be able to take naively the direct sum of the two complexes).

We proceed then by induction. Assume that we have constructed a map of the form

$$\begin{pmatrix} \partial'_n & \theta_n \\ 0 & \partial''_n \end{pmatrix} : P'_n \oplus P''_n \longrightarrow P'_{n-1} \oplus P''_{n-1}.$$

We want to find $\theta_{n+1} : P''_{n+1} \to P'_n$ to define the next boundary operation in the same fashion. To have a complex, we see by multiplying two matrices that we must have $\partial' \circ \theta_{n+1} = -\theta_n \circ \partial''$. A little diagram chase shows that $\partial' \circ \theta_n \circ \partial'' = 0$, so θ_{n+1} can be chosen, satisfying our condition, by exactness of (P', ∂') and the fact that P''_{n+1} is projective. More diagram chasing establishes that all the other requirements are met (exactness at P_n, and commutativity of the required diagram). □

Theorem 9.45 *Let*

$$0 \longrightarrow A \xrightarrow{\iota} B \xrightarrow{\pi} C \longrightarrow 0$$

be a short exact sequence of R-modules, and let N be another module. Then there is a long exact sequence

$$0 \longrightarrow \mathrm{Hom}_R(C,N) \xrightarrow{\pi^0} \mathrm{Hom}_R(B,N) \xrightarrow{\iota^0} \mathrm{Hom}_R(A,N)$$

$$\xrightarrow{\delta^0} \mathrm{Ext}^1_R(C,N) \xrightarrow{\pi^1} \mathrm{Ext}^1_R(B,N) \xrightarrow{\iota^1} \mathrm{Ext}^1_R(A,N)$$

$$\xrightarrow{\delta^1} \mathrm{Ext}^2_R(C,N) \xrightarrow{\pi^2} \mathrm{Ext}^2_R(B,N) \xrightarrow{\iota^2} \mathrm{Ext}^2_R(A,N) \longrightarrow \cdots$$

Proof. We let P'_* be a resolution of A and P''_* be a resolution of C, and we use the horseshoe lemma to find a compatible resolution P_* of B. The sequences

$$0 \longrightarrow \mathrm{Hom}_R(P'_n,N) \longrightarrow \mathrm{Hom}_R(P_n,N) \longrightarrow \mathrm{Hom}_R(P''_n,N) \longrightarrow 0$$

are exact in virtue of Lemma 9.32, since P''_n is projective. The long exact sequence is then produced by the zig-zag lemma. □

As an exercise, the reader can investigate the long exact sequence obtained by applying $\mathrm{Hom}_{\mathbb{Z}}(-,N)$ to

$$0 \longrightarrow \mathbb{Z} \xrightarrow{x \mapsto mx} \mathbb{Z} \longrightarrow \mathbb{Z}/m\mathbb{Z} \longrightarrow 0,$$

and recover the calculation of $\mathrm{Ext}(\mathbb{Z}/m\mathbb{Z},N)$. When this is done, one may ponder the fact that the long exact sequence just introduced, together with the observation that $\mathrm{Ext}^n_R(P,N) = 0$ when P is projective and $n > 0$, are enough to characterize $\mathrm{Ext}^n_R(M,N)$ in full generality. That is, the definition we have given, using projective resolutions, is forced upon us.

The Tor groups

What we have done with $\mathrm{Hom}_R(-,N)$ can be done with $M \otimes_R -$, with minor modifications, as we proceed to explain. We assume that the reader is familiar with tensor products *over a commutative ring*. Since we will need the general case, when R is any ring at all, we offer the following statements. The proofs are not provided, as they are very easy, and also can be adapted immediately from your favorite source on the commutative case (for example, the Appendix, or chapter XVI in [Lan02]). Our point is solely to recap what remains true over a non-commutative ring.

Proposition 9.46 *Let A be a* right *R-module, and let B be a* left *R-module. Then there exists an abelian group M with the following properties.*

There is a bilinear map of abelian groups $\varphi\colon A \times B \to M$ with the property that $\varphi(a \cdot \lambda, b) = \varphi(a, \lambda \cdot b)$ for any $\lambda \in R$, $a \in A$, $b \in B$; and for any map $f\colon A \times B \to C$ with the analogous properties, where C is some abelian group, we have $f = \tilde{f} \circ \varphi$ for some unique homomorphism $\tilde{f}\colon M \to C$.

It follows that M is unique, up to canonical isomorphism. It is denoted by $A \otimes_R B$, and the element $\varphi(a,b)$ is denoted by $a \otimes b$.

When R is commutative, or more generally when A is an $R-R^{op}$-bimodule, then $A \otimes_R B$ is a left R-module, with $\lambda \cdot (a \otimes b) = (\lambda \cdot a) \otimes b$. In this case φ is R-linear, as is $\tilde{f}$ whenever f is R-linear. □

In this chapter we will always view $A \otimes_R B$ as an abelian group. From the "universal property" of the tensor product (that is, from the proposition), one establishes the "adjunction" below.

Proposition 9.47 *Let A be a right R-module, let B be a left R-module, and let C be an abelian group. There is a canonical isomorphism of abelian groups*

$$\mathrm{Hom}(A \otimes_R B, C) \cong \mathrm{Hom}_R(B, \mathrm{Hom}(A, C)).$$

Here $\mathrm{Hom}(A,C)$ *is viewed as an R-module, via $(\lambda \cdot f)(a) = f(a \cdot \lambda)$, in obvious notation.* □

Definition 9.48 When $f\colon N_1 \to N_2$ is a homomorphism of R-modules, there is an induced map

$$f_0 = f_0^M : M \otimes_R N_1 \longrightarrow M \otimes_R N_2,$$

for any right module M, satisfying $f_0(m \otimes n) = m \otimes f(n)$. It is sometimes also written $\mathrm{id} \otimes f$, or $M \otimes f$. ■

Lemma 9.49 *The construction of f_0 from f enjoys the properties described in Lemma 9.29. As a result, $M \otimes_R (N_1 \oplus N_2)$ is naturally isomorphic to $(M \otimes_R N_1) \oplus (M \otimes_R N_2)$.* □

Lemma 9.50 *Suppose*

$$0 \longrightarrow A \xrightarrow{\iota} B \xrightarrow{\pi} C \longrightarrow 0$$

is an exact sequence of R-modules. Then there exists an exact sequence

$$M \otimes_R A \xrightarrow{\iota_0} M \otimes_R B \xrightarrow{\pi_0} M \otimes_R C \longrightarrow 0.$$

When C is projective, there is an exact sequence

$$0 \longrightarrow M \otimes_R A \xrightarrow{\iota_0} M \otimes_R B \xrightarrow{\pi_0} M \otimes_R C \longrightarrow 0.$$

Proof. Exactness at $M \otimes_R C$ is obvious, as is the fact that $\mathrm{Im}(\iota_0) \subset \ker(\pi_0)$. To prove equality, first write the exact sequence

$$0 \longrightarrow \mathrm{Hom}_R(C,N) \longrightarrow \mathrm{Hom}_R(B,N) \longrightarrow \mathrm{Hom}_R(A,N),$$

for any module N, by Lemma 9.31. Applied to $N = \mathrm{Hom}(M,X)$ where X is any R-module, and using the adjunction above, this gives the exactness of

$$0 \longrightarrow \mathrm{Hom}(M \otimes_R C,X) \longrightarrow \mathrm{Hom}(M \otimes_R B,X) \longrightarrow \mathrm{Hom}(M \otimes A,X).$$

Finally, pick $X = M \otimes_R B/\mathrm{Im}(\iota_0)$. The natural quotient map $M \otimes_R B \to X$ gives an element of $\mathrm{Hom}(M \otimes_R B,X)$ which, by exactness, must factor through π_0. This shows $\ker(\pi_0) \subset \mathrm{Im}(\iota_0)$.

When C is projective, the original sequence is split, so the induced sequence is also split and exact, by the arguments in the proof of Lemma 9.32. □

Definition 9.51 Let P_* be a projective resolution of N. We put

$$\mathrm{Tor}_n^R(M,N) = \mathrm{H}_n(M \otimes_R P_*),$$

for any $n \geq 0$. ■

Lemma 9.52 *This definition is independent of the choice of projective resolution, and the notation makes sense. In degree* 0, *we have* $\mathrm{Tor}_0^R(M,N) = M \otimes_R N$. *If* $f\colon N_1 \to N_2$, *there are induced maps*

$$f_n\colon \mathrm{Tor}_n^R(M,N_1) \longrightarrow \mathrm{Tor}_n^R(M,N_2),$$

extending the definition of f_0, *for any* $n \geq 0$, *and enjoying the properties described in Lemma 9.29. As a result, for each* $n \geq 0$ *the group* $\mathrm{Tor}(M,N_1 \oplus N_2)$ *is naturally isomorphic to* $\mathrm{Tor}_n^R(M,N_1) \oplus \mathrm{Tor}_n^R(M,N_2)$.

Similarly, there are induced maps on the first factor, and isomorphisms $\mathrm{Tor}_n^R(M_1 \oplus M_2,N) \cong \mathrm{Tor}_n^R(M_1,N) \oplus \mathrm{Tor}_n^R(M_2,N)$. □

Here, Lemma 9.50 is crucial to identify Tor in degree 0 as just the tensor product. Using the horseshoe lemma and the zig-zag lemma, we then get:

Theorem 9.53 *Suppose*

$$0 \longrightarrow A \xrightarrow{\iota} B \xrightarrow{\pi} C \longrightarrow 0$$

is an exact sequence of R-modules, and let M be any right R-module. Then there is a long exact sequence

$$\cdots \longrightarrow \mathrm{Tor}_2^R(M,A) \xrightarrow{\iota_2} \mathrm{Tor}_2^R(M,B) \xrightarrow{\pi_2} \mathrm{Tor}_2^R(M,C)$$

$$\longrightarrow \mathrm{Tor}_1^R(M,A) \xrightarrow{\iota_1} \mathrm{Tor}_1^R(M,B) \xrightarrow{\pi_1} \mathrm{Tor}_1^R(M,C)$$

$$\longrightarrow M \otimes_R A \xrightarrow{\iota_0} M \otimes_R B \xrightarrow{\pi_0} M \otimes_R C \longrightarrow 0.$$

Example 9.54 We pick $R = \mathbb{Z}$. As any module possesses a projective resolution with just two nonzero terms, we see that $\mathrm{Tor}_n^{\mathbb{Z}}(M,N) = 0$ for any abelian groups M,N whenever $n > 1$. We usually write $\mathrm{Tor}(M,N) = \mathrm{Tor}_1^{\mathbb{Z}}(M,N)$.

Certainly $\mathrm{Tor}(M,\mathbb{Z}) = 0$ as $\mathbb{Z}$ is projective, and to describe the Tor groups when N is finitely-generated, we only need to compute $\mathrm{Tor}(M,\mathbb{Z}/m\mathbb{Z})$ for any M and any m. We start from the exact sequence

$$0 \longrightarrow \mathbb{Z} \xrightarrow{x \mapsto mx} \mathbb{Z} \longrightarrow \mathbb{Z}/m\mathbb{Z} \longrightarrow 0,$$

and apply $M \otimes -$, yielding the six term exact sequence

$$0 \longrightarrow \mathrm{Tor}(M,\mathbb{Z}) \longrightarrow \mathrm{Tor}(M,\mathbb{Z}) \longrightarrow \mathrm{Tor}(M,\mathbb{Z}/m\mathbb{Z})$$

$$\longrightarrow M \otimes_{\mathbb{Z}} \mathbb{Z} \longrightarrow M \otimes_{\mathbb{Z}} \mathbb{Z} \longrightarrow M \otimes_{\mathbb{Z}} \mathbb{Z}/m\mathbb{Z} \longrightarrow 0.$$

Here, $\mathrm{Tor}(M,\mathbb{Z}) = 0$ as just observed. On the other hand, $M \otimes_{\mathbb{Z}} \mathbb{Z}$ can be identified with M itself, and the map between the two copies of M in this sequence is $x \mapsto mx$, by inspection. It follows that

$$\mathrm{Tor}(M,\mathbb{Z}/m\mathbb{Z}) \cong \{x \in M : mx = 0\},$$

that is, $\mathrm{Tor}(M,\mathbb{Z}/m\mathbb{Z})$ is the m-torsion in M. The notation Tor comes from this!

For example, as there is no torsion in the abelian group $\mathbb{Q}$, we see that $\mathrm{Tor}(\mathbb{Q},N) = 0$ for all N finitely generated (and one could generalize to any N), so that "tensoring with $\mathbb{Q}$ preserves the exact sequences".

We finish our treatment of Tor, and the chapter, with an answer to the natural question: Can you reverse the roles of M and N in the discussion above? The next proposition gives a qualified "yes".

Proposition 9.55 *One can compute $\mathrm{Tor}_n^R(M,N)$ by taking a resolution P_* of M by (right!) projective modules, tensoring it with N, and taking the homology groups of $P_* \otimes_R N$.*

Some comments are in order before we embark on a proof. Let R^{op} be the opposite ring, as in Definition 5.9. Then it is immediate that $M \otimes_R N$ can be identified with $N \otimes_{R^{op}} M$. The groups mentioned in the proposition are then $\mathrm{Tor}_n^{R^{op}}(N,M)$, by definition, and so it is claimed that

$$\mathrm{Tor}_n^{R^{op}}(N,M) \cong \mathrm{Tor}_n^R(M,N)$$

for all $n \geq 0$, having the case $n = 0$ settled. The question will not go away by use of an abstract argument – there is something to prove. We merely want to mention, in passing, that the operation $M \rightsquigarrow M \otimes_R N$ has the same formal properties as $N \rightsquigarrow M \otimes_R N$; in particular, there are long exact sequences, involving Tor groups, which the proposition identifies. To be more precise, suppose

$$0 \longrightarrow A \longrightarrow B \longrightarrow C \longrightarrow 0$$

is an exact sequence of *right* R-modules, and let N be any left R-module. Then there is a long exact sequence

$$\cdots \longrightarrow \mathrm{Tor}_2^R(A,N) \longrightarrow \mathrm{Tor}_2^R(B,N) \longrightarrow \mathrm{Tor}_2^R(C,N)$$

$$\longrightarrow \mathrm{Tor}_1^R(A,N) \longrightarrow \mathrm{Tor}_1^R(B,N) \longrightarrow \mathrm{Tor}_1^R(C,N)$$

$$\longrightarrow A \otimes_R N \longrightarrow B \otimes_R N \longrightarrow C \otimes_R N \longrightarrow 0.$$

Turning to the proof, we start with a particular case: If the proposition is to be true, then the next lemma must hold.

Lemma 9.56 *If P is a right projective R-module, then* $\mathrm{Tor}_n^R(P,N) = 0$ *for any module N, and any $n > 0$. It follows that the operation $P \otimes_R -$ preserves exact sequences.*

Proof. One has $R \otimes_R N = N$ for any module N (the canonical identifications are $r \otimes n \mapsto r \cdot n$ and $n \mapsto 1 \otimes n$). As a result, $R^k \otimes_R N = N^k$ for any integer k, and more generally, if F is a free R-module, on the right, then $F \otimes_R N$ is a direct sum of copies of N, indexed by a basis for F. Thus the operation $F \otimes_R -$ preserves exact sequences, by inspection.

If Q_* is a projective resolution of N, then the chain complex $F \otimes_R Q_*$ is exact in degrees > 0, so $\mathrm{Tor}_n^R(F,N) = 0$ for $n > 0$.

We turn to the case when we consider a projective R-module P. Let P' be such that $P \oplus P' = F$ is free. Then $0 = \mathrm{Tor}_n^R(F,N) \cong \mathrm{Tor}_n^R(P,N) \oplus \mathrm{Tor}_n^R(P',N)$, and so certainly $\mathrm{Tor}_n^R(P,N) = 0$.

The (usual) long exact sequence of Tor groups shows that $P \otimes_R -$ then preserves short exact sequences. It is a general fact, left to the reader (and not related to tensor products in particular), that this implies that the same operation preserves all exact sequences, short or not. □

Proof of Proposition 9.55. Let P_* be a projective resolution of the right module M, and let Q_* be a projective resolution of the left module N. We let $K_1 = \ker(Q_0 \to N)$, so that there is a short exact sequence

$$0 \longrightarrow K_1 \longrightarrow Q_0 \longrightarrow N \longrightarrow 0.$$

Applying $M \otimes_R -$, we get a long exact sequence of abelian groups, a portion of which is

$$0 \longrightarrow \mathrm{Tor}_1^R(M,N) \longrightarrow M \otimes_R K_1 \longrightarrow M \otimes_R Q_0 . \qquad (*)$$

Here we use the fact that $\mathrm{Tor}_1^R(M,Q_0) = 0$ since Q_0 is projective.

On the other hand, by the lemma, the sequences

$$0 \longrightarrow P_n \otimes_R K_1 \longrightarrow P_n \otimes_R Q_0 \longrightarrow P_n \otimes_R N \longrightarrow 0 \qquad (*)$$

are all exact, for all $n \geq 0$. Together they form an exact sequence of complexes, to which we apply the zig-zag lemma. Note that the homology groups of the complex $P_* \otimes_R N$ are $\mathrm{Tor}_n^{R^{op}}(N,M)$, by definition, and similarly for the other two complexes. The long exact sequence obtained includes the following terms:

$$0 \longrightarrow \mathrm{Tor}_1^{R^{op}}(N,M) \longrightarrow M \otimes_R K_1 \longrightarrow M \otimes_R Q_0 . \qquad (**)$$

We have used the fact that $\mathrm{Tor}_0^{R^{op}}(A,B) = A \otimes_{R^{op}} B = B \otimes_R A$, and also that $\mathrm{Tor}_1^{R^{op}}(Q_0,M) = 0$: This follows from the last lemma, applied to Q_0, which is a projective R^{op}-module on the right.

Together, the sequences (*) and (**) show that the groups $\mathrm{Tor}_1^R(M,N)$ and $\mathrm{Tor}_1^{R^{op}}(N,M)$ are isomorphic: indeed, they are both isomorphic to the kernel of $M \otimes_R K_1 \to M \otimes_R Q_0$.

This settles the case $n = 1$ of the proposition. The reader will continue along the same lines with the higher Tor groups. □

Example 9.57 When R is commutative, the proposition gives isomorphisms $\mathrm{Tor}_n^R(M,N) \cong \mathrm{Tor}_n^R(N,M)$. For $R = \mathbb{Z}$, we see for example

$$\begin{aligned} \mathrm{Tor}(\mathbb{Z}/n\mathbb{Z}, \mathbb{Z}/m\mathbb{Z}) &\cong \text{the } m\text{-torsion in } \mathbb{Z}/n\mathbb{Z} \\ &\cong \mathbb{Z}/\gcd(n,m)\mathbb{Z} \\ &\cong \text{the } n\text{-torsion in } \mathbb{Z}/m\mathbb{Z} \\ &\cong \mathrm{Tor}(\mathbb{Z}/m\mathbb{Z}, \mathbb{Z}/n\mathbb{Z}) . \end{aligned}$$

■

Remark 9.58 It may feel overwhelming to have so many different options for calculating a Tor group: Not only can we pick any projective resolution, but we can also choose either factor. We have proved the existence of isomorphisms, sometimes canonical ones, between the various groups thus obtained, but we have not answered all the questions. For example, if we systematically compute the groups $\mathrm{Tor}_n^R(M,N)$ using a resolution P_* of M, considering the complex $P_* \otimes_R N$, then we can define induced maps $\mathrm{Tor}_n^R(M,N_1) \to \mathrm{Tor}_n^R(M,N_2)$ from a map $N_1 \to N_2$; we have not looked into the question of checking whether these new maps agree with the previous ones (mentioned in Lemma 9.52).

Similarly, in the proof of the proposition we have seen a new way of constructing the long exact sequence of Theorem 9.53. Namely, consider the short exact sequence of complexes

$$0 \longrightarrow P_* \otimes_R A \longrightarrow P_* \otimes_R B \longrightarrow P_* \otimes C \longrightarrow 0,$$

where P_* is again a resolution of M, and apply the zig-zag lemma (the last proposition identifies the terms of the sequence as appropriate Tor groups). Does this sequence agree with the one previously constructed? (This includes the previous question, and extends it to the "connecting homomorphisms".)

One can answer all these questions, but it does require some sustained effort. The easy way out is to make definite choices. In the next chapter, when we study the (co)homology of groups, we will do just that.

Remark 9.59 The reader may similarly wonder whether the roles of M and N can be exchanged, in the discussion of $\mathrm{Ext}^n_R(M,N)$. The answer is again "yes", but we will never use this, and the discussion would take some time. The reason is fundamentally that the arrows are going in the "wrong direction", so projective modules are not the right objects to look at anymore – one needs "injective modules". Just as projective modules can be defined as those modules for which $\mathrm{Hom}_R(M,-)$ preserves exact sequences, injective modules are such that $\mathrm{Hom}_R(-,N)$ preserves exact sequences, by definition. We will not mention these any more (except in the exercises).

Problems

9.1. Compute the "homology of the tetrahedron", as sketched in Example 9.19.

9.2. Prove Lemma 9.31.

9.3. Prove the zig-zag lemma.

9.4. Finish the proof of the horseshoe lemma.

9.5. *In this long problem we define and study injective modules. In the next, we show that, when $R = F[G]$ where F is a field and G is a finite group, injective modules are the same as projective modules!*

Let R be a ring. An R-module I is said to be *injective* when $\mathrm{Hom}_R(-,I)$ preserves exact sequences.

1. Rewrite this plainly, as in the definition of "projective module", and make it clear that it is "the same definition, with the arrows reversed".
2. In this question, we take $R = \mathbb{Z}$. Show that a module I is injective if and only if it is *divisible*, that is, for any $x \in I$ and any integer $n \geq 1$, there exists $x' \in I$ such that $x = nx'$.

   Hints: The "only if" is easy (look at multiplication by n on $\mathbb{Z}$). For the converse, suppose $A \subset M \subset B$ is an inclusion of modules, and suppose $f_M : M \to I$ extends $f : A \to I$, where I is divisible; show that M can be enlarged to $M + \mathbb{Z}b$ for any $b \in B$, and use Zorn's lemma to extend to B.

In particular, with this question we see that $\mathbb{Q}$ and $\mathbb{Q}/\mathbb{Z}$ are injective abelian groups.

3. Let R be any ring again, and let I be an injective abelian group. Consider the R-module $M = \mathrm{Hom}(R, I)$, as in Proposition 9.47. Use this proposition to show that M is an injective R-module.
4. Let M be any R-module, and let $x \in M$. Show that there exists $\varphi: M \to \mathrm{Hom}(R, \mathbb{Q}/\mathbb{Z})$ such that $\varphi(x) \neq 0$.

   Hint: Find a nonzero map of abelian groups $\mathbb{Z}x \to \mathbb{Q}/\mathbb{Z}$, use that the latter is injective to extend to the abelian group M, and finally use the adjunction of Proposition 9.47 again.
5. Show that a product of injective modules, even over an infinite indexing set, is itself injective; deduce from this and the previous question that any module embeds into an injective module.
6. Give the definition of an injective resolution. Prove a statement of existence and uniqueness up to homotopy.
7. Prove that $\mathrm{Ext}^*_R(M,N)$ can be computed using injective resolutions of N.

9.6. Let $R = F[G]$, the group algebra of the finite group G over the field F. When M is an R-module, we put $M^* = \mathrm{Hom}_F(M, F)$, the dual of M, which is also an R-module, with $(g \cdot \varphi)(x) = \varphi(g^{-1}x)$ (in notation which is meant to be obvious). Note that $f: M \to N$ induces $f^*: N^* \to M^*$.

In this problem, all $F[G]$-modules considered are assumed to be finite-dimensional over F, in order for us to use the canonical isomorphism between M and M^{**}. We point out that the definitions of "projective" and "injective" are now understood to involve only finite-dimensional modules. (Generally, one may speak of projective or injective objects in a category, if you know what that is.)

1. Show that M is projective if and only if M^* is injective, and vice versa.
2. Show that $F[G]^1$ is isomorphic to its own dual.

   Hints: Let $\varepsilon: F[G] \to F$ be defined by $\varepsilon\left(\sum_g \alpha_g g\right) = \alpha_1$, and consider the bilinear form $F[G]^1 \times F[G]^1 \to F$ defined by $x, y \mapsto \langle x, y\rangle = \varepsilon(xy)$. Define $\varphi: F[G]^1 \to \left(F[G]^1\right)^*$ by $\varphi(x) = \langle x, -\rangle$.
3. Conclude that M is projective if and only if it is injective.

   One says that $F[G]$ is a Frobenius algebra to indicate the presence of the bilinear form used in the previous question, which prominently implies that projective modules and injective modules coincide.
4. Deduce a new proof for the fact that any module embeds in an injective module, in this case.

10 Group cohomology

We have finally arrived at the definition of the cohomology and homology groups of a finite group, with coefficients in a $\mathbb{Z}[G]$-module, as Ext and Tor groups respectively. We will describe some particular projective resolutions which are special to the rings at hand (namely, group algebras). These will allow us to make definite choices, with which the naturality of many constructions emerges. We will also be able to define cohomology groups for profinite groups, rather than merely finite ones.

Definition of group (co)homology

We apply the machinery developed in the previous chapter to the ring $R = \mathbb{Z}[G]$, where G is a finite group (any group would do, really, for the basic definitions). Recall that a $\mathbb{Z}[G]$-module is an abelian group M endowed with an action of G by linear maps. Of special importance is the *trivial module*, that is, the group $\mathbb{Z}$ on which every element of G acts as the identity. It can be seen as a left or right module over $\mathbb{Z}[G]$.

cohomology groups

Definition 10.1 The **cohomology groups** of G with coefficients in M are

$$\mathrm{H}^n(G,M) := \mathrm{Ext}^n_{\mathbb{Z}[G]}(\mathbb{Z},M),$$

homology groups

for $n \geq 0$. The **homology groups** of G with coefficients in M are

$$\mathrm{H}_n(G,M) = \mathrm{Tor}_n^{\mathbb{Z}[G]}(\mathbb{Z},M),$$

for $n \geq 0$. ∎

Of course this seems to clash with our previous definition of $\mathrm{H}^2(G,M)$ (see Definition 7.14). However, we will prove shortly that the two groups are actually isomorphic.

As the title of this chapter suggests, most of the time, cohomology groups will be more important to us than homology groups. In any case, there are theorems, which we are not going to prove here, asserting that homology groups can be computed from cohomology groups and vice versa. It is true that we will go on to define *Tate cohomology* shortly, which combines homology and cohomology, and in due time we will discover that it can be a good idea to consider all of these together at once. In all honesty though, the main application of Tate cohomology will be class field theory in the next part of the book, and if we did not have this in mind, then we would have skipped homology entirely.

Let us summarize some facts from the previous chapter. To compute the cohomology groups, we need to take a projective resolution of $\mathbb{Z}$ by $\mathbb{Z}[G]$-modules, say P_*, and consider the cochain complex $\mathrm{Hom}_{\mathbb{Z}[G]}(P_*, M)$: Its cohomology groups are the various $\mathrm{H}^n(G,M)$. Any projective resolution will do: If we pick another one, say Q_*, the groups obtained are not only isomorphic to those obtained with P_*, but they are *canonically* isomorphic. Still, in the next section we propose a choice of "standard" resolution.

To deal with the homology groups, following the initial definition we need a projective resolution Q_* of M, and consider the homology groups of the complex $\mathbb{Z} \otimes_{\mathbb{Z}[G]} Q_*$. However, by Proposition 9.55, we can also consider a projective resolution P_* of $\mathbb{Z}$ (by right $\mathbb{Z}[G]$-modules), and consider the complex $P_* \otimes_{\mathbb{Z}[G]} M$. We make this our official choice, and again, we will soon propose a resolution.

The fundamental facts to keep are in mind are, to start with, that an exact sequence

$$0 \longrightarrow A \xrightarrow{\iota} B \xrightarrow{\pi} C \longrightarrow 0$$

yields a long exact sequence

$$0 \longrightarrow \mathrm{H}^0(G,A) \xrightarrow{\iota_0} \mathrm{H}^0(G,B) \xrightarrow{\pi_0} \mathrm{H}^0(G,C) \xrightarrow{\delta_0} \mathrm{H}^1(G,A)$$

$$\xrightarrow{\iota_1} \mathrm{H}^1(G,B) \xrightarrow{\pi_1} \mathrm{H}^1(G,C) \xrightarrow{\delta_1} \mathrm{H}^2(G,A) \longrightarrow \cdots \qquad (10.1)$$

In homology, the same short exact sequence induces

$$\cdots \longrightarrow \mathrm{H}_2(G,C) \longrightarrow \mathrm{H}_1(G,A) \xrightarrow{\iota_1} \mathrm{H}_1(G,B) \xrightarrow{\pi_1} \mathrm{H}_1(G,C)$$

$$\longrightarrow \mathrm{H}_0(G,A) \xrightarrow{\iota_0} \mathrm{H}_0(G,B) \xrightarrow{\pi_0} \mathrm{H}_0(G,C) \longrightarrow 0.$$

(Notice how the notation ι_j is used in two different ways, as is π_j, though in practice this causes no confusion.) The existence of such sequences follows from the material in the previous chapter. Yet again, we will soon propose a standard construction.

Also highly important, of course, is the value of these groups in degree 0. In cohomology, we have

$$\mathrm{H}^0(G,M) = \mathrm{Hom}_{\mathbb{Z}[G]}(\mathbb{Z}, M) = M^G,$$

where the right-hand side denotes $\{m \in M : g \cdot m = m \text{ for all } g \in G\}$ (to see this, associate to $\varphi\colon \mathbb{Z} \to M$ the element $\varphi(1)$). Thus a first way of thinking about the cohomology groups is that they measure the extent to which the "fixed points" operation, taking M to M^G, fails to preserve exact sequences.

Turning to homology, in degree 0 we have

$$\mathrm{H}_0(G,M) = \mathbb{Z} \otimes_{\mathbb{Z}[G]} M.$$

We claim that this is isomorphic to $M_G := M/M'$ where $M' = \langle m - g\cdot m : m \in M, g \in G\rangle$. Indeed, the map $M \to \mathbb{Z}\otimes_{\mathbb{Z}[G]} M$ defined by $m \mapsto 1\otimes m$ factors through M_G. The resulting map is an isomorphism, with inverse $\mathbb{Z}\otimes_{\mathbb{Z}[G]} M \to M_G$ defined by $n\otimes m \mapsto n\overline{m}$ where $\overline{m} \in M_G$ is the class of $m \in M$.

The elements of M^G are sometimes called the *invariant* elements, and those of M_G are sometimes called the *coinvariant* elements.

Example 10.2 Here is a silly example, but we certainly need to understand it first: Consider the trivial group $G = \{1\}$. Then $\mathbb{Z}[G] = \mathbb{Z}$. The trivial module $\mathbb{Z}$ is now projective as a $\mathbb{Z}[G]$-module, exceptionally, so $\mathrm{H}^n(G,M) = \mathrm{Ext}^n_{\mathbb{Z}}(\mathbb{Z},M) = 0$ for all $n > 0$ and for any M (Example 9.36). Similarly, one has $\mathrm{H}_n(G,M) = 0$ for $n > 0$. ∎

Example 10.3 Let $G = C_r = \{1, T, \ldots, T^{r-1}\}$ be a cyclic group of order r. In Example 9.13, we have constructed a projective resolution of the trivial module $\mathbb{Z}$. This is precisely what we need to compute both the cohomology and the homology groups.

If this resolution is P_*, then the cohomology groups of G with coefficients in M are the cohomology groups of the cochain complex $\mathrm{Hom}_{\mathbb{Z}[G]}(P_*,M)$. In any degree n, we have $P_n = \mathbb{Z}[C_r]^1$, so $\mathrm{Hom}_{\mathbb{Z}[G]}(P_n,M)$ can be identified with M itself (by considering the image of $1 \in \mathbb{Z}[C_r]^1$). The cochain complex looks like

$$0 \longrightarrow M \xrightarrow{1-T} M \xrightarrow{N} M \xrightarrow{1-T} M \longrightarrow \cdots$$

where $N = 1 + T + \cdots + T^{m-1}$ is the norm element, and the maps are really given by the action of $1 - T$ or N on the module M. It follows for $n > 0$ that

$$\mathrm{H}^{2n}(C_r,M) \cong \frac{M^G}{\mathrm{N}(M)},$$

with $\mathrm{N}(M) = \{N\cdot m : m \in M\}$, and for $n \geq 0$ that

$$\mathrm{H}^{2n+1}(C_r,M) \cong \frac{{}_N M}{M'},$$

where ${}_N M = \{m \in M : N\cdot m = 0\}$, and as in the previous example, $M' = \{m - g\cdot m : m \in M, g \in G\}$. There is a little exception in degree 0, where of course we recover

$$\mathrm{H}^0(C_r,M) = M^{C_r}.$$

Note that, as promised, in degree 2 this agrees with our previous computation (Proposition 7.18).

If M is a module with trivial G-action (for example $M = \mathbb{Z}$), then the formulae above give

$$\mathrm{H}^{2n}(C_r,M) \cong M/rM$$

for $n > 0$, and

$$\mathrm{H}^{2n+1}(C_r,M) \cong \{m \in M : rm = 0\},$$

the r-torsion in M, for $n \geq 0$. (For $M = \mathbb{Z}$, the cohomology groups are thus 0 in odd degrees.)

We turn to the homology groups. Since $\mathbb{Z}[C_r]$ is a commutative ring, there is no difference between left and right modules, so we only need to consider the homology groups of the complex $P_* \otimes_{\mathbb{Z}[G]} M$. In each degree we have $P_n \otimes_{\mathbb{Z}[G]} M \cong M$ since $P_n = \mathbb{Z}[G]^1$ (consider $m \mapsto 1 \otimes m$). The chain complex looks like

$$\cdots \longrightarrow M \xrightarrow{1-T} M \xrightarrow{N} M \xrightarrow{1-T} M \longrightarrow 0$$

So in degree 0 we recover $\mathrm{H}_0(C_r, M) \cong M_G = M/M'$. Otherwise we have

$$\mathrm{H}_{2n}(C_r, M) \cong \frac{{}_N M}{M'}$$

for $n > 0$, and

$$\mathrm{H}_{2n+1}(C_r, M) \cong \frac{M^G}{\mathrm{N}(M)}.$$

■

The similarities between homology and cohomology groups in the case of a cyclic group is the (original) motivation for the next definition.

Tate cohomology groups

Definition 10.4 Let M be a $\mathbb{Z}[G]$-module. The **Tate cohomology groups** of G with coefficients in M are denoted by $\widehat{\mathrm{H}}^n(G,M)$, for $n \in \mathbb{Z}$, and are defined by

$$\widehat{\mathrm{H}}^n(G,M) = \begin{cases} \mathrm{H}^n(G,M) & \text{if } n \geq 1, \\ M^G/\mathrm{N}(M) & \text{if } n = 0, \\ {}_N M/M' & \text{if } n = -1, \\ \mathrm{H}_{-(n+1)}(G,M) & \text{if } n \leq -2. \end{cases}$$

Here $\mathrm{N}(M) = \{N \cdot m : m \in M\}$ and ${}_N M = \{m \in M : N \cdot m = 0\}$, where $N = \sum_{\sigma \in G} \sigma \in \mathbb{Z}[G]$. As above, $M^G = \{m \in M : \sigma \cdot m = m \text{ for all } \sigma \in G\}$, and $M' = \langle m - \sigma \cdot m : m \in M, \sigma \in G \rangle$. ■

Example 10.5 When G is a cyclic group, we have the utterly simple formulae

$$\widehat{\mathrm{H}}^{2n}(G,M) = M^G/\mathrm{N}(M), \qquad \widehat{\mathrm{H}}^{2n+1}(G,M) = {}_N M/M',$$

for all $n \in \mathbb{Z}$. ■

We conclude with a word of warning. The material in the previous chapter shows that, if M is a projective $\mathbb{Z}[G]$-module, then $\mathrm{H}_n(G,M) = 0$ for $n > 0$. However, we have *not* included any result indicating that the same should be true for cohomology. It turns out, nonetheless, that when M is projective, we have $\mathrm{H}^n(G,M) = 0$ for $n > 0$. This is a nontrivial fact, which we will derive as a consequence of *Shapiro's lemma* a little later. The formulae above show, at least, that this is true when G is cyclic. In fact we will have the easy-to-state result that $\widehat{\mathrm{H}}^n(G,M) = 0$ for all $n \in \mathbb{Z}$, when M is projective.

We will also prove the existence of long exact sequences in Tate cohomology, but this, too, will have to wait.

The standard resolution

It is possible to write down a projective resolution of $\mathbb{Z}$ as a $\mathbb{Z}[G]$-module which is "the same" for all groups G. This is useful to clarify the definition of the (co)homology groups and, quite importantly, the various maps between them.

We end up with a simple "formula" defining $\mathrm{H}^n(G,M)$. Of course, we could have used this formula as our starting point, bypassing Ext and Tor entirely. The trade-off is that it becomes awkward to prove the existence of the long exact sequences in (co)homology, among other things. Also, consider Example 10.3, where we computed the cohomology of a cyclic group: We were able to use a resolution P_* for which the modules P_n are all free of rank one. By contrast, the "standard resolution", to be introduced next, involves free modules whose rank grows exponentially with n.

The (great) benefits of this resolution are thus theoretical, not practical. We will discover that a map $H \to G$ induces a homomorphism $\mathrm{H}^n(G,M) \to \mathrm{H}^n(H,M)$, and that there is a "multiplication" $\mathrm{H}^m(G,M) \otimes \mathrm{H}^n(G,N) \to \mathrm{H}^{m+n}(G,M \otimes_{\mathbb{Z}} N)$, to name just these two facts. One could obtain the same conclusions in greater generality, by pursuing the homological algebra approach; for example, we could study how $\mathrm{Ext}^n_R(M,N)$ varies when R varies, and so on. While it is easy to appreciate that this would be more conceptual, and probably more satisfying, the extra work needed would be overwhelming for now, given our purposes.

Let $R = \mathbb{Z}[G]$, where G is a finite group. We start with a construction which could easily be performed with any algebra over any commutative ring. We consider $R \otimes R \otimes \cdots \otimes R = R^{\otimes n}$, where $\otimes = \otimes_{\mathbb{Z}}$ (*not* over R). Here we see R, and each $R^{\otimes n}$, as a mere abelian group (we will explicitly mention the right moment to look at these as $\mathbb{Z}[G]$-modules). We construct a map

$$\partial_n \colon R^{\otimes n+1} \longrightarrow R^{\otimes n},$$

to be also written ∂ when subscripts become too much of a burden. We first define

$$\partial^{(i)} = \partial^{(i)}_n \colon R^{\otimes n+1} \longrightarrow R^{\otimes n},$$

for $0 \leq i \leq n$, by

$$\partial^{(i)}(\sigma_0 \otimes \cdots \otimes \sigma_n) = \sigma_0 \otimes \cdots \otimes \widehat{\sigma_i} \otimes \cdots \otimes \sigma_n,$$

where $\sigma_i \in G$; here the notation means that the ith factor is to be omitted entirely. Since $R^{\otimes n+1}$ is a free abelian group on the basis given by all such tensors, this defines $\partial^{(i)}$ unambiguously. Then we put

$$\partial = \sum_{i=0}^{n} (-1)^i \partial^{(i)}.$$

Remark 10.6 This formula is inspired by the following informal, geometric considerations. Suppose you have a triangle with vertices v_0, v_1, v_2, with an "orientation" from v_0 to v_1 to v_2, back to v_0. The "boundary" of the triangle, intuitively, is made of the oriented segment v_0v_1, then v_1v_2, and then v_2v_0; the latter we denote by $-v_0v_2$ (still informally). So this boundary is $v_0v_1 - v_0v_2 + v_1v_2$, much like, for $n = 2$, we have

$$\partial(\sigma_0 \otimes \sigma_1 \otimes \sigma_2) = \sigma_0 \otimes \sigma_1 - \sigma_0 \otimes \sigma_2 + \sigma_1 \otimes \sigma_2 .$$

Similar analogies may be made in any dimension. ■

Lemma 10.7 *We have $\partial_{n-1} \circ \partial_n = 0$. In other words, $(R^{\otimes *+1}, \partial_*)$ is a chain complex.*

Proof. First note the *simplicial identities*:

$$\partial^{(i)}\partial^{(j)} = \partial^{(j-1)}\partial^{(i)} \quad \text{when} \quad i<j .$$

Then compute

$$\begin{aligned}
\partial\partial &= \sum_{i=0}^{n-1}\sum_{j=0}^{n}(-1)^i(-1)^j\partial^{(i)}\partial^{(j)} \\
&= \sum_{i=0}^{n-1}\sum_{0\le j\le i}(-1)^{i+j}\partial^{(i)}\partial^{(j)} + \sum_{i=0}^{n-1}\sum_{i<j\le n}(-1)^{i+j}\partial^{(j-1)}\partial^{(i)} \\
&= \sum_{0\le j\le i\le n-1}(-1)^{i+j}\partial^{(i)}\partial^{(j)} - \sum_{0\le i\le j\le n-1}(-1)^{i+j}\partial^{(j)}\partial^{(i)} \\
&= 0 .
\end{aligned}$$

So $\partial\partial = 0$ as claimed. □

Next, consider the chain complex (of abelian groups) which is $\mathbb{Z}$ in degree 0, and zero otherwise, and call it $\mathbb{Z}_*$. We want to construct a weak equivalence $\varepsilon_* \colon R^{\otimes *+1} \to \mathbb{Z}_*$, and for this it is enough to specify a single map $\varepsilon_0 \colon R \to \mathbb{Z}$, in degree 0. We take it to be the natural "augmentation" of $R = \mathbb{Z}[G]$, that is, the homomorphism satisfying $\varepsilon_0(\sigma) = 1$ for all $\sigma \in G$.

Lemma 10.8 *This morphism is a weak equivalence. In other words, the sequence*

$$\cdots \xrightarrow{\ \partial\ } R\otimes R \xrightarrow{\ \partial\ } R \xrightarrow{\ \varepsilon\ } \mathbb{Z} \longrightarrow 0$$

is exact.

Proof. We have seen the equivalence of the two statements during our discussion of weak equivalences. First, we construct a homomorphism $\iota_* \colon \mathbb{Z}_* \to R^{\otimes *+1}$, by specifying it in degree 0 to be the homomorphism $\iota_0 \colon \mathbb{Z} \to R$ mapping $1 \in \mathbb{Z}$ to $1 \in G$. We show that ι_* and ε_* are homotopy inverses of each other.

The composition $\varepsilon_* \circ \iota_*$ is the identity of $\mathbb{Z}_*$, and the nontrivial work happens with $\iota_* \circ \varepsilon_*$; in degree $n > 0$ this is the zero map, and in degree 0 it takes each $\sigma \in G$ to $1 \in G$. Consider now the map $s_n : R^{\otimes n+1} \to R^{\otimes n+2}$ defined by $s_n(\sigma_0 \otimes \cdots \otimes \sigma_n) = 1 \otimes \sigma_0 \otimes \cdots \otimes \sigma_n$. We let the reader compute that, for $n \geq 1$,

$$s_{n-1} \circ \partial_n + \partial_{n+1} \circ s_n = \mathrm{id},$$

using another "simplicial identity", namely

$$s \circ \partial^{(i)} = \partial^{(i+1)} \circ s$$

for those values of i that make sense. In degree 0 we have

$$\partial_1 \circ s_0(\sigma) = \partial_1(1 \otimes \sigma) = \sigma - 1 = \sigma - \varepsilon_0(\sigma).$$

In the end, we have established that s_* is a chain homotopy between the identity and $\iota_* \circ \varepsilon_*$ (definition 9.23).

So ε_* is a homotopy equivalence, and in particular, a weak equivalence (Lemma 9.24). □

Now we use a little more seriously the fact that $R = \mathbb{Z}[G]$. Namely, each $R^{\otimes n+1}$ can be seen as an R-module, with action of $\sigma \in G$ given by

$$\sigma \cdot (r_0 \otimes \cdots \otimes r_n) = \sigma r_0 \otimes \cdots \otimes \sigma r_n.$$

(In general, the ring R would have to be a "Hopf algebra" for something like this to hold.) The maps ∂_n above are maps of R-modules, and $(R^{\otimes *+1}, \partial_*)$ is a chain complex of R-modules. Moreover, the morphism ε_0 is also a map of R-modules, where $\mathbb{Z}$ is the "trivial" $\mathbb{Z}[G]$-module (in general, R would have to be an "augmented algebra" here).

The sequence mentioned in the statement of the last lemma is of course still exact when regarded as a sequence of maps of R-modules. However, there is a subtle point here: the homotopy constructed in the proof is, on the contrary, not R-linear! Conducting the argument with abelian groups was very helpful here.

The bottom line is that we have a weak equivalence $R^{\otimes *+1} \to \mathbb{Z}_*$ between chain complexes of R-modules. Perhaps unsurprisingly, we move on to show that each $R^{\otimes n}$ is projective as an R-module, and indeed, free.

Let us introduce the notation

$$[\sigma_1|\sigma_2|\cdots|\sigma_n] := 1 \otimes \sigma_1 \otimes \sigma_1\sigma_2 \otimes \cdots \otimes \sigma_1\sigma_2\cdots\sigma_n \in R^{\otimes n+1},$$

where $\sigma_i \in G$.

Lemma 10.9 *The elements $[\sigma_1|\sigma_2|\cdots|\sigma_n]$ form a basis for the R-module $R^{\otimes n+1}$, which is thus free of rank $|G|^n$.*

Proof. As an abelian group, $R^{\otimes n+1}$ is free with a basis consisting of all the elements of the form $\sigma_0 \otimes \cdots \otimes \sigma_n$, with $\sigma_i \in G$. However, note that

$$\sigma_0 \otimes \sigma_1 \otimes \cdots \otimes \sigma_n = \sigma_0 \cdot [\sigma_0^{-1}\sigma_1 | \sigma_1^{-1}\sigma_2 | \cdots | \sigma_{n-1}^{-1}\sigma_n] .$$

The result follows. □

Thus we have constructed a projective resolution of the trivial $\mathbb{Z}[G]$-module $\mathbb{Z}$, to be called henceforth the *standard resolution*. From the previous chapter, we know that the cohomology groups of G with coefficients in the $\mathbb{Z}[G]$-module M can be computed as the cohomology groups of the cochain complex $\mathrm{Hom}_R(R^{\otimes *+1}, M)$. Likewise, the homology groups of G with coefficients in M may be obtained from the chain complex $R^{\otimes *+1} \otimes_R M$. At least for cohomology, this leads to pretty simple formulae.

Definition 10.10 Let M be a $\mathbb{Z}[G]$-module. For $n \geq 0$, we define $C^n(G,M)$ to be the group of maps (of sets)

$$f \colon G^{n+1} \longrightarrow M$$

satisfying

$$f(\sigma\sigma_0, \dots, \sigma\sigma_n) = \sigma \cdot f(\sigma_0, \dots, \sigma_n) .$$

homogeneous cochains

The elements of $C^n(G,M)$ are called **homogeneous cochains** for G, of degree n, with values in M.

We also define $d^n \colon C^n(G,M) \to C^{n+1}(G,M)$ by

$$d^n(f)(\sigma_0, \dots, \sigma_{n+1}) = \sum_{i=0}^{n+1} (-1)^i f(\sigma_0, \dots, \widehat{\sigma_i}, \dots, \sigma_{n+1}) .$$

We put $Z^n(G,M) = \ker(d^n)$ and $B^n(G,M) = \mathrm{Im}(d^{n-1})$, the groups of cocycles and coboundaries respectively. ■

Lemma 10.11 *The quotient $Z^n(G,M)/B^n(G,M)$ is isomorphic to $\mathrm{H}^n(G,M)$.*

Proof. When we compute the cohomology groups using the standard resolution, this is exactly what we get. We use the fact that a map of abelian groups $f \colon R^{\otimes n+1} \to M$ is entirely determined by the values $f(\sigma_0 \otimes \cdots \otimes \sigma_n)$ where $\sigma_i \in G$, which we write $f(\sigma_0, \dots, \sigma_n)$. □

For definiteness, we make this our official definition of the cohomology groups:

Definition 10.12 From this point in the book, the cohomology groups will be computed using the standard resolution unless stated otherwise. In other words, we define

$$\mathrm{H}^n(G,M) := Z^n(G,M)/B^n(G,M) .$$

When $h \colon M_1 \to M_2$ is a homomorphism of $\mathbb{Z}[G]$-modules, the induced map

$$C^n(G,M_1) \longrightarrow C^n(G,M_2) ,$$

defined by composing the cocycles with h, is compatible with the boundary operators; it induces

$$h_n\colon \mathrm{H}^n(G,M_1) \longrightarrow \mathrm{H}^n(G,M_2).$$

If $\varphi\colon H \to G$ is a homomorphism of groups, the induced map

$$C^n(G,M) \longrightarrow C^n(H,M),$$

defined by precomposing the cocycles with φ, is also compatible with the boundary operators, and it induces

$$\varphi^n\colon \mathrm{H}^n(G,M) \longrightarrow \mathrm{H}^n(H,M).$$

Here on the right-hand side, the module M is seen as an H-module via φ.

In the presence of a short exact sequence

$$0 \longrightarrow A \xrightarrow{\ \iota\ } B \xrightarrow{\ \pi\ } C \longrightarrow 0,$$

the long exact sequence (10.1) is obtained by considering the short exact sequence of complexes

$$0 \longrightarrow \mathrm{Hom}_R(P_*,A) \xrightarrow{\ \iota_0\ } \mathrm{Hom}_R(P_*,B) \xrightarrow{\ \pi_0\ } \mathrm{Hom}_R(P_*,C) \longrightarrow 0,$$

where $P_* = R^{\otimes *+1}$ is the standard resolution, and applying the zig-zag lemma, as in the proof of Theorem 9.43. ∎

(We could be even more careful, and state that the "connecting homomorphisms" in the conclusion of the zig-zag lemma are just those which we described in the proof of that lemma, since no uniqueness statement was made.)

Another popular way of writing this uses fewer variables, and more complicated boundary maps.

Definition 10.13 Let M be a $\mathbb{Z}[G]$-module. For $n \geq 0$, we define $\mathscr{C}^n(G,M)$ to be the group of maps (of sets)

$$f\colon G^n \longrightarrow M.$$

(For $n = 0$, this means $\mathscr{C}^0(G,M) = M$.) The elements of $\mathscr{C}^n(G,M)$ are called **inhomogeneous cochains** for G, of degree n, with values in M.

inhomogeneous cochains

We define $d^0\colon \mathscr{C}^0(G,M) \to \mathscr{C}^1(G,M)$ by $d^0(m)(\sigma) = \sigma \cdot m - m$, as well as $d^1\colon \mathscr{C}^1(G,M) \to \mathscr{C}^2(G,M)$ by

$$d^1(f)(\sigma,\tau) = f(\sigma) + \sigma \cdot f(\tau) - f(\sigma\tau).$$

For $n \geq 2$ we define $d^n\colon \mathscr{C}^n(G,M) \to \mathscr{C}^{n+1}(G,M)$ by

$$\begin{aligned} d^n(f)(\sigma_1,\ldots,\sigma_{n+1}) = {} & \sigma_1 \cdot f(\sigma_2,\ldots,\sigma_{n+1}) \\ & + \sum_{i=1}^{n} (-1)^i f(\sigma_1,\ldots,\sigma_{i-1},\sigma_i\sigma_{i+1},\sigma_{i+2},\ldots,\sigma_{n+1}) \\ & + (-1)^{n+1} f(\sigma_1,\ldots,\sigma_n). \end{aligned}$$

We put $\mathscr{Z}^n(G,M) = \ker(d^n)$ and $\mathscr{B}^n(G,M) = \mathrm{Im}(d^{n-1})$, the groups of cocycles and coboundaries respectively. ∎

Lemma 10.14 *The quotient $\mathscr{Z}^n(G,M)/\mathscr{B}^n(G,M)$ is isomorphic to $\mathrm{H}^n(G,M)$.*

Proof. Again, let us compute the cohomology using the standard resolution. This time, we use Lemma 10.9 to see that a map of $\mathbb{Z}[G]$-modules $f\colon R^{\otimes n+1} \to M$ is entirely determined by the values $f([\sigma_1|\ldots|\sigma_n])$, which may be arbitrary. We write these values $f(\sigma_1,\ldots,\sigma_n) \in M$.

It remains to identify the coboundaries. The easiest way is probably to use the isomorphisms directly given by the proof of Lemma 10.9, that is,

$$C^n(G,M) \longrightarrow \mathscr{C}^n(G,M), \quad f \mapsto f',$$

given in degree 0 by $f' = f(1)$ and for $n \geq 1$ by

$$f'(\sigma_1,\ldots,\sigma_n) = f(1,\sigma_1,\sigma_1\sigma_2,\ldots,\sigma_1\cdots\sigma_n).$$

The inverse is given by $f' \mapsto f$ with

$$f(\sigma_0,\ldots,\sigma_n) = \sigma_0 \cdot f'(\sigma_0^{-1}\sigma_1, \sigma_1^{-1}\sigma_2,\ldots,\sigma_{n-1}^{-1}\sigma_n).$$

The translation is then straightforward, and left to the reader. □

We encourage the reader to think of $\mathscr{C}^n(G,M)$ and $C^n(G,M)$ as "the same groups, in different coordinates". In other words, be ready to work with either of these, keeping the isomorphism just given in mind if necessary.

We turn to homology. Here we first notice that $R^{\otimes n+1}$ can be seen as a right $\mathbb{Z}[G]$-module by defining

$$(r_0 \otimes \cdots \otimes r_n) \cdot \sigma := \sigma^{-1} r_0 \otimes \cdots \otimes \sigma^{-1} r_n.$$

Clearly, it is then a free module, with the same basis as before. Again, we make the use of the standard resolution official:

Definition 10.15 From this point in the book, the homology groups of G with coefficients in the $\mathbb{Z}[G]$-module M are taken to be the homology groups of the complex $R^{\otimes *+1} \otimes_R M$. A map $f\colon M_1 \to M_2$ induces maps

$$f_n\colon \mathrm{H}_n(G,M_1) \longrightarrow \mathrm{H}_n(G,M_2)$$

by considering $\mathrm{id} \otimes f$, which is compatible with the boundary operators. A homomorphism $\varphi\colon H \to G$ induces maps

$$\varphi_n\colon \mathrm{H}_n(H,M) \longrightarrow \mathrm{H}_n(G,M)$$

by considering first $\varphi'\colon S = \mathbb{Z}[H] \to R = \mathbb{Z}[G]$, which extends φ, and then

$$S^{\otimes n+1} \otimes_S M \longrightarrow R^{\otimes n+1} \otimes_R M.$$

Here, M is viewed as an H-module via φ. A short exact sequence

$$0 \longrightarrow A \xrightarrow{\iota} B \xrightarrow{\pi} C \longrightarrow 0$$

induces the long exact sequence (10) by considering the short exact sequence of complexes

$$0 \longrightarrow R^{\otimes *+1} \otimes_R A \xrightarrow{\iota} R^{\otimes *+1} \otimes_R B \xrightarrow{\pi} R^{\otimes *+1} \otimes_R C \longrightarrow 0,$$

and appealing to the zig-zag lemma, as in Remark 9.58. ∎

Having made these definite choices, some arguments become routine. For example, the next lemma is left to the reader.

Lemma 10.16 *Let $f\colon M_1 \to M_2$ be a homomorphism of G-modules, and let $\varphi\colon H \to G$ be a homomorphism of groups. Then for all $n \geq 0$ we have*

$$f_n \circ \varphi_n = \varphi_n \circ f_n, \qquad f_n \circ \varphi^n = \varphi^n \circ f_n.$$

What is more, if we have a commutative diagram with exact rows

$$\begin{array}{ccccccccc} 0 & \longrightarrow & A & \longrightarrow & B & \longrightarrow & C & \longrightarrow & 0 \\ & & \downarrow{\scriptstyle f} & & \downarrow{\scriptstyle g} & & \downarrow{\scriptstyle h} & & \\ 0 & \longrightarrow & A' & \longrightarrow & B' & \longrightarrow & C' & \longrightarrow & 0 \end{array}$$

then we have a commutative diagram

$$\begin{array}{ccc} \mathrm{H}^n(G,C) & \xrightarrow{\delta_n} & \mathrm{H}^{n+1}(G,A) \\ \downarrow{\scriptstyle h_n} & & \downarrow{\scriptstyle f_{n+1}} \\ \mathrm{H}^n\left(G,C'\right) & \xrightarrow{\delta_n} & \mathrm{H}^{n+1}\left(G,A'\right). \end{array}$$

(Recall that δ_n is our notation for the connecting homomorphisms in the long exact sequences.)

As a result, we obtain an infinite commutative diagram, between the long exact sequences associated to the two short exact sequences. A similar statement holds in homology.

Finally, there is also a commutative diagram

$$\begin{array}{ccc} \mathrm{H}^n(G,C) & \xrightarrow{\delta_n} & \mathrm{H}^{n+1}(G,A) \\ \downarrow{\scriptstyle \varphi^n} & & \downarrow{\scriptstyle \varphi^{n+1}} \\ \mathrm{H}^n(H,C) & \xrightarrow{\delta_n} & \mathrm{H}^{n+1}(H,A). \end{array}$$

Thus we have an infinite commutative diagram between long exact sequences, and a similar statement in homology. □

Low degrees

Using the standard resolution, we can give a "uniform" description of the groups $\mathrm{H}^n(G,M)$ for small values of n. It seems now urgent, for example, to finally establish that:

Lemma 10.17 *The definition of* $\mathrm{H}^2(G,M)$ *given in Definition 10.1 agrees with that given in Definition 7.14.*

Proof. We can work with inhomogeneous cocycles, and compute $\mathrm{H}^2(G,M)$ as $\mathscr{Z}^2(G,M)/\mathscr{B}^2(G,M)$; this is according to Definition 10.1 as in the rest of this chapter, via Lemma 10.14. The only difference, if we look back at Definition 7.14, is that our initial description used the subgroup $\overline{\mathscr{Z}}^2(G,M) \subset \mathscr{Z}^2(G,M)$ of those cocycles which are *normalized*, that is, satisfy $c(\sigma,1) = c(1,\tau) = 0$ for all $\sigma,\tau \in G$. These were considered modulo the subgroup $\overline{\mathscr{B}}^2(G,M) \subset \mathscr{B}^2(G,M)$ of cocycles of the form $d^1(f)$ with the extra condition $f(1) = 0$. However, consider the natural map

$$\varphi\colon \frac{\overline{\mathscr{Z}}^2(G,M)}{\overline{\mathscr{B}}^2(G,M)} \longrightarrow \frac{\mathscr{Z}^2(G,M)}{\mathscr{B}^2(G,M)}.$$

To see that it is surjective, let $c \in \mathscr{Z}^2(G,M)$. By definition, we have $d^2(c) = 0$, which is the familiar condition

$$\sigma \cdot c(\tau,\rho) + c(\sigma,\tau\rho) = c(\sigma\tau,\rho) + c(\sigma,\tau).$$

Applied with $\tau = \rho = 1$, this gives $c(\sigma,1) = \sigma \cdot c(1,1)$; with $\sigma = \tau = 1$, we draw $c(1,\rho) = c(1,1)$. So a cocycle such as c is normalized as soon as $c(1,1) = 0$.

However, if $f\colon G \to M$ is the constant function taking the value $c(1,1)$, then $d^1(f)(1,1) = c(1,1)$, so $c' = c - d^1(f)$ is normalized. Of course c and c' define the same cohomology class, so the map φ above is indeed surjective.

That φ is injective is equivalent to $\overline{\mathscr{B}}^2(G,M) = \mathscr{B}^2(G,M) \cap \overline{\mathscr{Z}}^2(G,M)$, or that $d^1(f)$ is normalized if and only if $f(1) = 0$. However, $d^1(f)(1,1) = f(1)$, so this is now clear. □

In degree 1, we merely paraphrase the definitions, and examine a particular case.

crossed homomorphism

Definition 10.18 Let M be a $\mathbb{Z}[G]$-module. A **crossed homomorphism** from G to M is a map $f\colon G \to M$ such that

$$f(\sigma\tau) = f(\sigma) + \sigma \cdot f(\tau).$$

A crossed homomorphism is one of the form $\sigma \mapsto \sigma \cdot m - m$, for some $m \in M$.

Lemma 10.19 *The group* $\mathrm{H}^1(G,M)$ *is isomorphic to the group of crossed homomorphisms from G to M, modulo the subgroup of trivial ones. In particular, if M has a trivial G-action, we have*

$$\mathrm{H}^1(G,M) = \mathrm{Hom}(G,M),$$

the group of homomorphisms.

Proof. Immediate with inhomogeneous cocycles. □

We have already examined the situation in degrees 0 and -1 (in the numbering of Tate cohomology). A nice example in degree -2, that is in homological degree 1, is the description of $\mathrm{H}_1(G,\mathbb{Z})$. Recall that if $\sigma,\tau \in G$, the notation $[\sigma,\tau]$ denotes the commutator $\sigma\tau\sigma^{-1}\tau^{-1}$, while $[G,G]$ is the subgroup of G generated by all the commutators. It is a characteristic subgroup of G, visibly. The quotient $G/[G,G]$ is written G^{ab} and called the *abelianization* of G: The name comes from the fact that G^{ab} is the largest abelian quotient of G. Indeed, if A is abelian and there is a homomorphism $G \to A$, then it clearly factors through G^{ab}.

Lemma 10.20 *There is an isomorphism*

$$\mathrm{H}_1(G,\mathbb{Z}) \cong G^{ab}.$$

Proof. We must examine a portion of the chain complex below, obtained from the standard resolution:

$$R^{\otimes 3} \otimes_R \mathbb{Z} \xrightarrow{\partial_2} R^{\otimes 2} \otimes_R \mathbb{Z} \xrightarrow{\partial_1} R \otimes_R \mathbb{Z}.$$

(As always, $R = \mathbb{Z}[G]$.) Keeping in mind the structure of the right module on $R^{\otimes n}$, we can identify $R^{\otimes n+1} \otimes_R \mathbb{Z}$ with the abelian group $R^{\otimes n}$: Consider $\sigma_1 \otimes \cdots \otimes \sigma_n \mapsto \sigma_1 \otimes \cdots \otimes \sigma_n \otimes 1 \otimes 1$, and $\sigma_0 \otimes \cdots \otimes \sigma_n \otimes 1 \mapsto \sigma_n^{-1}\sigma_0 \otimes \cdots \otimes \sigma_n^{-1}\sigma_{n-1}$. So we are left with

$$R \otimes R \xrightarrow{\partial_2} R \xrightarrow{\partial_1} \mathbb{Z}.$$

Examining the definition of the boundary maps in the standard resolution, we find that ∂_1 is the zero map, and that $\partial_2(\sigma \otimes \tau) = \tau^{-1}\sigma - (\sigma - \tau)$.

For $\sigma \in G$, let $\overline{\sigma} \in G^{ab}$ denote the image of σ in the quotient. We have a homomorphism of abelian groups $\mathbb{Z}[G] \to G^{ab}$, taking σ to $\overline{\sigma}$, and it factors through $\mathbb{Z}[G]/\mathrm{Im}(\partial_2)$; also, the inclusion of sets $G \to \mathbb{Z}[G]$, followed by $\mathbb{Z}[G] \to \mathbb{Z}[G]/\mathrm{Im}(\partial_2)$, is a homomorphism, by inspection, so it factors through G^{ab}. These two maps between $\mathbb{Z}[G]/\mathrm{Im}(\partial_2)$ and G^{ab} are inverses of each other. □

The proof of the next proposition involves mainly a close look at the groups in low degrees, so it seems fitting here.

Proposition 10.21 *Suppose*

$$0 \longrightarrow A \xrightarrow{\iota} B \xrightarrow{\pi} C \longrightarrow 0,$$

is a short exact sequence. Then there is a long exact sequence

$$\cdots \longrightarrow \widehat{\mathrm{H}}^n(G,A) \longrightarrow \widehat{\mathrm{H}}^n(G,B) \longrightarrow \widehat{\mathrm{H}}^n(G,C) \longrightarrow$$

$$\widehat{\mathrm{H}}^{n+1}(G,A) \longrightarrow \widehat{\mathrm{H}}^{n+1}(G,B) \longrightarrow \widehat{\mathrm{H}}^{n+1}(G,C) \longrightarrow \cdots$$

A commutative diagram between short exact sequences gives a commutative diagram between the corresponding long exact sequences.

Proof. When n is large and positive, this is the usual exact sequence in cohomology; when n is large and negative, this is the exact sequence in homology. It remains to examine the groups in low degrees, and the reader will help complete the following sketch.

Part of the cohomology sequence is

$$0 \longrightarrow A^G \longrightarrow B^G \longrightarrow C^G \xrightarrow{\delta_0} \mathrm{H}^1(G,A).$$

One starts by proving the following general fact. Pick $c \in C^G$, and choose $b \in B$ mapping to c. The association $\sigma \mapsto \sigma \cdot b - b$ is a crossed homomorphism with values in A, that is an element of $\mathrm{H}^1(G,A)$, and it is precisely $\delta_0(c)$. From this, one proves the exactness of

$$\widehat{\mathrm{H}}^0(G,A) \longrightarrow \widehat{\mathrm{H}}^0(G,B) \longrightarrow \widehat{\mathrm{H}}^0(G,C) \longrightarrow \mathrm{H}^1(G,A).$$

Here is also the definition of the map $\delta_{-1}\colon \widehat{\mathrm{H}}^{-1}(G,C) = {}_N C/C' \to \widehat{\mathrm{H}}^0(G,A) = A^G/\mathrm{N}(A)$. Choose $c \in C$ such that $N \cdot c = 0$, and select $b \in B$ mapping to c. Then $b - N \cdot b$ is an element of A^G, whose image in $A^G/\mathrm{N}(A)$ we define to be $\delta_{-1}(c)$. The reader will finish the argument easily. □

Profinite groups and Galois cohomology

Another virtue of the standard resolution is that we can extend our definitions from finite to profinite groups. We start with an exercise for the reader.

Lemma 10.22 *Let G be a profinite group, and let M be a $\mathbb{Z}[G]$-module. The following conditions are equivalent.*

1. *The map $G \times M \longrightarrow M$, mapping (σ, m) to $\sigma \cdot m$, is continuous, where the discrete topology is used on M (and the product topology on $G \times M$).*
2. *The stabilizer of any $m \in M$ is open in G.*
3. *$M = \bigcup_U M^U$, where U runs through the open subgroups of G, and $M^U = \{m \in M : \sigma \cdot m = m$ for all $\sigma \in U\}$.*

discrete G-module

In this situation, we say that M is a **discrete G-module**. □

Definition 10.23 Let G be a profinite group, and let M be a discrete G-module. We define $C^n(G,M)$ to be the abelian group of all *continuous* maps $f\colon G^{n+1} \to M$, where the discrete topology is used on M, satisfying

$$f(\sigma\sigma_0, \ldots, \sigma\sigma_n) = \sigma \cdot f(\sigma_0, \ldots, \sigma_n).$$

We define boundary maps by using the formulae from Definition 10.10, and the cohomology groups $\mathrm{H}^n(G,M)$ as in Definition 10.12. Induced maps h_n and φ^n are defined using the same recipe, when $\varphi\colon H \to G$ is a continuous homomorphism.

Similarly, the group $\mathscr{C}^n(G,M)$ is defined to be the abelian group of all continuous maps $G^n \to M$. There is an isomorphism between $\mathscr{C}^n(G,M)$ and $C^n(G,M)$, as in the case of finite groups, and the same formulae for the boundaries can be drawn.

Example 10.24 The group $\mathrm{H}^1(G,M)$ consists of *continuous* crossed homomorphisms, modulo trivial ones; if M has a trivial G-action, then $\mathrm{H}^1(G,M) = \mathrm{Hom}(G,M)$, the group of *continuous* homomorphisms.

Lemma 10.25 *When G is profinite and M is a discrete G-module, we have isomorphisms*

$$\mathrm{H}^n(G,M) \cong \mathrm{colim}_U \, \mathrm{H}^n\left(G/U, M^U\right),$$

where the colimit is taken over all open subgroups U.

Proof. Let $\pi : G \to G/U$ be the quotient map, and consider the composition

$$C^n\left(G/U, M^U\right) \xrightarrow{\pi^n} C^n\left(G, M^U\right) \longrightarrow C^n(G,M).$$

These maps, as U varies, can be combined into

$$\mathrm{colim}_U \, C^n\left(G/U, M^U\right) \longrightarrow C^n(G,M).$$

A cochain $f\colon G^{n+1} \to M$ is assumed to be continuous, so it is locally constant as M is discrete; thus there is an open subgroup $U \subset G$ such that f is constant on U^{n+1}, since G is profinite. It follows that f factors through π^n, and that the map above is surjective. It is evidently injective (by Lemma 7.30, if you will).

We conclude that $C^n(G,M) \cong \mathrm{colim}_U \, C^n(G/U, M^U)$. This isomorphism is obviously compatible with the boundary operators, so the lemma follows. □

Lemma 10.26 *An exact sequence of discrete G-modules induces a long exact sequence in cohomology as in* (10.1).

Proof. Same proof as for finite groups: Consider the exact sequence of complexes

$$0 \longrightarrow C^*(G,A) \longrightarrow C^*(G,B) \longrightarrow C^*(G,C) \longrightarrow 0$$

and appeal to the zig-zag lemma. □

We have finally arrived at the definition of Galois cohomology (or at any rate, a first definition, to be amended in Remark 12.28).

Definition 10.27 Let F be a field, and let $\overline{F}$ be a separable closure of F. For any discrete $\mathrm{Gal}(\overline{F}/F)$-module M, we put

$$\mathrm{H}^n(F,M) := \mathrm{H}^n\left(\mathrm{Gal}\left(\overline{F}/F\right), M\right),$$

Galois cohomology group

and call it the **Galois cohomology group** of F, with coefficients in M, in degree $n \geq 0$.

(Recall that a separable closure is the same thing as an algebraic closure when F is perfect, for example finite or of characteristic 0, and this is the only case of importance in this book.)

Remark 10.28 The separable closure of F is unique up to isomorphism, but there is no canonical isomorphism between two candidates. So $\mathrm{H}^n(F,M)$ is only defined up to isomorphism, and the notation may seem a little surprising. This will not matter much. Usually we can fix a choice for $\overline{F}$ at the beginning of any discussion surrounding the Galois cohomology of F. More profound is the fact, discussed in Remark 12.28, that $\mathrm{H}^n(F,M)$ is really defined up to *canonical* isomorphism, despite appearances. ■

Example 10.29 Here is an example of discrete G-module: Take $M = \overline{F}$, or $M = \overline{F}^\times$. In either case, since $\overline{F}$ is a union of finite extensions of F, we see that M is discrete. We then have

$$\mathrm{H}^n\left(F,\overline{F}^\times\right) = \mathrm{colim}_E\, \mathrm{H}^n(\mathrm{Gal}(E/F),E^\times),$$

where E/F runs through the finite extensions of F contained in $\overline{F}$, by Lemma 10.25. In degree 2, we recover the cohomological Brauer group as in Definition 7.34. Also note that $\mathrm{H}^0(F,\overline{F}^\times) = F^\times$, by Galois theory.

Another common discrete module is given by $\mu_N(\overline{F})$, the group of Nth roots of unity contained in $\overline{F}$. We have

$$\mathrm{H}^n\left(F,\mu_N\left(\overline{F}\right)\right) = \mathrm{colim}_E\, \mathrm{H}^n(\mathrm{Gal}(E/F),\mu_N(E))$$

in this case.

Modules with a trivial G-action are also important, and the cohomology groups $\mathrm{H}^n(F,\mathbb{F}_p)$, where p is a prime, contain much information about F. ■

Example 10.30 Consider $F = \mathbb{R}$, the field of real numbers, so that $\mathrm{Gal}(\overline{F}/F)$ has order two: The nontrivial element is, of course, complex conjugation. The cohomology group $\mathrm{H}^n(\mathbb{R},M)$ is then given by the formulae of Example 10.3. For instance, the group $\mathrm{H}^n(\mathbb{R},\mathbb{F}_p)$ is trivial when p is odd and $n > 0$, while $\mathrm{H}^n(\mathbb{R},\mathbb{F}_2) = \mathbb{F}_2$ for all $n \geq 0$. ■

Remark 10.31 It is a common shorthand to write simply $\mathrm{H}^n(F,\mu_N)$ instead of $\mathrm{H}^n(F,\mu_N(\overline{F}))$. Likewise, let us write $\mathbb{G}_m(A) := A^\times$, whenever A is a ring. We have $\mathbb{G}_m(\overline{F})^{\mathrm{Gal}(\overline{F}/E)} = \mathbb{G}_m(E)$, whenever E/F is a finite extension, by Galois theory. If we agree to write $\mathrm{H}^n(F,\mathbb{G}_m)$ instead of $\mathrm{H}^n(F,\mathbb{G}_m(\overline{F}))$, then

$$\mathrm{H}^n(F,\mathbb{G}_m) = \mathrm{colim}_E\, \mathrm{H}^n(\mathrm{Gal}(E/F),\mathbb{G}_m(E)).$$

In the same fashion, if we can associate to any finite E/F a group $T(E)$ in a reasonable way, then we may decide to use the notation

$$\mathrm{H}^n(F,T) := \mathrm{colim}_E\, \mathrm{H}^n(\mathrm{Gal}(E/F),T(E)).$$

Again, this will be discussed a little more in Remark 12.28. In this book, however, only the shorthands $T = \mu_N$ and $T = \mathbb{G}_m$ will be considered (besides cohomology with coefficients in a trivial module M such as $\mathbb{Z}$ or $\mathbb{F}_p$). ■

We conclude with a comment on how $\mathrm{H}^n(F,M)$ varies with F. Suppose $\iota\colon F \to E$ is a homomorphism, necessarily injective since F is a field. Pick a separable closure $\overline{E}$ of E, and let $\overline{F}$ be a separable closure of F, and extend ι to a homomorphism from $\overline{F}$ into $\overline{E}$. Then there is a map $\mathrm{Gal}(\overline{E}/E) \to \mathrm{Gal}(\overline{F}/F)$, and when we pass to cohomology we have a map

$$\mathrm{H}^n(F,M) \longrightarrow \mathrm{H}^n(E,M)$$

in all degrees, for any $\mathrm{Gal}(\overline{F}/F)$-module M. In this book this will typically be applied when $F \subset E \subset \overline{F} = \overline{E}$, in which case there is no need to choose an extension of ι – choosing the algebraic closure of F once and for all is enough. The map $\mathrm{Gal}(\overline{F}/E) \to \mathrm{Gal}(\overline{F}/F)$ is an inclusion in this case, and the induced map $\mathrm{H}^n(F,M) \to \mathrm{H}^n(E,M)$ is called the *restriction* map (because the action on M is restricted from $\mathrm{Gal}(\overline{F}/F)$ to its subgroup $\mathrm{Gal}(\overline{F}/E)$).

Problems

10.1. Prove Lemma 10.16.

10.2. Complete the proof of Proposition 10.21.

11 Hilbert 90

We interrupt the flow of generalities, and turn to a few applications of group cohomology to Galois theory. They all depend crucially on one result, a modern version of a property observed and proved by Hilbert as the 90th in a string of theorems collected together in his *Zahlbericht*, a report on number theory. Mathematicians frequently write colloquially "by Hilbert 90, we have ...", dropping the word "theorem" altogether. It appears that Kummer was aware of the result earlier, and that the general version which we will present is due to Noether.

Hilbert's Theorem 90 in Galois cohomology

It is the following result, of remarkable generality.

Theorem 11.1 (Hilbert 90) *Let K/F be a finite Galois extension. Then*

$$\mathrm{H}^1(\mathrm{Gal}(K/F), K^\times) = 0.$$

For any field F, we have also

$$\mathrm{H}^1(F, \mathbb{G}_m) = 0.$$

Proof. The second result follows from the first, since $\mathrm{H}^1(F, \mathbb{G}_m)$ is a colimit of groups of the form $\mathrm{H}^1(\mathrm{Gal}(K/F), K^\times)$ with K/F finite.

We turn to the first statement, and let $G = \mathrm{Gal}(K/F)$. We pick a crossed homomorphism $\varphi \in \mathscr{Z}^1(G, K^\times)$, so that $\varphi(\tau\sigma) = \varphi(\tau)\tau(\varphi(\sigma))$ for $\tau, \sigma \in G$, and we wish to prove that $\varphi \in \mathscr{B}^1(G, K^\times)$, that is, that there is some $a \in K^\times$ such that $\varphi(\tau) = \tau(a)a^{-1}$.

The key is Dedekind's lemma on the independence on characters, which guarantees that $\sum_\sigma \varphi(\sigma)\sigma \neq 0$. In other words, there is $c \in K^\times$ such that

$$b := \sum_{\sigma \in G} \varphi(\sigma)\sigma(c) \neq 0.$$

We have, for any $\tau \in G$:

$$\varphi(\tau)\tau(b) = \sum_{\sigma \in G} \varphi(\tau)\,\tau(\varphi(\sigma))\tau\sigma(c) = \sum_{\sigma \in G} \varphi(\tau\sigma)\tau\sigma(c) = b.$$

It follows that $\varphi(\tau) = b\tau(b)^{-1}$, or $\varphi(\tau) = \tau(a)a^{-1}$ with $a = b^{-1}$. □

The result actually stated by Hilbert was, historically:

Corollary 11.2 *Let K/F be a Galois extension with finite cyclic Galois group, generated by σ. Suppose $x \in K$ is an element such that $\mathrm{N}_{K/F}(x) = 1$. Then there exists $a \in K^\times$ such that $x = \sigma(a)a^{-1}$.*

Note that the converse is obvious: An element of the form $\sigma(a)a^{-1}$ has norm 1.

Proof. On the one hand, the theorem asserts that $\mathrm{H}^1(\mathrm{Gal}(K/F), K^\times) = 0$. On the other hand, we have seen with Example 10.3 that whenever G is a cyclic group generated by an element σ and M is a $\mathbb{Z}[G]$-module in multiplicative notation, we have

$$\mathrm{H}^1(G,M) = \frac{\{x : \prod_{\tau \in G} \tau \cdot x = 1\}}{\{\sigma(a)a^{-1} : a \in M\}}.$$

Thus the result is obvious. □

Example 11.3 If F contains an nth root of unity ω, where $n = [K : F]$, then $\mathrm{N}_{K/F}(\omega) = \omega^n = 1$, so there is $a \in K^\times$ such that $\sigma(a) = \omega a$ – a fact we had proved directly in the very first chapter of this book (see Lemma 1.6).

The corollary is given, of course, in a special situation. It is possible to derive a similar corollary whenever we have a presentation of $\mathrm{Gal}(K/F)$ by generators and relations. Rather than present the general case, we offer a second corollary illustrating the method.

Corollary 11.4 *Suppose K/F is Galois, with $\mathrm{Gal}(K/F)$ generated by two elements σ_1 and σ_2 satisfying $\sigma_1^2 = \sigma_2^2 = [\sigma_1, \sigma_2] = 1$ (so $\mathrm{Gal}(K/F) \cong C_2 \times C_2$ and $[K : F] = 4$).*

Given two elements $a_1, a_2 \in K^\times$ such that $a_i\sigma_i(a_i) = 1$ for $i = 1,2$, and

$$\frac{\sigma_2(a_1)}{a_1} = \frac{\sigma_1(a_2)}{a_2},$$

there exists $w \in K^\times$ such that

$$a_i = \frac{\sigma_i(w)}{w} \quad \textit{for } i = 1,2 \textit{ simultaneously}.$$

(Recall that $[a,b] = aba^{-1}b^{-1}$.) This corollary is reminiscent of the situation, familiar in analysis, when we have two differentiable maps f_1 and f_2 defined on an open set U in the plane, satisfying

$$\frac{\partial f_2}{\partial x_1} = \frac{\partial f_1}{\partial x_2},$$

and we wonder whether we can find a single function f with

$$f_i = \frac{\partial f}{\partial x_i} \quad \text{for } i = 1,2 \text{ simultaneously}.$$

The reader probably knows that such an f may be found when U is simply connected; more to the point, this happens precisely when $\mathrm{H}^1_{deR}(U) = 0$, the first de Rham cohomology group. To prove the corollary, we replace this with Hilbert 90, the vanishing of another cohomology group in degree 1.

Proof. Since we are dealing with a group of order 4, we could shorten the argument, but we try to get a sense of the general case. Let us work backwards, first. Let $\varphi \in \mathscr{Z}^1(G, K^\times)$ be a crossed homomorphism, where $G = \mathrm{Gal}(K/F)$. The definition of "crossed homomorphism" makes it clear that φ is entirely determined by $a_1 = \varphi(\sigma_1)$ and $a_2 = \varphi(\sigma_2)$, since σ_1, σ_2 generate G. Note that $\varphi(1) = \varphi(1 \cdot 1) = \varphi(1)\varphi(1)$ so $\varphi(1) = 1$.

Each relation in G gives information on φ. First, since $\sigma_1^2 = 1$, we must have $\varphi(\sigma_1^2) = a_1\sigma_1(a_1) = 1$, and likewise $a_2\varphi(a_2) = 1$. Second, since we have $\sigma_1\sigma_2\sigma_1\sigma_2 = 1$ (taking into account $\sigma_i = \sigma_i^{-1}$), we apply the defining property of crossed homomorphisms repeatedly and find

$$\varphi(\sigma_1\sigma_2\sigma_1\sigma_2) = a_1\sigma_1(a_2)\sigma_1\sigma_2(a_1)\sigma_1\sigma_2\sigma_1(a_2) = 1.$$

Taking into account that $\sigma_1\sigma_2\sigma_1 = \sigma_2$, and $\sigma_i(a_i) = a_i^{-1}$, this can be rewritten as

$$\frac{\sigma_2(a_1)}{a_1} = \frac{\sigma_1(a_2)}{a_2}.$$

Having made this analysis, the point is to observe, conversely, that two elements a_1, a_2 satisfying the conditions above define a unique crossed homomorphism φ with $\varphi(\sigma_i) = a_i$. (The reader can just work out the four values of φ and work from there, and hopefully the general argument for more complicated groups can be envisaged, too.) By Hilbert 90, this φ must be trivial, that is, for some $w \in K^\times$ we have $\varphi(\tau) = \tau(w)w^{-1}$ for all $\tau \in G$. □

Remark 11.5 We can also use Hilbert 90 to give the most conceptual proof (bordering on silly) of Wedderburn's theorem. Indeed, let F be a finite field, and let K/F be a finite extension. We have $\mathrm{H}^1(\mathrm{Gal}(K/F), K^\times) = 0$ by Hilbert 90. However, the group $\mathrm{Gal}(K/F)$ is cyclic, and here is an amusing exercise for the reader: For any cyclic group G, and any *finite* $\mathbb{Z}[G]$-module M, the order of $\mathrm{H}^n(G,M)$ is the same for all $n \geq 1$. So $\mathrm{H}^n(\mathrm{Gal}(K/F), K^\times) = 0$ for all $n \geq 1$, and in particular for $n = 2$. So all the groups $\mathrm{Br}(K/F)$ are trivial, as is their union $\mathrm{Br}(F)$. This means that a finite skewfield must be equal to its center, so it must be commutative.

More Kummer theory

We collect a number of results which complement those of Chapter 1 nicely. Here, F is any field, and $\overline{F}$ is a separable closure of F.

Lemma 11.6 *Suppose that either F is perfect, or the integer n is prime to the characteristic of F. Then there is a short exact sequence of discrete $\mathrm{Gal}(\overline{F}/F)$-modules*

$$1 \longrightarrow \mu_n(\overline{F}) \longrightarrow \overline{F}^\times \xrightarrow{x \mapsto x^n} \overline{F}^\times \longrightarrow 1.$$

Kummer exact sequence

It is called the **Kummer exact sequence**.

Proof. That $\mu_n(\overline{F})$ is the kernel of $x \mapsto x^n$ is tautological. When F is perfect, then $\overline{F}$ is an algebraic closure of F, so $x^n - a = 0$ always has solutions in $\overline{F}$, for any $a \in \overline{F}^\times$. Likewise, this equation has no multiple roots when n is prime to the characteristic of F, and we come to the same conclusion. □

Theorem 11.7 *Suppose that either F is perfect, or the integer n is prime to the characteristic of F. Then there is an isomorphism*

$$\mathrm{H}^1(F, \mu_n) \cong F^\times / F^{\times n}.$$

Also, there are isomorphisms

$$\begin{aligned} \mathrm{H}^2(F, \mu_n) &\cong \text{the } n\text{-torsion in } \mathrm{H}^2(F, \mathbb{G}_m) \\ &\cong \mathrm{Br}_n(F). \end{aligned}$$

If F contains a primitive nth root of unity, there is an isomorphism of discrete $\mathrm{Gal}(\overline{F}/F)$-modules $\mu_n \cong \mathbb{Z}/n\mathbb{Z}$, so $\mathrm{H}^*(F, \mu_n) \cong \mathrm{H}^*(F, \mathbb{Z}/n\mathbb{Z})$ for $* \geq 0$.

In particular, there is an isomorphism

$$F^\times / F^{\times n} \cong \mathrm{H}^1(F, \mathbb{Z}/n\mathbb{Z}) = \mathrm{Hom}(\mathrm{Gal}\left(\overline{F}/F\right), \mathbb{Z}/n\mathbb{Z}).$$

Explicitly, if ω is a primitive root, the continuous homomorphism χ_a associated to $a \in F^\times$ is characterized by

$$\frac{\sigma\left(\sqrt[n]{a}\right)}{\sqrt[n]{a}} = \omega^{\chi_a(\sigma)}$$

for all $\sigma \in \mathrm{Gal}\left(\overline{F}/F\right)$, and any choice of nth root $\sqrt[n]{a}$ of a.

Proof. Everything will follow from examining the long exact sequence in cohomology associated to the Kummer exact sequence. A portion of it is

$$\mathrm{H}^0(F, \mathbb{G}_m) \longrightarrow \mathrm{H}^0(F, \mathbb{G}_m) \longrightarrow \mathrm{H}^1(F, \mu_n) \longrightarrow \mathrm{H}^1(F, \mathbb{G}_m).$$

By Hilbert 90, we have $\mathrm{H}^1(F, \mathbb{G}_m) = 0$. On the other hand, $\mathrm{H}^0(F, \mathbb{G}_m) = F^\times$. So the exact sequence is really

$$F^\times \xrightarrow{x \mapsto x^n} F^\times \longrightarrow \mathrm{H}^1(F, \mu_n) \longrightarrow 0.$$

Hence the isomorphism $H^1(F,\mu_n) \cong F^\times/F^{\times n}$.

Another portion of the same sequence is

$$H^1(F,\mathbb{G}_m) \longrightarrow H^2(F,\mu_n) \longrightarrow H^2(F,\mathbb{G}_m) \xrightarrow{x\mapsto x^n} H^2(F,\mathbb{G}_m).$$

Again, the first group is 0 by Hilbert 90, so that $H^2(F,\mu_n)$ is indeed isomorphic to the n-torsion in $H^2(F,\mathbb{G}_m)$. The latter is isomorphic to the Brauer group $\mathrm{Br}(F)$, and $\mathrm{Br}_n(F)$ is our notation for the n-torsion in $\mathrm{Br}(F)$.

In the presence of a primitive nth root of unity $\omega \in F$, we have an isomorphism of discrete $\mathrm{Gal}(\overline{F}/F)$-modules $\mathbb{Z}/n\mathbb{Z} \to \mu_n$ given by $k \mapsto \omega^k$ (and in particular, this depends on the choice of ω). So μ_n can be replaced by $\mathbb{Z}/n\mathbb{Z}$, with trivial Galois action, in the isomorphisms above.

It remains to establish the explicit formula, which we leave to the reader. The explicit description of various maps must be examined – in particular, choosing a root $\sqrt[n]{a}$ is one of the moves made in the proof of the zig-zag lemma. □

Example 11.8 As an illustration, suppose F is a finite field which contains the nth roots of unity. On the one hand, the group $F^\times$ is cyclic, and its order is a multiple of n, so $F^\times/F^{\times n}$ is cyclic of order n; on the other hand, the absolute Galois group $\mathrm{Gal}(\overline{F}/F)$ is $\widehat{\mathbb{Z}}$, which contains $\mathbb{Z}$ as a dense subgroup (see Example 3.24), so a continuous $\chi : \widehat{\mathbb{Z}} \to \mathbb{Z}/n\mathbb{Z}$ is determined by $\chi(1) \in \mathbb{Z}/n\mathbb{Z}$, and we recover that $H^1(F,\mathbb{Z}/n\mathbb{Z})$ is cyclic of order n, as was expected.

Also, the group $H^2(F,\mathbb{Z}/n\mathbb{Z})$ is isomorphic to the n-torsion in the cohomological Brauer group $H^2(F,\mathbb{G}_m)$, which is trivial by Wedderburn, so $H^2(F,\mathbb{Z}/n\mathbb{Z}) = 0$. It is possible to prove, in fact, that $H^*(F,\mathbb{Z}/n\mathbb{Z}) = 0$ for $* \geq 2$ (a key ingredient being the case $* = 2$ just treated). This is quite far from our objectives in this book, so we will not go into this.

Example 11.9 Now consider a local number field F, again containing the nth roots of unity. The group $H^1(F,\mathbb{Z}/n\mathbb{Z}) \cong F^\times/F^{\times n}$ is entirely described by Lemma 4.10 (in particular, it is finite). Also, $H^2(F,\mathbb{G}_m) \cong \mathrm{Br}(F) \cong \mathbb{Q}/\mathbb{Z}$, the major result in the previous part of the book, so the n-torsion subgroup is $\frac{1}{n}\mathbb{Z}/\mathbb{Z}$, which is cyclic of order n. Thus $H^2(F,\mathbb{Z}/n\mathbb{Z}) \cong \mathbb{Z}/n\mathbb{Z}$. In fact, one has $H^*(F,\mathbb{Z}/n\mathbb{Z}) = 0$ for $* \geq 3$, so we have entirely described the cohomology of such a local number field (we will not go into the proof of this; one uses the result about finite fields mentioned in the previous example, and a general procedure to relate the cohomology of a local field and that of its residue field).

The Hilbert symbol

Throughout this section, let F be a field, and let $\overline{F}$ be a separable closure. We will first define a "symbol" (χ,b) when $b \in F^\times$, and $\chi \in H^1(F,\mathbb{Q}/\mathbb{Z})$, that is, when χ is a continuous homomorphism $G := \mathrm{Gal}(\overline{F}/F) \longrightarrow \mathbb{Q}/\mathbb{Z}$.

The kernel of χ is open (Lemma 3.3), so it is of the form $\mathrm{Gal}(\overline{F}/E)$ for some finite Galois extension E/F, by infinite Galois theory. The image of χ has finite order $d = [E:F]$, so $\chi(G) = \frac{1}{d}\mathbb{Z}/\mathbb{Z}$. Thus the Galois group $\mathrm{Gal}(E/F)$ is cyclic, but we have more: A canonical generator is at our disposal, namely that $\rho \in \mathrm{Gal}(E/F)$ such that $\chi(\rho) = \frac{1}{d}$.

We have the data required to apply Proposition 7.18. For $M = E^\times$, this provides us with an *explicit* isomorphism

$$\varphi_\rho \colon \mathrm{H}^2\left(\mathrm{Gal}(E/F), E^\times\right) \longrightarrow F^\times/\mathrm{N}_{E/F}\left(E^\times\right).$$

Definition 11.10 The element of $\mathrm{H}^2(\mathrm{Gal}(E/F), E^\times)$ mapping to the class of b via φ_ρ is written (χ, b). ■

Recall that $\mathrm{H}^2(\mathrm{Gal}(E/F), E^\times)$ can be seen as a subgroup of $\mathrm{H}^2(F, \mathbb{G}_m)$ (Theorem 7.35), so we may view (χ, b) as an element of $\mathrm{H}^2(F, \mathbb{G}_m)$ if we wish. Also note that $b \mapsto (\chi, b)$ is a homomorphism, and it is tautological that its kernel is $\mathrm{N}_{E/F}(E^\times)$.

More delicate will be the proof that $\chi \mapsto (\chi, b)$ is also a homomorphism (so that the operation is in fact bilinear). We start with a general lemma.

Lemma 11.11 *Let C be a cyclic group of order d, let $\chi : C \to \mathbb{Q}/\mathbb{Z}$ be an isomorphism between C and $\frac{1}{d}\mathbb{Z}/\mathbb{Z}$, and let φ be the element of C with $\chi(\rho) = \frac{1}{d}$. Consider the short exact sequence (of trivial $\mathbb{Z}[C]$-modules)*

$$0 \longrightarrow \mathbb{Z} \longrightarrow \mathbb{Q} \longrightarrow \mathbb{Q}/\mathbb{Z} \longrightarrow 0,$$

and its corresponding connecting homomorphism $\delta \colon \mathrm{H}^1(C, \mathbb{Q}/\mathbb{Z}) \to \mathrm{H}^2(C, \mathbb{Z})$, so that $\delta(\chi) \in \mathrm{H}^2(C, \mathbb{Z})$.

Finally, let M be any $\mathbb{Z}[G]$-module, and construct a map

$$\mathrm{H}^0(C, M) \times \mathrm{H}^2(C, \mathbb{Z}) \longrightarrow \mathrm{H}^2(C, M)$$

written $(m, x) \mapsto m \smile x$, by the following recipe. The element $m \in \mathrm{H}^0(C, M) = M^C$ defines $\lambda \colon \mathbb{Z} \to M$ by $n \mapsto n \cdot m$, and $m \smile x = \lambda_2(x)$, the map induced on cohomology in degree 2.

Then

$$\varphi_\rho(m \smile \delta(\chi)) = m \in M^C/\mathrm{N}(M),$$

with the notation of Proposition 7.18.

Proof. We examine the definition of δ by having a look at the proof of the zig-zag lemma. We have $\chi \in \mathscr{Z}^1(C, \mathbb{Q}/\mathbb{Z})$, and we pick a lift $\widetilde{\chi} \in \mathscr{C}^1(C, \mathbb{Q})$, so $\widetilde{\chi}(\sigma) = \chi(\sigma) \in \mathbb{Q}/\mathbb{Z}$, for all $\sigma \in C$. We can arrange to have $\widetilde{\chi}(\rho) = \frac{1}{d}$ (part of the proof of the zig-zag lemma asserts that the choices are irrelevant). Then the cocycle $c = d^1(\widetilde{\chi})$, for which

$$c(\sigma, \tau) = \widetilde{\chi}(\sigma) + \widetilde{\chi}(\tau) - \widetilde{\chi}(\sigma\tau),$$

takes its values in $\mathbb{Z}$, and when we see it as an element of $\mathscr{C}^2(C,\mathbb{Z})$, it is actually a cocycle which represents $\delta(\chi)$.

The cohomology class $m \smile \delta(\chi)$ is thus represented by the cocycle $c' \in \mathscr{Z}^2(C,M)$ with

$$c'(\sigma,\tau) = c(\sigma,\tau)\cdot m.$$

Looking back at the definition of φ_ρ in Proposition 7.18, we have

$$\varphi_\rho(c') = \left[\sum_{\tau\in C} \widetilde{\chi}(\tau) + \widetilde{\chi}(\rho) - \widetilde{\chi}(\tau\rho)\right]\cdot m.$$

In the brackets on the right-hand side, we have the cancellation

$$\sum_\tau \widetilde{\chi}(\tau) = \sum_\tau \widetilde{\chi}(\tau\rho),$$

while

$$\widetilde{\chi}(\rho) = \frac{1}{d} = \frac{1}{|C|},$$

so $\varphi_\rho(c') = m$. □

Corollary 11.12 *For $\chi,\chi' \colon G \longrightarrow \mathbb{Q}/\mathbb{Z}$ and $b \in F^\times$, we have $(\chi+\chi',b) = (\chi,b)+(\chi',b) \in \mathrm{H}^2(F,\mathbb{G}_m)$. In other words, the symbol (χ,b) is bilinear.*

Proof. First, we apply the last lemma to $C = \mathrm{Gal}(E/F)$ where $\mathrm{Gal}(\overline{F}/E)$ is the kernel of $\chi \colon G \to \mathbb{Q}/\mathbb{Z}$, and with $M = E^\times$. Note that $\mathrm{H}^0(C,M) = F^\times$ here, so we can see $b \in F^\times$ as an element of $\mathrm{H}^0(C,M)$. The conclusion is that

$$(\chi,b) = b\smile\delta(\chi) \in \mathrm{H}^2(\mathrm{Gal}(E/F),E^\times). \tag{*}$$

This is the crux of the matter, but we have to go through a tedious verification that $\mathrm{Gal}(E/F)$ can be replaced by $\mathrm{Gal}(\overline{F}/F)$. So, mimicking the lemma, first define an operation

$$\mathrm{H}^0(F,M)\times\mathrm{H}^2(F,\mathbb{Z}) \longrightarrow \mathrm{H}^2(F,M),$$

written $(m,x)\mapsto m\smile x$, using the exact same procedure. (In the next chapter, we shall study these *cup-products* systematically.) We apply this to $M = \overline{F}^\times$, or rather, with our usual notation, $M = \mathbb{G}_m$. If χ is viewed as an element of $\mathrm{H}^1(F,\mathbb{Q}/\mathbb{Z})$, we can consider $\delta(\chi) \in \mathrm{H}^2(F,\mathbb{Z})$, where δ is the obvious connecting homomorphism. Finally, we have $\mathrm{H}^0(F,\mathbb{G}_m) = F^\times$, so $b \in F^\times$ can be seen as an element of $\mathrm{H}^0(F,\mathbb{G}_m)$.

We claim now that (χ,b), once we map it to $\mathrm{H}^2(F,\mathbb{G}_m)$, is none other than $b\smile\delta(\chi)$. In other words, we have again $(\chi,b) = b\smile\delta(\chi)$, but on the right-hand side, each letter has a slightly different meaning this time, compared to (*) (for example δ is not quite the *same* connecting homomorphism). The routine verification of the claim consists in checking that the operations $(m,x)\mapsto m\smile x$ commute with the homomorphism induced in cohomology by the projection $\mathrm{Gal}(\overline{F}/F) \to \mathrm{Gal}(E/F)$, and then invoking Lemma 10.16.

Since the expression $b \smile \delta(\chi)$ is clearly bilinear, we are done. □

From now on, we choose an integer $n \geq 1$ and we assume that F contains a primitive nth root of unity. In fact, we pick one, call it ω, and we accept that everything to follow will depend on this choice. (Except for $n = 2$.)

By "Kummer theory", that is by Theorem 11.7, there is an (explicit) identification of $F^\times/F^{\times n}$ with $\mathrm{Hom}(G, \mathbb{Z}/n\mathbb{Z})$, and we continue to write χ_a for the homomorphism corresponding to $a \in F^\times$, so that χ_a only depends on a modulo nth powers. We turn our attention to (χ_a, b).

Lemma 11.13 *Let $a, b \in F^\times$.*

1. *The fixed field of $\ker(\chi_a)$ is $F[\sqrt[n]{a}]$. It follows that $(\chi_a, b) = 0$ happens precisely when b is a norm from $F[\sqrt[n]{a}]$.*
2. *The element $(\chi_a, b) \in \mathrm{H}^2(F, \mathbb{G}_m)$ is n-torsion.*
3. *(χ_a, b) depends only on the classes of a and b in $F^\times/F^{\times n}$.*

Proof. (1) The expression given in Theorem 11.7 makes it clear that $\chi_a(\sigma) = 0$ if and only if σ fixes $\sqrt[n]{a}$, proving the first statement. The kernel of $b \mapsto (\chi, b)$ is $\mathrm{N}_{E/F}(E^\times)$, where E is the fixed field of $\ker(\chi)$, so the second statement is obvious.

(2) We have $n(\chi_a, b) = (\chi_a, b^n)$ by linearity. However, since the degree $m := [F[\sqrt[n]{a}] : F]$ divides n (by (1) of Proposition 1.7), we have $b^n = \mathrm{N}_{F[\sqrt[n]{a}]/F}(b^{\frac{n}{m}})$. By (1), we draw $n(\chi_a, b) = 0$.

(3) This is now obvious. □

We have seen that the inclusion $\mu_n(\overline{F}) \to \mathbb{G}_m(\overline{F})$ induces an identification of $\mathrm{H}^2(F, \mu_n)$ with the n-torsion in $\mathrm{H}^2(F, \mathbb{G}_m)$ (Theorem 11.7 again). Having also picked the root ω, we have an isomorphism $\mathbb{Z}/n\mathbb{Z} \to \mu_n$. Thus it makes sense to introduce, using (2) of the last lemma:

Hilbert symbol

Definition 11.14 The element of $\mathrm{H}^2(F, \mathbb{Z}/n\mathbb{Z})$ corresponding to (χ_a, b) under the above identifications is written (a, b) or $(a, b)_F$, and is called the **Hilbert symbol** of a and b.

We think of the Hilbert symbol as a bilinear map

$$F^\times/F^{\times n} \times F^\times/F^{\times n} \longrightarrow \mathrm{H}^2(F, \mathbb{Z}/n\mathbb{Z}).$$

We may replace $F^\times/F^{\times n}$ by $\mathrm{H}^1(F, \mathbb{Z}/n\mathbb{Z})$, using our identifications, and sometimes this is more natural. Note that some authors reserve the term "Hilbert symbol" for the case $n = 2$, speaking merely of "the symbol $(a, b)_F$" in the general case. Also, the notation $(a, b)_\omega$, highlighting the dependence on the choice of ω, is sometimes encountered.

Remark 11.15 Earlier in this book we used the notation $(a,b)_F$ for quaternion algebras. However, it follows from Problem 7.7, as the reader will establish, that the element $(a,b)_F$ just defined coincides with the element of $\mathrm{Br}_2(F) \cong \mathrm{H}^2(F,\mathbb{Z}/2\mathbb{Z})$ defined by the quaternion algebra $(a,b)_F$, viewed as an element of order 2 in the Brauer group (as explained in Example 6.23). Notice how (1) of Lemma 11.13 is consistent with Problem 5.3.

Example 11.16 When F is a local number field, we have given an isomorphism $\mathrm{H}^2(F,\mathbb{Z}/n\mathbb{Z}) \cong \mathbb{Z}/n\mathbb{Z}$; for $n=2$, this is even canonical, and we fix $n=2$ for the next comments. We can then see $(a,b)_F \in \mathbb{F}_2$ as

$$(a,b)_F = \begin{cases} 0 \text{ if } b \text{ is a norm from } F[\sqrt{a}] \\ 1 \text{ otherwise}. \end{cases}$$

This is the classical definition of the Hilbert symbol. For $F = \mathbb{Q}_p$, there are various recipes to compute it, and they will be explored in the exercises. Note that, when $a,b \in \mathbb{Q}^\times$, one frequently encounters the notation $(a,b)_p$ for $(a,b)_{\mathbb{Q}_p}$.

We also point out that $(a,b)_F = 0$ exactly when the equation $x^2 - ay^2 = b$ has a solution with $x,y \in F$. This works even when a is already a square in F, as a moment's thought reveals (in fact, if $a = u^2$, try $x = (b+1)/2$ and $y = (1-b)/2u$). Further inspection brings us to realize that the condition is symmetric in a and b, so that $(a,b) = (b,a) \in \mathbb{F}_2$ in this case. This may come as a surprise, as the definition of (a,b) seems to use a and b very differently. However, there is a general statement here, as we proceed to show, after considering one more example.

Example 11.17 Let $F = \mathbb{R}$, and let us consider $(a,b)_\mathbb{R}$ for $a,b \in \mathbb{R}^\times$. Here $n=2$, as there are not enough roots of unity in $\mathbb{R}$ for us to deal with other cases anyway. If either a or b is a square, then certainly $(a,b)_\mathbb{R} = 0$. It remains to compute $(-1,-1)_\mathbb{R}$. Here, we see that -1 is not a norm from $\mathbb{R}[\sqrt{-1}] = \mathbb{C}$, for it would be positive if it were. As a result, we conclude that $(-1,-1)_\mathbb{R}$ is the nonzero element of $\mathrm{H}^2(\mathbb{R},\mathbb{F}_2) = \mathbb{F}_2$ (Example 10.30). If we believe the claim of Remark 11.15, then we see that this is consistent with the fact that the classical quaternion algebra $(-1,-1)_\mathbb{R} = \mathbb{H}$ is a field, so it is not split (=Brauer equivalent to $\mathbb{R}$).

Lemma 11.18 ([Ser79], chapter XIV, section 2, proposition 4.) *Let F be any field with a chosen primitive nth root of unity.*

1. *Suppose $x,a \in F$, with $a \neq 0$ and $x^n - a \neq 0$. Then $(a, x^n - a) = 0$. In particular, we have $(a,-a) = 0$ and $(a, 1-a) = 0$.*
2. *$(b,a) = -(a,b)$.*

Proof. (1) Let α be an nth root of a, and let $K = F[\alpha]$. From Proposition 1.7, the extension K/F is cyclic, of degree m dividing n; if $n = dm$, we have moreover, for any $\sigma \in \mathrm{Gal}(K/F)$, the expression $\sigma(\alpha) = \omega^{kd}\alpha$ for some $k \in \mathbb{Z}$, where $\omega \in F$ is our favorite nth root of unity. So $\sigma(x - \omega^r\alpha) = x - \omega^{kd+r}\alpha$, and as σ runs through $\mathrm{Gal}(K/F)$, all the integers k with $0 \leq k < m$ occur.

Now write

$$x^n - a = \prod_{i=0}^{n-1}\left(x - \omega^i\alpha\right) = \prod_{r=0}^{d-1}\prod_{k=0}^{m-1}\left(x - \omega^{kd+r}\alpha\right) = \prod_{r=0}^{d-1} \mathrm{N}_{K/F}\left(x - \omega^r\alpha\right).$$

So $x^n - a$ is a norm from K, and (by (1) of the previous lemma), we have $(a, x^n - a) = 0$. For $x = 0$ and $x = 1$, we deduce the two particular cases announced.

For (2) we use that $(ab, -ab) = 0$ by (1), and we note

$$(ab,-ab) = (a,(-a)b)+(b,(-b)a) = (a,-a)+(a,b)+(b,-b)+(b,a) = (a,b)+(b,a).$$

So $(b,a) = -(a,b)$, as promised. □

Problems

11.1. Provide the details for the discussion outlined in Remark 11.15.

11.2. Here is a formula for $(a,b)_p = (a,b)_{\mathbb{Q}_p}$ for $a,b \in \mathbb{Q}_p$ (and for $n = 2$). It is given assuming that $\mathrm{H}^2(\mathbb{Q}_p, \mathbb{F}_2) \cong \mathbb{F}_2$ is identified with $\{\pm 1\}$. Write $a = p^\alpha u$ and $b = p^\beta v$ with $u, v \in \mathbb{Z}_p^\times$. Then for $p > 2$ one has

$$(a,b)_p = (-1)^{\alpha\beta\varepsilon(p)}\left(\frac{u}{p}\right)^\beta\left(\frac{v}{p}\right)^\alpha.$$

In this expression, the *Legendre symbol* $\left(\frac{u}{p}\right)$ is 1 if u is a square mod p and -1 otherwise; and $\varepsilon(p) = (p-1)/2 \bmod p$. When $p = 2$ one has

$$(a,b)_2 = (-1)^{\varepsilon(u)\varepsilon(v)+\alpha\omega(v)+\beta\omega(u)}.$$

Here $\varepsilon(u) = (u-1)/2 \bmod 2$ as above, while $\omega(u) = (u^2-1)/8 \bmod 2$. (Notice that an element $u \in \mathbb{Z}_2^\times$ must be "odd", that is of the form $u = 2k+1$ with $k \in \mathbb{Z}_2$, and it follows immediately that $u^2 - 1$ is divisible by 8).

In this problem, we ask you to write a computer program which, given $a, b \in \mathbb{Q}^\times$, computes the (finite!) set of prime numbers p such that $(a,b)_p$ is nontrivial. Then, assume the following consequence of the Hasse–Minkowski theorem: The symbol $(a,b)_\mathbb{Q}$ is trivial if and only if all the symbols $(a,b)_p$ are trivial (for each prime p) as well as the symbol $(a,b)_\mathbb{R}$. Based in this, write a program which decides whether $(a,b)_\mathbb{Q}$ is trivial.

A proof of the above formulae is given in [Ser73], as well as a proof of the Hasse–Minkowski theorem. The reader can consult this reference now. Another argument for the formulae is given in [Ser79], but it relies on material to be introduced in Chapter 13.

12 Finer structure

We return to generalities about group cohomology, and present several complements. These will all be used in the next part of the book, and besides, they are also classical facts.

Shapiro's isomorphism

Our setup involves a finite group G and a subgroup H. We will typically write M for a $\mathbb{Z}[G]$-module, and A for a $\mathbb{Z}[H]$-module. It will often be convenient to choose a *transversal set* for H in G, that is a subset $T \subset G$ such that G is the disjoint union of the cosets Ht for $t \in T$. Such a T is obviously not unique, so whenever we make use of it, we shall need to make sure that our constructions are independent of all choices. For example, the statement of the next lemma does not involve T, while the proof does.

Lemma 12.1 *The free $\mathbb{Z}[G]$-module of rank 1, that is $\mathbb{Z}[G]^1$, is also free as a $\mathbb{Z}[H]$-module, of rank $[G : H]$ (the index of H in G). As a result, any free $\mathbb{Z}[G]$-module is free as a $\mathbb{Z}[H]$-module.*

Proof. The elements $t \in \mathbb{Z}[G]^1$, for $t \in T$ where T is a transversal set as above, clearly form a basis for $\mathbb{Z}[G]^1$ as a $\mathbb{Z}[H]$-module. The second statement is then obvious. □

induced module

Definition 12.2 Let A be a $\mathbb{Z}[H]$-module. The corresponding **induced** $\mathbb{Z}[G]$-module, written $\mathrm{Ind}_H^G(A)$ or simply $\mathrm{Ind}(A)$, is $\mathrm{Hom}_{\mathbb{Z}[H]}(\mathbb{Z}[G]^1, A)$. It is viewed as a $\mathbb{Z}[G]$-module, where the action of $\sigma \in G$ on $f \in \mathrm{Ind}(A)$ is given by $(\sigma \cdot f)(x) = f(x\sigma)$.

There is an alternative description of $\mathrm{Ind}(A)$ which is often employed:

Lemma 12.3 *There is an isomorphism $\mathrm{Ind}_H^G(A) \cong \mathbb{Z}[G] \otimes_{\mathbb{Z}[H]} A$, with G-action given by $\sigma \cdot (g \otimes a) = \sigma g \otimes a$.*

Proof. Observe that the proposed tensor product is the direct sum of the abelian groups $t^{-1} \otimes A$ for $t \in T$, a transversal set. (It is useful to realize that G is the disjoint union of the cosets $t^{-1}H$, for $t \in T$.) If we associate to $t^{-1} \otimes a$, for $a \in A$, the unique element $f_{t,a} \in \mathrm{Hom}_{\mathbb{Z}[H]}(\mathbb{Z}[G]^1, A)$ for which $f_{t,a}(t) = a$ and $f_{t,a}(s) = 0$ for $s \in T \setminus \{t\}$, we obtain the desired isomorphism. The reader will check the crucial (but easy) point, namely that this is G-linear. □

Remark 12.4 In the more general setting when G is not finite, and the index of H in G is not finite either, the two modules may not be isomorphic. The module in Definition 12.2 is sometimes called the *coinduced* module, written $\mathrm{Coind}_H^G(A)$. ■

The statement of the next proposition is known as "Shapiro's lemma". Later, we will work to make things more explicit, but it is helpful to know from the outset that the isomorphisms presented exist.

Proposition 12.5 *For each $n \geq 0$, there is an isomorphism*

$$\mathrm{H}^n\left(G, \mathrm{Ind}_H^G(A)\right) \cong \mathrm{H}^n(H, A),$$

and an isomorphism

$$\mathrm{H}_n\left(G, \mathrm{Ind}_H^G(A)\right) \cong \mathrm{H}_n(H, A).$$

Proof. For any $\mathbb{Z}[G]$-module P, there is an isomorphism

$$\mathrm{Hom}_{\mathbb{Z}[G]}(P, \mathrm{Ind}(A)) \cong \mathrm{Hom}_{\mathbb{Z}[H]}(P, A).$$

Indeed, in one direction this takes $f\colon P \to \mathrm{Hom}_{\mathbb{Z}[H]}(\mathbb{Z}[G]^1, A)$ to the map $p \mapsto f(p)(1)$; in other words, we compose f with the map

$$\mathrm{ev}\colon \mathrm{Hom}_{\mathbb{Z}[H]}\left(\mathbb{Z}[G]^1, A\right) \longrightarrow A,$$

which evaluates at $1 \in G$. In the other direction, having at our disposal $h \in \mathrm{Hom}_{\mathbb{Z}[H]}(P, A)$, we define $f\colon P \to \mathrm{Hom}_{\mathbb{Z}[H]}(\mathbb{Z}[G]^1, A)$ by $f(p)(\sigma) = h(\sigma p)$ for $\sigma \in G$, extending by linearity. The details are readily checked.

This isomorphism is "natural", so that if we pick a free resolution P_* of $\mathbb{Z}$ as a $\mathbb{Z}[G]$-module, we end up with an isomorphism of cochain complexes

$$\mathrm{Hom}_{\mathbb{Z}[G]}(P_*, \mathrm{Ind}(A)) \cong \mathrm{Hom}_{\mathbb{Z}[H]}(P_*, A). \tag{*}$$

The left-hand-side, by definition, has cohomology equal to $\mathrm{H}^*(G, \mathrm{Ind}(A))$. By Lemma 12.1, we can view P_* as a free resolution of $\mathbb{Z}$ as a $\mathbb{Z}[H]$-module, so that the cohomology of the cochain complex on the right-hand side of (*) is isomorphic $\mathrm{H}^*(H, A)$.

The case of homology groups is established similarly, and rests on the existence of an isomorphism

$$P_* \otimes_{\mathbb{Z}[G]} \left(\mathbb{Z}[G] \otimes_{\mathbb{Z}[H]} A\right) \cong P_* \otimes_{\mathbb{Z}[H]} A$$

of cochain complexes, and on Lemma 12.3. □

Corollary 12.6 *Let M be a $\mathbb{Z}[G]$-module which is induced from the trivial subgroup of G, that is $M \cong \mathrm{Ind}_{\{1\}}^G(A)$ for some abelian group A. Then $\widehat{\mathrm{H}}^n(G,M) = 0$ for all $n \in \mathbb{Z}$.*

Proof. Shapiro's lemma does the hard work: For $n > 0$, we have $\mathrm{H}^n(G,M) \cong \mathrm{H}^n(\{1\},A) = 0$ (the cohomology groups of the trivial group are trivial in positive degrees, cf. Example 10.2), and likewise $\mathrm{H}_n(G,M) = 0$. This shows that $\widehat{\mathrm{H}}^n(G,M) = 0$ for all n except $n = 0$ and $n = -1$, two cases which we may as well treat directly.

An element of M^G is a homomorphism $f\colon \mathbb{Z}[G]^1 \to A$ which is constant on G. If the value that it takes on G is $a \in A$, then we let φ denote the "Dirac function at $1 \in G$", that is $\varphi(1) = a$ and $\varphi(\sigma) = 0$ for $\sigma \neq 1$. We can view φ as a homomorphism $\mathbb{Z}[G]^1 \to A$ (extending by linearity), and $\sum_{\sigma\in G} \sigma \cdot \varphi = f$. So f is in $\mathrm{N}(M)$, and $\widehat{\mathrm{H}}^0(G,M) = M^G/\mathrm{N}(M) = 0$.

Finally, we show that $\widehat{\mathrm{H}}^{-1}(G,M) = {}_N M/M' = 0$. For any $f\colon \mathbb{Z}[G]^1 \to A$, let $\lfloor f \rfloor$ be the number of elements $\sigma \in G$ with $f(\sigma) \neq 0$. Starting with such an f, we let $f' \in M'$ be such that $\lfloor f - f' \rfloor$ is minimal, and we put $f'' = f - f'$. If $f \in {}_N M$, then $f'' \in {}_N M$ as well, and this means precisely that $\sum_{\sigma\in G} f''(\sigma) = 0$. It follows that we cannot have $\lfloor f'' \rfloor = 1$, clearly. Suppose we had $\lfloor f'' \rfloor \geq 2$, so let σ, τ be distinct elements of G with $f''(\sigma) \neq 0$ and $f''(\tau) \neq 0$. Let φ be the "Dirac" function on G, taking the value $f''(\sigma)$ at σ and 0 elsewhere, and consider $\psi = \varphi - \sigma\tau^{-1} \cdot \varphi$. Then ψ is an element of M', with $\psi(x) = 0$ except when we have either $x = \sigma$, in which case $\psi(\sigma) = f''(\sigma)$, or when $x = \tau$. As a result, we have $\lfloor f'' - \psi \rfloor < \lfloor f'' \rfloor$, a contradiction. This absurd conclusion shows $\lfloor f'' \rfloor = 0$, so $f'' = 0$ and $f \in M'$. □

Corollary 12.7 *Suppose G is a finite group and P is a projective $\mathbb{Z}[G]$-module. Then $\widehat{\mathrm{H}}^n(G,P) = 0$ for all $n \in \mathbb{Z}$.*

Proof. There exists a module Q such that $P \oplus Q$ is free, and $\widehat{\mathrm{H}}^*(G, P \oplus Q) = \widehat{\mathrm{H}}^*(G,P) \oplus \widehat{\mathrm{H}}^*(G,Q)$, so it is enough to prove the corollary when P is free. Also, we reduce immediately to the case $P = \mathbb{Z}[G]^1$.

Now, consider the previous lemma with $A = \mathbb{Z}$. In this case $\mathrm{Ind}(\mathbb{Z}) = \mathrm{Hom}(\mathbb{Z}[G]^1, \mathbb{Z})$, which is usually written $\mathbb{Z}[G]^*$ and called the *dual* of $\mathbb{Z}[G]$. We have an isomorphism $\mathbb{Z}[G]^1 \to \mathbb{Z}[G]^*$ by sending $1 \in G$ to the "Dirac function at 1", that is the homomorphism $f\colon \mathbb{Z}[G]^1 \to \mathbb{Z}$ with $f(1) = 1$ and $f(\sigma) = 0$ for $\sigma \in G$, $\sigma \neq 1$. Alternatively, use Lemma 12.3 to deduce that $\mathbb{Z}[G]^* \cong \mathbb{Z}[G] \otimes_{\mathbb{Z}} \mathbb{Z} = \mathbb{Z}[G]$. Either way, we see that $\mathbb{Z}[G]^1$ is induced from the trivial subgroup, so its Tate cohomology is zero. □

The next corollary will be the base of the technique of "dimension shifting", to be illustrated later in the chapter. We need a lemma first, whose immediate proof is left to the reader.

Lemma 12.8 *Let M be a $\mathbb{Z}[G]$-module, and let H be any subgroup of G. Then there is an injective homomorphism $M \to \mathrm{Ind}_H^G(M)$ mapping $m \in M$ to $f\colon \mathbb{Z}[G]^1 \to M$ with $f(\sigma) = \sigma \cdot m$.* □

Corollary 12.9 *Let M be any $\mathbb{Z}[G]$-module. Then we can both embed M as a submodule of a module M' with trivial Tate cohomology, and express M as a quotient of a module M'' with trivial Tate cohomology.*

As a result, if we are given any integer $k \in \mathbb{Z}$, then we can find a $\mathbb{Z}[G]$-module $M(k)$ such that

$$\widehat{\mathrm{H}}^n(G,M) \cong \widehat{\mathrm{H}}^{n+k}(G,M(k)).$$

Note that $M(k)$ is in no way uniquely determined by M and k alone; it is merely convenient to have a notation for one such module.

Proof. The module M' is given by Lemma 12.8 with $H = \{1\}$, given Corollary 12.6. As for M'', pick any free module mapping onto M, and use Corollary 12.7.

If we form the exact sequence

$$0 \longrightarrow M \longrightarrow M' \longrightarrow M'/M \longrightarrow 0$$

and consider the long exact sequence in Tate cohomology, we see that $\widehat{\mathrm{H}}^n(G,M) \cong \widehat{\mathrm{H}}^{n-1}(G,M'/M)$, for all $n \in \mathbb{Z}$. On the other hand, if we work with

$$0 \longrightarrow K \longrightarrow M'' \longrightarrow M \longrightarrow 0,$$

where K is the kernel of the given map $M'' \to M$, we obtain $\widehat{\mathrm{H}}^n(G,M) \cong \mathrm{H}^{n+1}(G,K)$, for all $n \in \mathbb{Z}$.

So we can shift indices to the left or the right. By doing this $|k|$ times, using either procedure depending on the sign of k, we find our module $M(k)$. □

The consequences are striking when you first discover them. We see that any (co)homology group $\widehat{\mathrm{H}}^n(G,M)$ can be computed as $\widehat{\mathrm{H}}^0(G,M(-n))$: The entire theory of both homology and cohomology groups is contained in the theory of $\widehat{\mathrm{H}}^0$! Of course we also have $\widehat{\mathrm{H}}^n(G,M) \cong \mathrm{H}^1(G,M(1-n))$, so that the theory of crossed homomorphisms contains everything else. And it is also true that $\widehat{\mathrm{H}}^n(G,M) \cong \mathrm{H}^2(G,M(2-n))$, so that everything in this chapter can *in principle* be reduced to the group extensions studied in the previous part of the book.

In practice, "translating everything in terms of H^2" (or H^1, or $\widehat{\mathrm{H}}^0$) is not a good strategy. Still, we will have several applications of "dimension shifting". Here is one.

Example 12.10 We have seen with Lemma 7.27 that $\mathrm{H}^2(G,M)$ is always $|G|$-torsion. By dimension shifting, it follows that $\widehat{\mathrm{H}}^n(G,M) \cong \mathrm{H}^2(G,M(2-n))$ is also $|G|$-torsion, for all $n \in \mathbb{Z}$. Starting with $\widehat{\mathrm{H}}^0$ instead of H^2 is even easier (and provides an alternative proof of Lemma 7.27).

A few explicit formulae

We wish to prove that the isomorphism in Shapiro's lemma can be constructed in a "natural" way – that is, that we can define specific isomorphisms $\mathrm{H}^n(G,\mathrm{Ind}_H^G(A)) \to \mathrm{H}^n(H,A)$, for all G,H,A,n, which are compatible with the various induced maps on modules, and also with the connecting homomorphisms. This will follow from the simple statement that we can choose a combination of induced maps, namely "restriction" and "evaluation at 1", to construct the isomorphisms. It will also follow that Shapiro's lemma holds for profinite groups.

The reason why this is not entirely trivial can be seen from the proof of Shapiro's lemma (Proposition 12.5): We have used the definition of cohomology as an Ext, and the possibility of choosing any projective resolution at all to conduct the calculations. By contrast, our definition of induced maps is in terms of the groups $C^*(G,M)$. We need to bridge the two approaches. We will also obtain an explicit description of the inverse of Shapiro's lemma, and this will be instrumental in defining the *corestriction* in the next section.

We keep the same setup. In this section, we focus on cohomology.

restriction

Definition 12.11 The homomorphism $\mathrm{H}^*(G,M) \to \mathrm{H}^*(H,M)$ induced by the inclusion map $H \to G$ is called the **restriction**, and is denoted $\mathrm{Res}_{G,H}$ or just Res. ■

This definition saves us the trouble of naming the inclusion map, say $\iota\colon H \to G$, and talking about ι^*.

We need to return briefly to the standard resolution in order to understand how it plays with the restriction. We write $R = \mathbb{Z}[G]$ and $S = \mathbb{Z}[H]$, so the standard projective resolutions are $R^{\otimes *+1}$ and $S^{\otimes *+1}$, with the boundary maps introduced in Chapter 10.

Lemma 12.12 *The inclusion $\iota\colon S \to R$, tensored with itself several times, yields a collection of maps $\iota_*\colon S^{\otimes *+1} \to R^{\otimes *+1}$ which is a morphism of chain complexes (of $\mathbb{Z}[H]$-modules).*

Let T be a transversal for H in G, and consider for $n \geq 0$ the map $R^{\otimes n+1} \to S^{\otimes n+1}$ defined by

$$h_0t_0 \otimes h_1t_1 \otimes \cdots \otimes h_nt_n \mapsto h_0 \otimes h_1 \otimes \cdots \otimes h_n\,,$$

where $h_i \in H$ and $t_i \in T$. Together, these form a morphism of chain complexes $\pi_*\colon R^{\otimes *+1} \to S^{\otimes *+1}$ (again, of $\mathbb{Z}[H]$-modules). Moreover, it is a homotopy inverse for the previous morphism. As a result, the homotopy class of π_* does not depend on the choice of T.

Proof. From the definition of the boundary maps, we see immediately that ι_* and π_* are morphisms of chain complexes over S.

Next, we observe that both ι_* and π_* induce the identity of $\mathbb{Z}$ in homology in degree 0. Since Lemma 12.1 tells us that $R^{\otimes *+1}$ and $S^{\otimes *+1}$ are both projective resolutions of $\mathbb{Z}$ as a $\mathbb{Z}[H]$-module, Proposition 9.25 applies (more precisely, the uniqueness statement in this proposition is what matters here). Thus ι_* and π_* are determined up to homotopy by this property. By the same token, the composition $\pi_* \circ \iota_*$ and $\iota_* \circ \pi_*$ must be homotopic to the identity. In other words, ι_* and π_* are homotopy inverses of each other. □

Lemma 12.13 *Identify $C^*(G,M)$ with $\mathrm{Hom}_{\mathbb{Z}[G]}(R^{\otimes *+1},M)$ and $C^*(H,M)$ with $\mathrm{Hom}_{\mathbb{Z}[H]}(S^{\otimes *+1},M)$ as usual (that is, as in the proof of Lemma 10.11). Then the restriction map $C^*(G,M) \to C^*(H,M)$ is identified with the map*

$$\mathrm{Hom}_{\mathbb{Z}[G]}\left(R^{\otimes *+1},M\right) \longrightarrow \mathrm{Hom}_{\mathbb{Z}[H]}\left(S^{\otimes *+1},M\right)$$

induced by ι_* from the previous lemma.

Proof. By inspection. □

Recall that $\mathrm{Ind}_H^G(A) = \mathrm{Hom}_{\mathbb{Z}[H]}(\mathbb{Z}[G]^1,A)$, and let us write $\mathrm{ev}\colon \mathrm{Ind}_H^G(A) \to A$ for the map, which evaluates at $1 \in G$. It is important to realize that ev is $\mathbb{Z}[H]$-linear, but *not* $\mathbb{Z}[G]$-linear (in fact, A is not assumed to be a $\mathbb{Z}[G]$-module). Thus the composition envisaged in the theorem below makes sense, but we could not compose the maps the other way around.

Theorem 12.14 *The composition of the homomorphism*

$$\mathrm{Res}_{G,H}\colon \mathrm{H}^*\left(G,\mathrm{Ind}_H^G(A)\right) \longrightarrow \mathrm{H}^*\left(H,\mathrm{Ind}_H^G(A)\right)$$

followed by

$$\mathrm{ev}_*\colon \mathrm{H}^*\left(H,\mathrm{Ind}_H^G(A)\right) \longrightarrow \mathrm{H}^*(H,A)$$

is an isomorphism, called Shapiro's isomorphism, *usually written* sh. *It follows that* Res *is injective here, and* ev_* *is surjective.*

The inverse of Shapiro's isomorphism is induced by the following operation $c \mapsto c'$ on cochains. Let T be a transversal set for H in G. Any $g \in G$ can be written uniquely $g = h_g t_g$ with $h_g \in H$ and $t_g \in T$. Now with the homogeneous cochain $c \in C^n(H,A)$, we associate $c' \in C^n(G,\mathrm{Ind}_H^G(A))$ given by

$$c'(\sigma_0,\sigma_1,\ldots,\sigma_n)(\sigma) = c(h_{\sigma\sigma_0},h_{\sigma\sigma_1}\ldots,h_{\sigma\sigma_n}).$$

Proof. Consider the following diagram, for $* \geq 0$:

$$\begin{array}{ccc}
\mathrm{Hom}_{\mathbb{Z}[G]}\left(R^{\otimes *+1},\mathrm{Ind}(A)\right) & \longrightarrow & \mathrm{Hom}_{\mathbb{Z}[H]}\left(R^{\otimes *+1},A\right) \\
\downarrow & & \downarrow \\
\mathrm{Hom}_{\mathbb{Z}[H]}\left(S^{\otimes *+1},\mathrm{Ind}(A)\right) & \longrightarrow & \mathrm{Hom}_{\mathbb{Z}[H]}\left(S^{\otimes *+1},A\right).
\end{array}$$

Both horizontal maps can be described as $f \mapsto \mathrm{ev}\circ f$. Both vertical maps can be described as $f \mapsto f \circ \iota_*$, where $\iota_*\colon S^{\otimes *+1} \to R^{\otimes *+1}$ is the inclusion $S \to R$ tensored with itself $*+1$ times. The diagram obviously commutes.

Let us start at the top-left corner of the diagram. If we go down and then right, we obtain a morphism of cochain complexes which induces $\mathrm{ev}_* \circ \mathrm{Res}$ in cohomology (using Lemma 12.13). We must show that it is a weak equivalence.

However, if we start at the top-left corner again, but move to the right, we encounter the isomorphism described in the proof of Proposition 12.5 (and its inverse was given explicitly); by going down afterwards, we use a morphism which has a homotopy inverse, as is deduced from Lemma 12.12 (and explicitly so). So we have a weak equivalence indeed. The formula for the inverse is worked out from 12.12 and 12.5 as suggested. □

Corollary 12.15 *Let $f\colon A_1 \to A_2$ be a homomorphism of $\mathbb{Z}[H]$-modules, and let $f'\colon \mathrm{Ind}_H^G(A_1) \to \mathrm{Ind}_H^G(A_2)$ be induced by f. Then the following diagram commutes:*

$$\begin{array}{ccc} \mathrm{H}^*(G,\mathrm{Ind}(A_1)) & \xrightarrow{f'_*} & \mathrm{H}^*(G,\mathrm{Ind}(A_2)) \\ \mathrm{sh}\downarrow & & \downarrow\mathrm{sh} \\ \mathrm{H}^*(H,A_1) & \xrightarrow{f_*} & \mathrm{H}^*(H,A_2)\,. \end{array}$$

If δ is the connecting homomorphism in the long exact sequence associated with the exact sequence of $\mathbb{Z}[H]$-modules

$$0 \longrightarrow A \longrightarrow B \longrightarrow C \longrightarrow 0,$$

and if δ' is the connecting homomorphism in the long exact sequence associated to

$$0 \longrightarrow \mathrm{Ind}_H^G(A) \longrightarrow \mathrm{Ind}_H^G(B) \longrightarrow \mathrm{Ind}_H^G(C) \longrightarrow 0,$$

then the following diagram commutes:

$$\begin{array}{ccc} \mathrm{H}^*(G,\mathrm{Ind}(C)) & \xrightarrow{\delta'} & \mathrm{H}^{*+1}(G,\mathrm{Ind}(A)) \\ \mathrm{sh}\downarrow & & \downarrow\mathrm{sh} \\ \mathrm{H}^*(H,C) & \xrightarrow{\delta} & \mathrm{H}^{*+1}(H,A). \end{array}$$

Finally, if $\pi\colon G' \to G$ is a surjective homomorphism between finite groups, and if $H' = \pi^{-1}(H)$, then the following diagram commutes:

$$\begin{array}{ccc} \mathrm{H}^*\left(G,\mathrm{Ind}_H^G(A)\right) & \xrightarrow{\pi^*} & \mathrm{H}^*\left(G',\mathrm{Ind}_{H'}^{G'}(A)\right) \\ \mathrm{sh}\downarrow & & \downarrow\mathrm{sh} \\ \mathrm{H}^*(H,A) & \xrightarrow{\pi^*} & \mathrm{H}^*(H',A). \end{array}$$

Given the explicit description of sh provided by the theorem, this corollary is a routine exercise for the reader. Use Lemma 10.16; prove that the operation $\mathrm{Ind}_H^G(-)$ preserves exact sequences; and observe that $\mathrm{Ind}_{H'}^{G'}(A)$ is none other than $\mathrm{Ind}_H^G(A)$ seen as a $\mathbb{Z}[G']$-module via π.

Remark 12.16 The discussion can now be extended to profinite groups. Briefly, suppose that H is an open subgroup of the profinite group G (necessarily of finite index). For any discrete H-module A, define $\mathrm{Ind}_H^G(A)$ to the the group of *continuous* functions $f\colon G \to A$ which satisfy $f(hx) = f(x)$ for all $x \in G$ and $h \in H$, with action of G given by $(\sigma \cdot f)(x) = f(x\sigma)$. Then $\mathrm{Ind}_H^G(A)$ is a discrete G-module.

The homomorphism $\mathrm{sh} := \mathrm{ev}_* \circ \mathrm{Res}$, as above, is shown to be an isomorphism, based on the finite case, and the commutativity of the last square in the corollary. Note that for any open subgroup $U \subset H$, we have $\mathrm{Ind}_H^G(A^U) = \mathrm{Ind}_H^G(A)^U$. ∎

The corestriction

In this section, G is a profinite group and H is an open subgroup, although our priority is the case when G is finite (thus we drop the word "discrete" when speaking of modules, and so on, for simplicity of phrasing).

Let us assume that M is a $\mathbb{Z}[G]$-module. Then the "evaluation at 1" map $\mathrm{ev}\colon \mathrm{Ind}_H^G(M) \to M$ is now $\mathbb{Z}[G]$-linear. As a result, Shapiro's isomorphism $\mathrm{H}^*(G, \mathrm{Ind}(M)) \to \mathrm{H}^*(G, M)$, which is given by $\mathrm{ev}_* \circ \mathrm{Res}$ by the last theorem, can also be described as $\mathrm{Res} \circ \mathrm{ev}_*$.

A first consequence is given in the next lemma. Recall that we have defined a map $M \to \mathrm{Ind}_H^G(M)$ in Lemma 12.8, to be called "the natural embedding" in what follows.

Lemma 12.17 *Let $\iota\colon M \to \mathrm{Ind}_H^G(M)$ be the natural embedding. Then the composition $\mathrm{sh} \circ \iota_*$ is the restriction $\mathrm{Res}_{G,H}\colon \mathrm{H}^*(G,M) \to \mathrm{H}^*(H,M)$.*

In other words, if we allow ourselves to think of Shapiro's isomorphism as the identity, then the restriction map is just the map induced by the natural embedding.

Proof. We have $\mathrm{sh} \circ \iota_* = \mathrm{ev}_* \circ \mathrm{Res} \circ \iota_* = \mathrm{Res} \circ (\mathrm{ev}_* \circ \iota_*) = \mathrm{Res} \circ (\mathrm{ev} \circ \iota)_* = \mathrm{Res}$, since $\mathrm{ev} \circ \iota$ is the identity of M. □

More importantly, we can now define a "wrong-way map":

Definition 12.18 Let M be a $\mathbb{Z}[G]$-module. Define a map

$$\mathrm{N}\colon \mathrm{Ind}_H^G(M) \longrightarrow M$$

by

$$\mathrm{N}(f) = \sum_{t \in T} t^{-1} \cdot f(t),$$

where a transversal set T has been chosen. (Recall that $\mathrm{Ind}_H^G(M)$ is by definition $\mathrm{Hom}_{\mathbb{Z}[H]}(\mathbb{Z}[G]^1, M)$.) Notice that N is independent of the choice of T.

The composition of

$$\mathrm{sh}^{-1}\colon \mathrm{H}^*(H,M) \longrightarrow \mathrm{H}^*\left(G, \mathrm{Ind}_H^G(M)\right)$$

followed by

$$\mathrm{N}_*\colon \mathrm{H}^*\left(G, \mathrm{Ind}_H^G(M)\right) \longrightarrow \mathrm{H}^*(G,M)$$

is called the **corestriction**, or **transfer**, and is written

$$\mathrm{Cor}_{G,H}\colon \mathrm{H}^*(H,M) \longrightarrow \mathrm{H}^*(G,M),$$

or simply Cor. ∎

Example 12.19 Let us explore this (rich) definition in degree 0, where the corestriction is a map $M^H \to M^G$. In fact, the isomorphism $\mathrm{H}^0(G,M) \cong M^G$ is given by $c \mapsto c(1)$, where $c \in Z^0(G,M) = \mathrm{H}^0(G,M)$ is a constant function $G \to M^G$. Likewise for H.

Now, the inverse of Shapiro's isomorphism, as per Theorem 12.14, produces from $c \in Z^0(H,M)$ the map $c' \colon G \to \mathrm{Ind}(M)$ given by $c'(\sigma_0)(\sigma) = c(h_{\sigma\sigma_0})$. Composing with N_0, we end up with $c'' \in Z^0(G,M)$ given by

$$c''(\sigma_0) = \mathrm{N}(c'(\sigma_0)) = \sum_{t\in T} t^{-1}c'(\sigma_0)(t) = \sum_{t\in T} t^{-1}c(h_{t\sigma_0}).$$

If c is constant with value $m \in M^H$, then c'' is constant with value

$$\mathrm{Cor}(m) = \sum_{t\in T} t^{-1} \cdot m \in M^G.$$

Note that the elements t^{-1} for $t \in T$ form a transversal set for the *right* cosets of H, that is, G is the disjoint union of the sets $t^{-1}H$. If H is normal in G (and perhaps also in general), then it makes a lot of sense to write

$$\mathrm{Cor}(m) = \sum_{t\in G/H} t \cdot m.$$

For example, assuming $G = \mathrm{Gal}(\overline{F}/F)$ and $H = \mathrm{Gal}(\overline{F}/E)$ where E/F is finite and Galois, we have $\mathrm{H}^0(F,\mathbb{G}_m) = F^\times$ and $\mathrm{H}^0(E,\mathbb{G}_m) = E^\times$, and the corestriction is precisely identified with the norm $\mathrm{N}_{E/F}$. The letter N was chosen because of this, of course.

Recall that M has a submodule $\mathrm{N}(M) = \mathrm{N}_G(M)$, spanned by elements of the form $\sum_{\sigma\in G}\sigma \cdot m$ for some $m \in M$. The explicit form of the corestriction just given makes it clear that it carries $\mathrm{N}_H(M)$ into $\mathrm{N}_G(M)$. In fact, the submodule $\mathrm{N}_G(M)$ is the image of the corestriction from the trivial subgroup, showing that the multiple uses of the letter N are quite consistent.

As another consequence, observe that there is also a map $\widehat{\mathrm{H}}^0(H,M) = M^H/\mathrm{N}_H(M) \to \widehat{\mathrm{H}}^0(G,M) = M^G/\mathrm{N}_G(M)$, obtained from the corestriction.

Finally, what happens if we restrict down to H, and then corestrict back up to G? Easy: The formula above for $\mathrm{Cor}(m)$, when $m \in M^G$, gives $\mathrm{Cor}(m) = [G : H]m$, where $[G : H] = |G|/|H|$ is the index of H in G. In other words, $\mathrm{Cor} \circ \mathrm{Res}$ is multiplication by $[G : H]$. ■

Using dimension shifting, we generalize the last property to all degrees at once:

Proposition 12.20 *Let $n \geq 0$. For any $x \in \mathrm{H}^n(G,M)$, we have*

$$\mathrm{Cor}_{G,H}(\mathrm{Res}_{G,H}(x)) = [G : H]x.$$

Proof. This is true when $n = 0$ by the example, both in ordinary and Tate cohomology. From the naturality of Shapiro's isomorphism (Corollary 12.15), the corestriction commutes with the dimension-shifting isomorphisms $\mathrm{H}^n(G,M) \cong \widehat{\mathrm{H}}^0(G,M(-n))$, as does the restriction. The result follows. □

Below we expand on the naturality of the corestriction, mentioned in the proof. Before we do that, we must mention some important consequences of the proposition.

Example 12.21 Taking H to be the trivial subgroup, we recover the fact that $\mathrm{H}^n(G,M)$ is $|G|$-torsion for $n \geq 1$, as in Example 12.10, since the cohomology of H is zero in degrees ≥ 1.

A classical application is given in Problem 12.1, which we urge the reader to consult (it is stated in terms of Tate cohomology, for which we define a corestriction below, but the case of positive degrees can be tackled now).

We need some notation for an important corollary of the proposition. Let p be a prime number. For any abelian group A, let $A_{(p)} = A \otimes_{\mathbb{Z}} \mathbb{Z}_{(p)}$ (recall that $\mathbb{Z}_{(p)}$ was defined in Example 2.16). For example, if $A = \mathbb{Z}/n\mathbb{Z}$, then $A_{(p)} = \mathbb{Z}/p^{v_p(n)}\mathbb{Z}$ (and so it is 0 if n is prime to p). We call $A_{(p)}$ the localization of A at p.

Corollary 12.22 *Let H be a subgroup of G of index prime to p, for example a p-Sylow subgroup of G. Then the homomorphism*

$$\mathrm{H}^n(G,M)_{(p)} \longrightarrow \mathrm{H}^n(H,M)_{(p)}$$

induced by the restriction is injective, for any $n \geq 0$ and any $\mathbb{Z}[G]$-module M. In particular, we have $\mathrm{H}^n(G,M)_{(p)} = 0$ for $n \geq 1$ if the order of G is prime to p.

Proof. If we compose with the corestriction, we get multiplication by $[G : H]$, a number prime to p; this is an isomorphism for any group of the form $A_{(p)}$, since $[G : H]$ is invertible in $\mathbb{Z}_{(p)}$. For the last statement, simply take $H = \{1\}$. □

Remark 12.23 First, an observation: Suppose A is an abelian group such that multiplication by N is an isomorphism on A whenever N is prime to p. Then A is naturally a $\mathbb{Z}_{(p)}$-module, hence a map $A \otimes_{\mathbb{Z}} \mathbb{Z}_{(p)} \to A$ which is an inverse for the obvious $A \to A_{(p)}$, and we see that $A = A_{(p)}$.

This applies to $A = \mathrm{H}^n(G,M)$, where M is a $\mathbb{Z}_{(p)}$-module. For such an M, the statement of the corollary holds without having to localize at p.

A typical example is $M = \mathbb{F}_p$. Then $\mathrm{H}^n(G,\mathbb{F}_p)$ is equal to its p-localization, as explained. By the corollary, we see that $\mathrm{H}^n(G,\mathbb{F}_p)$ injects into $\mathrm{H}^n(H,\mathbb{F}_p)$, where H is a p-Sylow subgroup of G (and it is zero if the order of G is prime to p). The cohomology groups $\mathrm{H}^n(G,\mathbb{F}_p)$ are of prime interest in algebraic topology, and in "modular" representation theory (that is, representation theory in positive characteristic). Hence the special role played by p-groups there.

However, the observation also applies to $A = \mathrm{H}^n(G,M)$ with arbitrary M, but assuming that G is a p-group, since A is then $|G|$-torsion. In particular, one can always state that $\mathrm{H}^n(G,M)_{(p)}$ injects into $\mathrm{H}^n(H,M)$ $(= \mathrm{H}^n(H,M)_{(p)})$ where H is a p-Sylow.

Finally, suppose that A is entirely N-torsion for some N – for example $A = \mathrm{H}^n(G,M)$, which is $|G|$-torsion. Then A splits off as a finite direct sum of various groups of the form $A^{[p]} := \{a \in A : p^k a = 0 \text{ for some } k \geq 1\}$, for finitely many primes p. We have obviously $A^{[p]}_{(\ell)} = 0$ if $\ell \neq p$, while $A^{[p]}_{(p)} = A^{[p]}$ because $A^{[p]}$ is a $\mathbb{Z}_{(p)}$-module. As a result, using that the process of tensoring with $\mathbb{Z}_{(p)}$ obviously commutes with finite direct sums, we have $A_{(p)} \cong A^{[p]}$. To summarize,

$$A \cong \bigoplus_p A_{(p)} .$$

(Keep in mind that this would fail for $A = \mathbb{Z}$, say.) This explains why $A_{(p)}$ is sometimes called the "p-primary part" of A, in this case. Besides, we see now that $\mathrm{H}^n(G,M)$ injects into the direct sum of the various $\mathrm{H}^n(G_p,M)$, where for each p a p-Sylow subgroup G_p has been chosen. ■

The naturality of the corestriction will be stated in the general setting of Tate cohomology. To extend the definition to negative degrees, we use the simplest possible construction.

corestriction

Definition 12.24 The homomorphism $\mathrm{H}_*(H,M) \to \mathrm{H}_*(G,M)$ induced by the inclusion map $H \to G$ is (also) called the **corestriction**, and is denoted $\mathrm{Cor}_{G,H}$ as above. ■

Example 12.25 Let us examine this in degree 0. Recall from the arguments right before Example 10.2 that $\mathrm{H}_0(G,M)$ is isomorphic to M/M' where $M' = M'_G = \langle m - \sigma \cdot m : m \in M, \sigma \in G \rangle$. In fact, let us refer back to the standard resolution, writing as ever $R = \mathbb{Z}[G]$ and $S = \mathbb{Z}[H]$; we have a commutative diagram:

$$\begin{array}{ccc} (S \otimes_{\mathbb{Z}} S) \otimes_S M & \longrightarrow & S \otimes_S M \\ \downarrow & & \downarrow \\ (R \otimes_{\mathbb{Z}} R) \otimes_R M & \longrightarrow & R \otimes_R M . \end{array}$$

The bottom line is a portion of the complex computing $\mathrm{H}_*(G,M)$, displayed in degrees 1 and 0, using the standard resolution. We can identify $R \otimes_R M$ with M itself, and the map is $r_0 \otimes r_1 \otimes m \mapsto r_0 m - r_1 m$. The top line is described in analogy, with H replacing G, and the vertical maps are the induced maps on homology – so here, the corestrictions, as we now call them.

The conclusion is that the corestriction, as a map $M/M'_H \to M/M'_G$, is the "obvious" map. It follows that this restricts to a map $M^H/M'_H \to M^G/M'_G$, that is, a map from $\widehat{\mathrm{H}}^{-1}(H,M)$ to $\widehat{\mathrm{H}}^{-1}(G,M)$. ■

All in all, we have defined

$$\mathrm{Cor} \colon \widehat{\mathrm{H}}^n(H,M) \longrightarrow \widehat{\mathrm{H}}^n(G,M)$$

for all $n \in \mathbb{Z}$.

Proposition 12.26 *The corestriction in Tate cohomology commutes with the morphisms induced by maps between modules, and with connecting homomorphisms.*

Proof. For positive degrees, this follows mostly from the naturality of Shapiro's isomorphism (Corollary 12.15), in degrees ≤ -2 this follows from Lemma 10.16, and in degrees -1 and 0 this requires a direct verification, left to the reader. □

It follows that Cor commutes with dimension shifting. As an application, we see that the statement of Proposition 12.20 extends to all $x \in \widehat{\mathrm{H}}^n(G,M)$ and all $n \in \mathbb{Z}$, and likewise for Corollary 12.22. More philosophically, we see that the corestriction is the unique "natural" extension of the "norm map" in degree 0 to all the groups of Tate cohomology (where "natural" means "with the properties stated in the proposition").

The conjugation action

Our plan is to investigate the special features of the above constructions when H is *normal* in G. However, we must start with some simple, general considerations about automorphisms.

When $\varphi\colon G_1 \to G_2$ is a homomorphism, we have defined the induced maps

$$\varphi^n\colon \mathrm{H}^n(G_2,M) \longrightarrow \mathrm{H}^n(G_1,M).$$

Here, we have abused notation. In fact, one should probably write

$$\varphi^n\colon \mathrm{H}^n(G_2,M) \longrightarrow \mathrm{H}^n(G_1,M_\varphi),$$

where M_φ is M viewed as a G_1-module via φ. It is easy to overlook this, especially when $G_1 = G_2$ and φ is an automorphism. *There is no action of* $\mathrm{Aut}(G)$ *on* $\mathrm{H}^*(G,M)$, *unless the action of G on M is trivial.* This is rarely emphasized in the literature, surprisingly enough.

There is, however, a particular situation in which we can remedy this. Suppose φ is the automorphism of G defined by $\varphi(g) = x^{-1}gx$, that is, φ is conjugation by x. In fact, for any normal subgroup H of G, we can see φ as an automorphism of H. Now, in this particular case, we do have an isomorphism

$$M_\varphi \longrightarrow M$$

given by $m \mapsto xm$. Please check that this is G-equivariant, so H-equivariant. We write

$$x_*\colon \mathrm{H}^*(H,M) \xrightarrow{\ \varphi^*\ } \mathrm{H}^*(H,M_\varphi) \longrightarrow \mathrm{H}^*(H,M),$$

where the second map is induced by our isomorphism $M_\varphi \cong M$.

Proposition 12.27 *The association $x \mapsto x_*$ is a well-defined homomorphism $G \to \mathrm{Aut}(\mathrm{H}^n(H,M))$, for any $n \ge 0$, or in other words we have a (left) action of G on $\mathrm{H}^n(H,M)$. Moreover, this action is trivial when $G = H$. As a result, there is an action of G/H on $\mathrm{H}^n(H,M)$.*

Proof. On cocycles, the operation x_* is given by

$$x_*(c)(\sigma_0,\dots,\sigma_n) = x\cdot c\left(\sigma_0^x,\dots,\sigma_n^x\right),$$

using the notation $a^b = b^{-1}ab$. From this we draw $(yx)_* = y_*\circ x_*$ readily, hence the first statement.

The second statement (which is a classic of algebraic topology when the action on M is trivial) only needs to be proved in degree 0, since the action is visibly compatible with dimension shifting. In degree 0, if $c\colon G\to M$ is constant with value $m\in M^G$, then the same can be said of $x_*(c)$ by the expression just given, so $x_*(c) = c$. So the action is indeed the identity when $G = H$ (when $G\neq H$, this argument breaks because c only has values in M^H).

By the previous point, the action of G on $\mathrm{H}^*(H,M)$ is trivial when restricted to H, so we may view it as an action of G/H. □

Remark 12.28 We can use the proposition to tie up a loose end. We will (very mildly) improve on the definition of Galois cohomology (Definition 10.27) to answer the objections raised in Remark 10.28. Recall that the issue is that we define $\mathrm{H}^n(F,M)$ to be $\mathrm{H}^n(\mathrm{Gal}(\overline{F}/F),M)$, where we have had to choose a separable closure $\overline{F}$ of F, and M is a discrete $\mathrm{Gal}(\overline{F}/F)$-module (a condition which depends on the choice of $\overline{F}$!).

First, we need to discuss the possible coefficients. Let us assume that we know how to associate to any separable closure K of F an abelian group $T(K)$, which is a discrete $\mathrm{Gal}(K/F)$-module. Moreover, assume that any isomorphism $\varphi\colon K\to K'$ induces an isomorphism $T(\varphi)\colon T(K)\to T(K')$ which is compatible with the Galois actions, in the obvious way. For example, we may take $T(K) = M$, a fixed abelian group with trivial Galois action, and $T(\varphi)$ is always the identity; or, we may take $T = \mu_N$ or $T = \mathbb{G}_m$, as in Remark 10.31.

In this situation we can combine $T(\varphi)$ with the isomorphism $\mathrm{Gal}(K/F)\cong\mathrm{Gal}(K'/F)$ (also obtained from φ) to obtain the isomorphism

$$\varphi_!\colon\ \mathrm{H}^*(\mathrm{Gal}(K/F),T(K))\longrightarrow \mathrm{H}^*(\mathrm{Gal}(K'/F),T(K')).$$

However, the proposition implies that $\varphi_!$ *does not depend on the specific choice of* φ. Indeed, φ can only be altered by postcomposing it with $\alpha\in\mathrm{Gal}(K'/F)$; then $(\alpha\circ\varphi)! = \alpha_*\circ\varphi_!$, but α_* is the identity by the proposition (with $G = H = \mathrm{Gal}(K'/F)$). We can now confidently define

$$\mathrm{H}^*(F,T) := \mathrm{H}^*\left(\mathrm{Gal}\left(\overline{F}/F\right),T\left(\overline{F}\right)\right)$$

for any separable closure $\overline{F}$ of F, and this definition makes sense up to *canonical* isomorphism (which is pretty good sense indeed). When T is "constant", with $T(K) = M$, we write this $\mathrm{H}^*(F,M)$.

All the sensible candidates for T will give a meaning to $T(E)$ for any extension E/F at all, and will satisfy $T(E) = T(\overline{F})^{\mathrm{Gal}(\overline{F}/E)}$, so that we may apply Lemma 10.25 to draw

$$\mathrm{H}^*(F,T) = \mathrm{colim}_E\, \mathrm{H}^*(\mathrm{Gal}(E/F), T(E)).$$

Here, E runs through the finite Galois extensions of F contained in $\overline{F}$, and up to canonical isomorphism, this does not depend on the choice of $\overline{F}$. ∎

The five-term exact sequence

We return to the case of a finite group G with a normal subgroup H. Our next goal is to define a homomorphism

$$\mathrm{tg}\colon \mathrm{H}^1(H,M)^{G/H} \longrightarrow \mathrm{H}^2\left(G/H, M^H\right),$$

called the *transgression*. Theorem 12.29 will state the existence of an exact sequence involving tg. In the sequel, however, we will only use a little part of that sequence, which does not feature the transgression, so the reader may decide to skip parts of this section. (See Theorem 12.29 and its proof.)

For the construction, we work with inhomogeneous cochains, and we let $c \in \mathscr{Z}^1(H,M)$ represent a cohomology class $[c] \in \mathrm{H}^1(H,M)$ which is fixed by the action of G/H. Recall that c is a function $H \to M$, and the condition means that for any $x \in G$ we have $x_*([c]) = [c]$, so that there exists $m = m_x \in M$ with the property

$$x \cdot c\left(\sigma^x\right) - c(\sigma) = \sigma \cdot m - m$$

for all $\sigma \in H$. We choose m_x for all $x \in G$, with $m_1 = 0$. Now let T denote a transversal set for the *right* cosets of H in G, that is, $G = \cup_{t \in T} tH$, and assume $1 \in T$. Finally, define

$$f\colon G \to M \qquad \text{by} \qquad f(th) = m_t + t \cdot c(h).$$

We will see that the coboundary $d^1(f) \in \mathscr{B}^2(G,M)$ factors through G/H, and takes values in M^H, giving a cocycle in $\mathscr{Z}^2(G/H, M^H)$ (which is not necessarily a coboundary), defining our class $\mathrm{tg}([c]) \in \mathrm{H}^2(G/H, M^H)$.

We first observe that the restriction of f to H is c itself. Next, routine calculations show that for $\sigma \in G$, $\tau \in H$, we have

$$f(\sigma\tau) = f(\sigma) + \sigma \cdot f(\tau), \qquad f(\tau\sigma) = f(\tau) + \tau \cdot f(\sigma).$$

Recalling that $d^1(f)(\sigma_1,\sigma_2) = \sigma_1 \cdot f(\sigma_2) + f(\sigma_1) - f(\sigma_1\sigma_2)$, one establishes, as claimed, that $d^1(f)(\sigma_1,\sigma_2\tau) = d^1(f)(\sigma_1,\sigma_2)$ for $\tau \in H$, and likewise, $d^1(f)(\sigma_1\tau,\sigma_2) = d^1(f)(\sigma_1,\sigma_2)$.

Now write that $d^2 d^1(f) = 0$ as always, and apply at (τ,σ_1,σ_2) with $\tau \in H$, to draw $\tau \cdot d^1(f)(\sigma_1,\sigma_2) = d^1(f)(\sigma_1,\sigma_2)$, or in other words, $d^1(f)$ takes its values in M^H. We have proved that $d^1(f)$ could be used to define an element of $\mathscr{Z}^2(G/H, M^H)$.

We have made some choices, but $d^1(f)$ is characterized easily. Indeed, suppose that $f'\colon G \to M$ is any cochain at all which restricts to c on H, and such that $d^1(f')$ factors through G/H, and takes values in M^H. Then $z = f - f'$ also factors through G/H and takes its values in M^H, as one checks easily; it follows that $d^1(f) - d^1(f') = d^1(z)$, where this time $z \in \mathscr{C}^1(G/H, M^H)$, so $d^1(f)$ and $d^1(f')$ differ by a coboundary. The common cohomology class defined by all such f is $\mathrm{tg}([c])$, by definition. The characterization implies easily that tg is a homomorphism.

Theorem 12.29 *When H is a normal subgroup of the finite group G, there is an exact sequence*

$$0 \longrightarrow \mathrm{H}^1\left(G/H, M^H\right) \xrightarrow{\mathrm{Inf}} \mathrm{H}^1(G,M) \xrightarrow{\mathrm{Res}} \mathrm{H}^1(H,M)^{G/H}$$

$$\xrightarrow{\mathrm{tg}} \mathrm{H}^2\left(G/H, M^H\right) \xrightarrow{\mathrm{Inf}} \mathrm{H}^2(G,M),$$

where the homomorphisms written Inf *are produced combining the projection $G \to G/H$ and the inclusion $M^H \to M$.*

This is usually called the "five-term exact sequence", presumably because 0 is not viewed as a term.

Proof. We will leave the hard work as a rather long exercise for the reader, which is also elementary (a solution is given in [NSW08], section 1.6.6). At least we will prove the exactness of the first line (which does not even involve the transgression), and as it turns out, this is all we use in the sequel.

The composition of the inclusion $H \to G$ and the projection $G \to G/H$ is the trivial homomorphism $H \to G/H$, which factors through the trivial group, and it follows that $\mathrm{Res} \circ \mathrm{Inf} = 0$. We need to show that Inf is injective, and that the kernel of Res is included in the image of Inf.

For the first statement, by Lemma 10.19, we need to consider a crossed homomorphism $f\colon G/H \to M^H$, such that $f'(\sigma) = \sigma \cdot m - m$ for some $m \in M$, where f' is f precomposed with $G \to G/H$. Since $f(1) = 0$, however, we must have $f'(h) = 0$ for $h \in H$, showing that $m \in M^H$. By the same lemma, this means precisely that f represents the zero cohomology class, so $\ker(\mathrm{Inf}) = \{0\}$, and we have the required injectivity.

Now suppose that the crossed homomorphism $f\colon G \to M$ restricts to a trivial crossed homomorphism on H, that is, we can find $m \in M$ such that $f(h) = h \cdot m - m$ for all $h \in H$. Replace f by $f' = f - f_1$ where $f_1(\sigma) = \sigma \cdot m - m$ for all $\sigma \in G$: Then f_1 is a trivial crossed homomorphism, so f and f' are in the same cohomology class, but f' restricts to 0 on H. For $\sigma \in G$, $\tau \in H$ we have thus

$$f'(\sigma\tau) = f'(\sigma) + \sigma \cdot f'(\tau) = f'(\sigma),$$

showing that f' can be viewed as a function on G/H. We also have $f'(\tau\sigma) = f'(\sigma\tau^\sigma) = f'(\sigma)$, and this is profitably compared with

$$f'(\tau\sigma) = f'(\tau) + \tau \cdot f'(\sigma) = \tau f'(\sigma).$$

In the end $f'(\sigma) = \tau \cdot f'(\sigma)$, so that f' takes its values in M^H. It follows that f' is "inflated" from $\mathrm{H}^1(G/H, M^H)$, as requested. □

The brave will even extend the theorem, and the definition of tg, to the case when G is profinite and H is open and normal (there are some minor continuity issues to deal with).

Example 12.30 For a very easy example, suppose M has a trivial G-action. Also suppose that, for some prime p, the abelian group M is in fact a $\mathbb{Z}_{(p)}$-module (for example, M may be an $\mathbb{F}_p$-vector space). Finally, assume that the order of G/H is prime to p (that is, H contains a p-Sylow of G). Then $\mathrm{H}^i(G/H, M^H) = 0$ for $i = 1,2$ by Corollary 12.22 (and the subsequent remark). The five-term exact sequence then shows that the restriction gives an isomorphism $\mathrm{Hom}(G,M) \cong \mathrm{Hom}(H,M)^{G/H}$. ■

Corollary 12.31 *Suppose* $\mathrm{H}^i(H,M) = 0$ *for* $1 \leq i < n$*, for some* $n \geq 1$*. Then there is an exact sequence*

$$0 \longrightarrow \mathrm{H}^n\left(G/H, M^H\right) \xrightarrow{\mathrm{Inf}} \mathrm{H}^n(G,M) \xrightarrow{\mathrm{Res}} \mathrm{H}^n(H,M)^{G/H}$$

$$\longrightarrow \mathrm{H}^n\left(G/H, M^H\right) \xrightarrow{\mathrm{Inf}} \mathrm{H}^n(G,M).$$

Proof. We proceed by induction on n, the theorem providing the base case $n = 1$.

Suppose the corollary holds for some n (and for all modules M!). Essentially, we use dimension shifting. More precisely, we consider the natural embedding $M \to \mathrm{Ind}_{\{1\}}^G(M)$ and we put $N = \mathrm{Ind}_{\{1\}}^G(M)/M$, so that we have an exact sequence of $\mathbb{Z}[G]$-modules

$$0 \longrightarrow M \longrightarrow \mathrm{Ind}_{\{1\}}^G(M) \longrightarrow N \longrightarrow 0. \tag{*}$$

By Shapiro and the long exact sequence in cohomology associated with (*), we have $\mathrm{H}^n(G,N) \cong \mathrm{H}^{n+1}(G,M)$. Also, we can view (*) as an exact sequence of $\mathbb{Z}[H]$-modules. We note that $\mathrm{Ind}_{\{1\}}^G(M)$ splits off, as a $\mathbb{Z}[H]$-module, as $[G:H]$ copies of $\mathrm{Ind}_{\{1\}}^H(M)$, so by Shapiro, it has trivial cohomology in degrees ≥ 1. As a result, we have $\mathrm{H}^i(H,N) \cong \mathrm{H}^{i+1}(H,M)$ for $i \geq 1$, and in particular we have $\mathrm{H}^i(H,N) = 0$ for $1 \leq i < n$ if we suppose (as we do) that $\mathrm{H}^i(H,M) = 0$ for $1 \leq i < n+1$.

Thus by induction, we have an exact sequence in degree n, with N replacing M. We have already identified two terms of the sequence as $\mathrm{H}^{n+1}(G,M)$, and another one as $\mathrm{H}^{n+1}(H,M)^{G/H}$ (the G/H action being compatible with the isomorphism $\mathrm{H}^n(H,N) \cong \mathrm{H}^{n+1}(H,M)$). Now consider the following portion of the long exact sequence in the cohomology of H associated with (*), taking into account that $\mathrm{H}^1(H,M) = 0$:

$$0 \longrightarrow M^H \longrightarrow \mathrm{Ind}_{\{1\}}^G(M)^H \longrightarrow N^H \longrightarrow 0. \tag{**}$$

We view this as an exact sequence of G/H-modules. Observe that $\mathrm{Ind}_{\{1\}}^G(M)^H$ can be identified with $\mathrm{Ind}_{\{1\}}^{G/H}(M)$, and so Shapiro applies again. The long exact sequence associated with (**) gives, this time, that $\mathrm{H}^n(G/H, N^H) \cong \mathrm{H}^{n+1}(G/H, M^H)$.

Thus we have indeed an exact sequence where the terms are as announced. That the homomorphisms are in fact Inf and Res follows from the fact that these are compatible with the connecting homomorphisms. □

For good measure, we also show the existence of an exact sequence in degree 0.

Lemma 12.32 *When H is normal in G, there is an exact sequence*

$$\widehat{\mathrm{H}}^0(H,M) \xrightarrow{\mathrm{Cor}} \widehat{\mathrm{H}}^0(G,M) \longrightarrow \widehat{\mathrm{H}}^0(G/H,M^H) \longrightarrow 0.$$

Proof. Looking at the definitions, we are looking for an exact sequence of the form

$$M^H/\mathrm{N}_H(M) \xrightarrow{\mathrm{Cor}} M^G/\mathrm{N}_G(M) \longrightarrow (M^H)^{G/H}/\mathrm{N}_{G/H}(M^H) \longrightarrow 0.$$

We note immediately that $(M^H)^{G/H} = M^G$, and we define the second map to be induced by the identity; it is obviously surjective. The explicit computations in Example 12.19 show that $\mathrm{N}_{G/H}(M^H) = \mathrm{Cor}(M^H)$, proving the exactness in the middle. □

Cup-products

We define new operations on the cohomology groups of a (pro)finite group. These are not needed in the sequel (although they can be used to clarify a few things), so we skip some of the proofs, especially when they are routine calculations.

As before, in this section G can be any profinite group, but for simplicity of language we leave the references to continuity implicit.

Definition 12.33 Suppose M_1 and M_2 are $\mathbb{Z}[G]$-modules, and let us view $M_1 \otimes_{\mathbb{Z}} M_2$ as a $\mathbb{Z}[G]$-module with action $\sigma \cdot (m_1 \otimes m_2) = \sigma \cdot m_1 \otimes \sigma \cdot m_2$. Define a bilinear operation

$$C^n(G,M_1) \times C^m(G,M_2) \longrightarrow C^{n+m}(G,M_1 \otimes M_2)$$

written $f_1,f_2 \mapsto f_1 \smile f_2$, by the formula

$$f_1 \smile f_2(\sigma_0,\ldots,\sigma_{n+m}) = f_1(\sigma_0,\ldots,\sigma_n) \otimes f_2(\sigma_n,\ldots,\sigma_{n+m}).$$

cup-product

We call $f_1 \smile f_2$ the **cup-product** of f_1 and f_2. Using inhomogeneous cochains instead, this gives a bilinear operation

$$\mathscr{C}^n(G,M_1) \times \mathscr{C}^m(G,M_2) \longrightarrow \mathscr{C}^{n+m}(G,M_1 \otimes M_2),$$

also written $f_1,f_2 \mapsto f_1 \smile f_2$ and also called the cup-product, given by

$$f_1 \smile f_2(\sigma_1,\ldots,\sigma_{n+m}) = f_1(\sigma_1,\ldots,\sigma_n) \otimes \sigma_1 \cdots \sigma_n \cdot f_2(\sigma_{n+1},\ldots,\sigma_{n+m})$$

when $n > 0$ and $m > 0$. When $n = 0$, recalling that $\mathscr{C}^0(G,M_1) = M_1$, we have

$$f_1 \smile f_2(\sigma_1,\ldots,\sigma_m) = f_1 \otimes f_2(\sigma_1,\ldots,\sigma_m);$$

when $m = 0$ we have

$$f_1 \smile f_2(\sigma_1,\ldots,\sigma_n) = f_1(\sigma_1,\ldots,\sigma_n) \otimes \sigma_1 \cdots \sigma_n f_2 .$$

When $n = m = 0$, we have $f_1 \smile f_2 = f_1 \otimes f_2$.

As usual, abstract properties will be established using homogeneous cochains, and concrete calculations will be carried out with inhomogeneous cochains. The next lemma must be given at once.

Lemma 12.34 *Let $f_1 \in C^n(G,M_1)$ and $f_2 \in C^m(G,M_2)$. Then*

$$d(f_1 \smile f_2) = d(f_1) \smile f_2 \; + \; (-1)^n f_1 \smile d(f_2),$$

or in full,

$$d^{n+m}(f_1 \smile f_2) = d^n(f_1) \smile f_2 \; + \; (-1)^n f_1 \smile d^m(f_2).$$

In particular, if f_1 and f_2 are both cocycles, then so is $f_1 \smile f_2$.

Proof. Direct calculation. □

As a result, the cup-products induce bilinear operations

$$\mathrm{H}^n(G,M_1) \times \mathrm{H}^m(G,M_2) \longrightarrow \mathrm{H}^{n+m}(G,M_1 \otimes M_2),$$

also written $x_1, x_2 \mapsto x_1 \smile x_2$ and also called cup-products.

We will now examine several examples, but before we do that, let us point out that typical cases include the situation $M_1 = M_2 = \mathbb{Z}$ or $M_1 = M_2 = \mathbb{F}_p$ for a prime p, with the trivial action. Then $M_1 \otimes M_2$ can be canonically identified with $\mathbb{Z}$ or $\mathbb{F}_p$, as the case may be (with 1 identified with $1 \otimes 1$). So the cup-products take the form

$$\mathrm{H}^n(G,\mathbb{Z}) \times \mathrm{H}^m(G,\mathbb{Z}) \longrightarrow \mathrm{H}^{n+m}(G,\mathbb{Z}),$$

or

$$\mathrm{H}^n(G,\mathbb{F}_p) \times \mathrm{H}^m(G,\mathbb{F}_p) \longrightarrow \mathrm{H}^{n+m}(G,\mathbb{F}_p).$$

Below we shall explain that these operations are associative, and thus define a multiplication on the direct sum of the groups $\mathrm{H}^n(G,\mathbb{Z})$ or $\mathrm{H}^n(G,\mathbb{F}_p)$, for all $n \geq 0$.

Example 12.35 In Lemma 11.11, we have used cup-products in a particular case (you should check this).

Example 12.36 In Example 7.10, we expressed the cohomology class of the extension

$$0 \longrightarrow \mathbb{F}_2 \longrightarrow D_8 \longrightarrow \mathbb{F}_2 \times \mathbb{F}_2 \longrightarrow 0$$

as a cup-product. It follows from Problem 7.5 that there exist two nonzero classes $x, y \in \mathrm{H}^1(D_8,\mathbb{F}_2)$ such that $x \smile y = 0$ (which ones?).

Example 12.37 Let p be a prime number, and let us choose $\varphi \in \mathrm{H}^1(G,\mathbb{F}_p) = \mathrm{Hom}(G,\mathbb{F}_p)$. Let V_φ denote $\mathbb{F}_p^2$ equipped with the action of G given by

$$\begin{pmatrix} 1 & \varphi(\sigma) \\ 0 & 1 \end{pmatrix}$$

for $\sigma \in G$. By construction, we have an exact sequence of $\mathbb{Z}[G]$-modules

$$0 \longrightarrow \mathbb{F}_p \longrightarrow V_\varphi \longrightarrow \mathbb{F}_p \longrightarrow 0.$$

Correspondingly, we have a long exact sequence in cohomology, and in particular connecting homomorphisms

$$\delta\colon \mathrm{H}^n(G,\mathbb{F}_p) \longrightarrow \mathrm{H}^{n+1}(G,\mathbb{F}_p).$$

We claim that in fact

$$\delta(x) = \varphi \smile x.$$

Indeed, suppose $c \in \mathscr{C}^n(G,\mathbb{F}_p)$ is an inhomogeneous cocycle representing x. Letting e_1, e_2 denote the canonical basis of V_φ, we put

$$f(\sigma_1,\dots,\sigma_n) = c(\sigma_1,\dots,\sigma_n)\,e_2 \in V_\varphi.$$

Thus $f \in \mathscr{C}^n(G,V_\varphi)$ is a lift of c. Following the zig-zag lemma, we look at $d^n(f)$; keeping in mind that $d^n(c) = 0$, we find

$$d^n(f)(\sigma_1,\dots,\sigma_{n+1}) = \varphi(\sigma_1)c(\sigma_2,\dots,\sigma_{n+1})\,e_1.$$

The class $\delta(x)$ is, according to the zig-zag lemma, represented by

$$\sigma_1,\dots,\sigma_{n+1} \mapsto \varphi(\sigma_1)c(\sigma_2,\dots,\sigma_{n+1}) \in \mathbb{F}_p,$$

and this is just $\varphi \smile c$. ■

When $p = 2$, this last example fits into a particularly nice result, called the *Arason exact sequence*, or sometimes the *Smith–Gysin sequence*. We digress briefly to discuss this.

Proposition 12.38 (Arason exact sequence) *Let G be a profinite group, let $\varphi \in \mathrm{H}^1(G,\mathbb{F}_2)$ be a nonzero class, and let $H = \ker(\varphi)$, an open subgroup of index 2 in G. Then there is a long exact sequence*

$$\cdots \xrightarrow{\mathrm{Cor}} \mathrm{H}^{n-1}(G,\mathbb{F}_2) \xrightarrow{x \mapsto \varphi \smile x}$$

$$\mathrm{H}^n(G,\mathbb{F}_2) \xrightarrow{\mathrm{Res}} \mathrm{H}^n(H,\mathbb{F}_2) \xrightarrow{\mathrm{Cor}} \mathrm{H}^n(G,\mathbb{F}_2) \xrightarrow{x \mapsto \varphi \smile x} \mathrm{H}^{n+1}(G,\mathbb{F}_2)$$

$$\xrightarrow{\mathrm{Res}} \mathrm{H}^{n+1}(H,\mathbb{F}_2) \longrightarrow \cdots$$

Proof. The module V_φ as in the previous example is now $\mathrm{Ind}_H^G(\mathbb{F}_2)$, and the map $V_\varphi \to \mathbb{F}_2$ is the map N from Definition 12.18. The exact sequence is the long exact sequence in cohomology associated with

$$0 \longrightarrow \mathbb{F}_2 \longrightarrow \mathrm{Ind}_H^G(\mathbb{F}_2) \longrightarrow \mathbb{F}_2 \longrightarrow 0.$$

We have already described the boundary operator as a cup-product. Then Shapiro's isomorphism identifies $\mathrm{H}^n(G, \mathrm{Ind}_H^G(\mathbb{F}_2))$ with $\mathrm{H}^n(H, \mathbb{F}_2)$. When we do that, Lemma 12.17 identifies certain maps in the sequence as restrictions. That the other maps are corestrictions is true by definition. □

Example 12.39 Let $G = C_2$, the group with two elements. We already know that $\mathrm{H}^n(C_2, \mathbb{F}_2) \cong \mathbb{F}_2$ for all $n \geq 0$ from Example 10.3, but now we know much more. Indeed, let $t \in \mathrm{H}^1(C_2, \mathbb{F}_2)$ be the identity (!), so that $\ker(t) = \{1\}$. We apply the Arason exact sequence, and deduce that the cup-product with t is an isomorphism between $\mathrm{H}^n(C_2, \mathbb{F}_2)$ and $\mathrm{H}^{n+1}(C_2, \mathbb{F}_2)$. So $\mathrm{H}^2(C_2, \mathbb{F}_2)$ is generated by $t \smile t$, which we write simply t^2; likewise, in obvious notation, the group $\mathrm{H}^n(C_2, \mathbb{F}_2)$ is generated by t^n. If we put

$$\mathrm{H}^*(C_2, \mathbb{F}_2) = \bigoplus_{n \geq 0} \mathrm{H}^n(C_2, \mathbb{F}_2)$$

(another example of the flexibility of the symbol $*$, see Remark 9.16), then we have

$$\mathrm{H}^*(C_2, \mathbb{F}_2) = \mathbb{F}_2[t],$$

a polynomial ring in one variable.

We now state an omnibus proposition listing the properties of cup-products. One of them is the compatibility with connecting homomorphisms; this is delicate, since an exact sequence

$$0 \longrightarrow M_1' \longrightarrow M_1 \longrightarrow M_1'' \longrightarrow 0$$

may not remain exact after we tensor it with M_2 (when M_2 is not free as a $\mathbb{Z}$-module). The usual way around this is to generalize slightly the cup-product operation. Suppose we have a homomorphism

$$M_1 \otimes M_2 \longrightarrow M_3$$

of $\mathbb{Z}[G]$-modules, which we will call a *pairing*. Then we can consider the composition

$$\mathrm{H}^n(G, M_1) \otimes \mathrm{H}^m(G, M_2) \longrightarrow \mathrm{H}^{n+m}(G, M_1 \otimes M_2) \longrightarrow \mathrm{H}^{n+m}(G, M_3),$$

where the first map is the cup-product defined so far, and the second one is induced by the pairing. The tradition is to also call the composition a cup-product, and to still write $x_1 \smile x_2 \in \mathrm{H}^{n+m}(G, M_3)$. The first point in Proposition 12.40 refers to these "generalized" cup-products.

Proposition 12.40 *The cup-products have the following properties.*

1. Compatibility with connecting homomorphisms. *Suppose*

$$0 \longrightarrow M_1' \longrightarrow M_1 \longrightarrow M_1'' \longrightarrow 0 \text{ and } 0 \longrightarrow M_3' \longrightarrow M_3 \longrightarrow M_3'' \longrightarrow 0$$

are exact sequences of $\mathbb{Z}[G]$-modules. Let M_2 be another $\mathbb{Z}[G]$-module, and suppose we have a pairing $M_1 \otimes M_2 \to M_3$ which induces pairings $M_1' \otimes M_2 \to M_3'$ and $M_1'' \otimes M_2 \to M_3''$. Then for $x_1'' \in \mathrm{H}^n(G, M_1'')$ and $x_2 \in \mathrm{H}^m(G, M_2)$ we have

$$\delta\left(x_1'' \smile x_2\right) = \delta\left(x_1''\right) \smile x_2 .$$

If

$$0 \longrightarrow M_2' \longrightarrow M_2 \longrightarrow M_2'' \longrightarrow 0$$

is another exact sequence of $\mathbb{Z}[G]$-modules, and if we have a pairing $M_1 \otimes M_2 \to M_3$ which induces pairings $M_1 \otimes M_2' \to M_3'$ and $M_1 \otimes M_2'' \to M_3''$, then for $x_1 \in \mathrm{H}^n(G, M_1)$ and $x_2'' \in \mathrm{H}^m(G, M_2'')$ we have

$$(-1)^n \delta\left(x_1 \smile x_2''\right) = x_1 \smile \delta\left(x_2''\right) .$$

2. Compatibility with homomorphisms of modules. *If $h_1 \colon M_1 \to M_1'$ and $h_2 \colon M_2 \to M_2'$ are homomorphisms of $\mathbb{Z}[G]$-modules, then*

$$(h_1 \otimes h_2)_*(x_1 \cup x_2) = (h_1)_*(x_1) \smile (h_2)_*(x_2) .$$

3. Compatibility with homomorphisms of groups. *If $\varphi \colon H \to G$ is a homomorphism of groups, we have*

$$\varphi^*(x_1 \smile x_2) = \varphi^*(x_1) \smile \varphi^*(x_2) .$$

4. Projection formula. *If H is an open subgroup of G, then*

$$\mathrm{Cor}(x \smile \mathrm{Res}(y)) = \mathrm{Cor}(x) \smile y,$$

for all $x \in \mathrm{H}^n(H, M_1)$ and $y \in \mathrm{H}^m(G, M_2)$.

5. Associativity. $(x_1 \smile x_2) \smile x_3 = x_1 \smile (x_2 \smile x_3)$.
6. Graded-commutativity. *Let $\mathrm{sw} \colon M_1 \otimes M_2 \to M_2 \otimes M_1$ be the switch (or swap) homomorphism defined by $\mathrm{sw}(m_1 \otimes m_2) = m_2 \otimes m_1$. Then for $x_1 \in \mathrm{H}^n(G, M_1)$ and $x_2 \in \mathrm{H}^m(G, M_2)$ we have*

$$x_2 \smile x_1 = (-1)^{nm} \mathrm{sw}_*(x_1 \smile x_2) .$$

Sketch. Properties (1), (2), (3), and (5) are all proved by direct computation. Note that the sign in (1) appears essentially because of the sign in Lemma 12.34. Also note that the associativity in (5) actually holds on the level of cochains.

One proves (6) by dimension shifting. The sign appears because of repeated uses of (1). Likewise, property (4) is established by dimension shifting. □

Let $M_1 = M_2 = \mathbb{Z}$, so $M_1 \otimes M_2$ is identified with $\mathbb{Z}$, and the switch homomorphism as in property (6) is the identity. It follows from the proposition that

$$\mathrm{H}^*(G,\mathbb{Z}) := \bigoplus_{n\geq 0} \mathrm{H}^n(G,\mathbb{Z})$$

is a graded-commutative ring. That is, it has a multiplication, given by the cup-products but usually just written $x, y \mapsto xy$, satisfying $yx = (-1)^{nm}xy$, when x has degree n and y has degree m.

Almost identical comments can be made about $\mathrm{H}^*(G,\mathbb{F}_p)$ where p is a prime (or with $\mathrm{H}^*(G,\mathbb{Z}/n\mathbb{Z})$ for any integer n, for that matter, although this is less typical). When $p = 2$, the signs disappear, and we see that $\mathrm{H}^*(G,\mathbb{F}_2)$ is always a commutative ring (as in Example 12.39).

Example 12.41 When p is an odd prime, graded-commutativity implies that the elements of odd degree in the ring $\mathrm{H}^*(G,\mathbb{F}_p)$ square to 0. So for $G = C_p$, we cannot expect the cohomology ring to be the exact analog of the one given for $p = 2$ in Example 12.39 – keeping in mind that $\mathrm{H}^n(C_p,\mathbb{F}_p) \cong \mathbb{F}_p$ for all $n \geq 0$, there are definitely some nonzero elements squaring to 0. In the exercises, we will prove the following. Let $t \in \mathrm{H}^1(C_p,\mathbb{F}_p)$ be nonzero, and let $u \in \mathrm{H}^2(C_p,\mathbb{F}_p)$ be nonzero. Then the group $\mathrm{H}^{2n}(C_p,\mathbb{F}_p)$ is generated by u^n, for all $n \geq 1$, while $\mathrm{H}^{2n+1}(C_p,\mathbb{F}_p)$ is generated by tu^n, for all $n \geq 0$. This describes the mod p cohomology ring entirely, given the relations $t^2 = 0$ and $tu = -ut$, which are guaranteed by graded-commutativity. ■

Let us give an application of cup-products to Galois cohomology. Let F be a field containing a primitive nth root of unity ω, where $n \geq 1$ is an integer. We have defined the symbol

$$(a,b) \in \mathrm{H}^2(F,\mathbb{Z}/n\mathbb{Z})$$

whenever $a, b \in F^\times$ (see Definition 11.14). On the other hand, by Theorem 11.7, there is an isomorphism

$$\mathrm{H}^1(F,\mathbb{Z}/n\mathbb{Z}) = \mathrm{Hom}(\mathrm{Gal}(\overline{F}/F),\mathbb{Z}/n\mathbb{Z}) \cong F^\times/F^{\times n}.$$

As before, we write $\chi_a \colon \mathrm{Gal}(\overline{F}/F) \to \mathbb{Z}/n\mathbb{Z}$ for the homomorphism associated with $a \in F^\times$. We prove now that the symbols are just cup-products.

Proposition 12.42 *For any $a, b \in F^\times$ we have*

$$(a,b) = \chi_a \smile \chi_b .$$

Proof. ([Ser79], chapter XIV, proposition 5.) We are comparing elements of $\mathrm{H}^2(F,\mathbb{Z}/n\mathbb{Z})$. Using our favorite root of unity ω, we consider

$$\iota \colon \mathbb{Z}/n\mathbb{Z} \longrightarrow \overline{F}^\times,$$

given by $\iota(k) = \omega^k$. Theorem 11.7 and its proof show that

$$\iota_*\colon \mathrm{H}^2(F, \mathbb{Z}/n\mathbb{Z}) \longrightarrow \mathrm{H}^2(F, \mathbb{G}_m)$$

is injective (and that its image is the subgroup of n-torsion in $\mathrm{H}^2(F, \mathbb{G}_m)$). We are going to show that $\iota_*(a,b) = \iota_*(\chi_a \smile \chi_b)$. Keep in mind that we will have to deal with cocycles in multiplicative notation.

By definition, we have $\iota_*(a,b) = (\chi_a, b)$, which by Lemma 11.11 is the cup-product $b \smile \delta(\chi_a')$, with the notation as follows. We view b as an element of $\mathrm{H}^0(F, \mathbb{G}_m)$; also, χ_a' is essentially χ_a itself, but viewed as taking its values in $\frac{1}{n}\mathbb{Z}/\mathbb{Z} \subset \mathbb{Q}/\mathbb{Z}$; and δ is the connecting homomorphism associated to the exact sequence

$$0 \longrightarrow \mathbb{Z} \longrightarrow \mathbb{Q} \longrightarrow \mathbb{Q}/\mathbb{Z} \longrightarrow 0.$$

(Incidentally, when we write $b \smile \delta(\chi_a')$, we use that any abelian group A can be identified with $A \otimes_{\mathbb{Z}} \mathbb{Z}$; here this is for $A = \overline{F}^\times$.) Concretely, let

$$\widetilde{\chi}_a\colon \mathrm{Gal}(\overline{F}/F) \longrightarrow \mathbb{Q}$$

be a lift of χ_a'; we have $\widetilde{\chi}_a(\sigma) + \widetilde{\chi}_a(\tau) - \widetilde{\chi}_a(\sigma\tau) \in \mathbb{Z}$ for all σ, τ in the Galois group of F. The class (χ_a, b) is represented by the (inhomogeneous) 2-cocycle

$$\sigma, \tau \mapsto b^{\widetilde{\chi}_a(\sigma) + \widetilde{\chi}_a(\tau) - \widetilde{\chi}_a(\sigma\tau)}.$$

It will be convenient to rewrite this slightly. Let $\sqrt[n]{b}$ be a choice of an nth root of b in $\overline{F}^\times$. Then the cocycle is

$$\sigma, \tau \mapsto \left(\sqrt[n]{b}\right)^{n[\widetilde{\chi}_a(\sigma) + \widetilde{\chi}_a(\tau) - \widetilde{\chi}_a(\sigma\tau)]} = \left(\sqrt[n]{b}\right)^{n\widetilde{\chi}_a(\sigma) + n\widetilde{\chi}_a(\tau) - n\widetilde{\chi}_a(\sigma\tau)}. \qquad (*)$$

We prefer to work with this expression, since $\widetilde{\chi}_a$ takes its values in $\frac{1}{n}\mathbb{Z}$, so $n\widetilde{\chi}_a$ takes its values in $\mathbb{Z}$; when we reduce $n\widetilde{\chi}_a$ mod n, we recover χ_a itself, rather than χ_a'.

Now we turn our attention to $\chi_a \smile \chi_b = -\chi_b \smile \chi_a$. By definition, the class $\iota_*(-\chi_b \smile \chi_a)$ is represented by the 2-cocycle

$$\sigma, \tau \mapsto \omega^{-\chi_b(\sigma)\chi_a(\tau)}.$$

By Theorem 11.7, we also have

$$\omega^{\chi_b(\sigma)} = \frac{\sigma\left(\sqrt[n]{b}\right)}{\sqrt[n]{b}},$$

so our cocycle can be rewritten

$$\sigma, \tau \mapsto \left(\frac{\sigma(\sqrt[n]{b})}{\sqrt[n]{b}}\right)^{-\chi_a(\tau)}. \qquad (**)$$

Now divide (*) by (**). We obtain

$$\sigma, \tau \mapsto \sigma\left(\sqrt[n]{b}\right)^{n\widetilde{\chi}_a(\tau)} \left(\sqrt[n]{b}\right)^{n\widetilde{\chi}_a(\sigma) - n\widetilde{\chi}_a(\sigma\tau)},$$

which is the coboundary of $\sigma \mapsto (\sqrt[n]{b})^{n\tilde{\chi}_a(\sigma)}$. Thus the cocycles given by (*) and (**) define the same cohomology class. □

One may wonder why we did not define (a,b) to be $\chi_a \smile \chi_b$ in the first place, which may appear simpler. However, while this would make the anti-commutativity of the symbol obvious, it would make it hard to prove that $(a,b)=0$ if and only if b is a norm from $F[\sqrt[n]{a}]$ (see (1) of Lemma 11.13).

Remark 12.43 It is possible to extend the definition of cup-products to Tate cohomology in all degrees. We will not do this here, but let us show that we have at least an induced product

$$\widehat{\mathrm{H}}^n(G,M_1)\times\widehat{\mathrm{H}}^m(G,M_2)\longrightarrow\widehat{\mathrm{H}}^{n+m}(G,M_1\otimes M_2)$$

for all $n,m\geq 0$. Indeed, we have $\widehat{\mathrm{H}}^n(G,M)=\mathrm{H}^n(G,M)/\mathrm{Cor}(\mathrm{H}^n(\{1\},M))$ for any module M and for all $n\geq 0$. Thus the projection formula

$$\mathrm{Cor}(x)\smile y=\mathrm{Cor}(x\smile\mathrm{Res}(y))$$

shows that $\mathrm{Cor}(x)\smile y$ is a corestriction, for $x\in\mathrm{H}^n(\{1\},M_1)$ and $y\in\mathrm{H}^n(G,M_2)$. So the product is well-defined. ■

Milnor and Bloch–Kato

We have just enough to state the remarkable conjectures by Milnor and Bloch–Kato, which are now theorems by Rost and Voevodsky. In a nutshell, the cohomology ring

$$\mathrm{H}^*\left(F,\mathbb{F}_p\right)=\bigoplus_{n\geq 0}\mathrm{H}^n\left(F,\mathbb{F}_p\right)$$

can be described completely, and is very simple.

So let F be a field containing a primitive pth root of unity, for a prime p. There are cohomology classes which we know: Indeed, by Theorem 11.7, we have

$$\mathrm{H}^1\left(F,\mathbb{F}_p\right)\cong F^\times/F^{\times p},$$

and thus we understand the classes of degree 1. Now that we have cup-products, we can create more elements of the cohomology ring by simply multiplying away.

What is more, certain simple relations must hold. Let us write χ_a for the element of $\mathrm{H}^1(F,\mathbb{F}_p)$ corresponding to $a\in F^\times$, as we have done so far. By (1) of Lemma 11.18, we have $(a,1-a)=0$; so by Proposition 12.42, we have

$$\chi_a\smile\chi_{1-a}=0. \tag{*}$$

Milnor has conjectured that $\mathrm{H}^*(F,\mathbb{F}_2)$ is generated as a ring by the elements of degree 1, that is by the various χ_a, and moreover, that all the relations between them are consequences of $(*)$. Bloch and Kato have made the analogous conjecture for p odd (and see the remark below for a refinement). This turns out to be true. Rephrasing slightly, we have:

Theorem 12.44 (Rost–Voevodsky) *Let F be a field containing a primitive pth root of unity. For each $a \in F^\times$, let $\ell(a)$ be a formal variable.*

If $p = 2$, then let A be the polynomial ring on the symbols $\ell(a)$, for all $a \in F^\times$; if p is odd, let A be the ring of non-commutative *polynomials on the same variables. Let I be the (two-sided) ideal of A generated by the elements*

$$\ell(ab) - \ell(a) - \ell(b) \quad \text{for } a,b \in F^\times,$$

$$\ell(a)\ell(1-a) \quad \text{for } a \in F \setminus \{0,1\}.$$

Then

$$\mathrm{H}^*\left(F, \mathbb{F}_p\right) \cong A/I.$$

The proof of this result is very difficult indeed, and is a spectacular achievement.

Remark 12.45 Let us write μ_p for $\mu_p(\overline{F})$, the group of pth roots of unity in $\overline{F}$. When we assume that F has a primitive root, we can identify μ_p and $\mathbb{F}_p$; the group $G := \mathrm{Gal}(\overline{F}/F)$ acts trivially on μ_p in this case, since $\mu_p \subset F$, so the isomorphism $\mu_p = \mathbb{F}_p$ is one of G-modules. However, if we do not make this assumption on F, then μ_p may have a nontrivial action of G, so it should not be confused with $\mathbb{F}_p$. What is more, for $i \geq 0$ write

$$\mu_p^{\otimes i} = \mu_p \otimes_{\mathbb{Z}} \mu_p \otimes_{\mathbb{Z}} \cdots \otimes_{\mathbb{Z}} \mu_p \ (p \text{ times}),$$

with $\mu_p^0 = \mathbb{F}_p$. Then $\mu_p^{\otimes i}$ and $\mu_p^{\otimes j}$ may not be isomorphic as G-modules, for some values of i,j.

There is a general version of the Bloch–Kato conjecture, also established by Rost–Voevodsky, which uses the ring

$$\mathrm{H}^*\left(F, \mu_p^{\otimes *}\right) := \bigoplus_{n \geq 0} \mathrm{H}^n\left(F, \mu_p^{\otimes n}\right).$$

The statement is essentially the same. Recall from Theorem 11.7 that we always have

$$\mathrm{H}^1\left(F, \mu_p\right) \cong F^\times / F^{\times p},$$

assuming merely that the characteristic of F is not p. ■

Problems

12.1.
1. Let G be a finite group, and let M be a $\mathbb{Z}[G]$-module such that multiplication by $|G|$ is an isomorphism on M (for example, M might be a $\mathbb{Q}$-vector space). Show that $\widehat{\mathrm{H}}^n(G,M) = 0$ for all $n \in \mathbb{Z}$.
2. Deduce that there are canonical isomorphisms

$$\widehat{\mathrm{H}}^{n+1}(G,\mathbb{Z}) \cong \widehat{\mathrm{H}}^n(G,\mathbb{Q}/\mathbb{Z}) \cong \widehat{\mathrm{H}}^n(G,\mathbb{C}^\times),$$

and in particular

$$\mathrm{H}^2(G,\mathbb{Z}) \cong \mathrm{Hom}(G,\mathbb{Q}/\mathbb{Z}) \cong \mathrm{Hom}(G,\mathbb{C}^\times).$$

3. Generalize to profinite groups.

12.2. Let F be a field of characteristic $\neq 2$, and let $a,b \in F^\times$. Use results scattered in the book to show (without much extra work, normally) the equivalence of the properties below:

1. The quaternion algebra $(a,b)_F$ is isomorphic to $\mathrm{M}_2(F)$.
2. The Hilbert symbol $(a,b)_F \in \mathrm{H}^2(F,\mathbb{F}_2)$ is zero.
3. b is a norm from $F[\sqrt{a}]$.
4. a is a norm from $F[\sqrt{b}]$.
5. There exist $x,y \in F$ such that $ax^2 + by^2 = 1$.
6. There exists a finite Galois extension L/F with $F[\sqrt{a},\sqrt{b}] \subset L$, and a homomorphism $\mathrm{Gal}(L/F) \to D_8$, such that the composition

$$\mathrm{Gal}(\overline{F}/F) \longrightarrow \mathrm{Gal}(L/F) \longrightarrow D_8 \longrightarrow \mathbb{F}_2 \times \mathbb{F}_2$$

is (χ_a, χ_b). (Here χ_a and χ_b correspond to the classes of a and b respectively when we use Kummer theory to identify $F^\times/F^{\times 2}$ with $\mathrm{Hom}(\mathrm{Gal}(\overline{F}/F),\mathbb{F}_2)$. Also D_8 is the dihedral group of order 8, and the map $\pi : D_8 \to \mathbb{F}_2 \times \mathbb{F}_2$ is that described in Example 7.10.)

Then, explain how to recover Proposition 1.28 as a particular case.

You should make sure that your arguments include the case when a or b is a square in F, or when a = b. Also note that we have used the equation $ax^2 + by^2 = 1$, which has a nice symmetry in a and b, but it is of course closely related to the equation $x^2 - ay^2 = b$, which appears elsewhere in the book.

12.3. Let p be an odd prime. In this problem, we show that the cohomology ring $\mathrm{H}^*(C_p,\mathbb{F}_p)$ is as described in Example 12.41. Let $t \in \mathrm{H}^1(C_p,\mathbb{F}_p)$ be a nonzero element.

1. Show that taking cup-products with t gives isomorphisms

$$\mathrm{H}^{2n}(C_p,\mathbb{F}_p) \longrightarrow \mathrm{H}^{2n+1}(C_p,\mathbb{F}_p)$$

for all $n \geq 0$.
Hint: Example 12.37.

2. For any (pro)finite group G, consider the exact sequence of $\mathbb{Z}[G]$-modules

$$0 \longrightarrow \mathbb{Z}/p\mathbb{Z} \longrightarrow \mathbb{Z}/p^2\mathbb{Z} \longrightarrow \mathbb{Z}/p\mathbb{Z} \longrightarrow 0,$$

where the G-actions are all trivial. (The first map is induced by multiplication by p.) The connecting homomorphisms are written

$$\beta_n \colon \mathrm{H}^n(G,\mathbb{F}_p) \longrightarrow \mathrm{H}^{n+1}(G,\mathbb{F}_p),$$

and called *Bockstein homomorphisms*. One usually considers "the" Bockstein homomorphism β which is defined on $\mathrm{H}^*(G,\mathbb{F}_p) = \bigoplus_n \mathrm{H}^n(G,\mathbb{F}_p)$ by using the various β_n.

In the case of $G = C_p$, show that the Bockstein yields an isomorphism between the groups $\mathrm{H}^{2n+1}(C_p, \mathbb{F}_p)$ and $\mathrm{H}^{2n+2}(C_p, \mathbb{F}_p)$, for all $n \geq 0$. On the other hand, show that the Bockstein between $\mathrm{H}^{2n}(G, \mathbb{F}_p)$ and $\mathrm{H}^{2n+1}(G, \mathbb{F}_p)$ is zero.

3. (Not used in the sequel.) For a general finite group G, show that $\beta \circ \beta = 0$.
Hint: Look at

$$\begin{array}{ccccccccc} 0 & \longrightarrow & \mathbb{Z} & \xrightarrow{x \mapsto px} & \mathbb{Z} & \longrightarrow & \mathbb{Z}/p\mathbb{Z} & \longrightarrow & 0 \\ & & \downarrow & & \downarrow & & \downarrow & & \\ 0 & \longrightarrow & \mathbb{Z}/p\mathbb{Z} & \longrightarrow & \mathbb{Z}/p^2\mathbb{Z} & \longrightarrow & \mathbb{Z}/p\mathbb{Z} & \longrightarrow & 0. \end{array}$$

Use a "naturality" property of connecting homomorphisms.

4. For $x \in \mathrm{H}^n(G, \mathbb{F}_p)$ and $y \in \mathrm{H}^m(G, \mathbb{F}_p)$, show that

$$\beta(xy) = \beta(x)y + (-1)^n x\beta(y).$$

One says that the Bockstein is a derivation "in the graded sense".
In the case of $G = C_p$, put $u = \beta(t)$, and compute $\beta(tu^n)$.

5. Conclude that $u^n \neq 0$ and $tu^n \neq 0$, for all $n \geq 0$.

12.4. Let K/F be an extension of local number fields. Comparing Proposition 12.20 from this chapter, and Theorem 8.10, prove that "corestriction is compatible with Hasse invariants"; that is, the following diagram commutes:

$$\begin{array}{ccc} \mathrm{H}^2(F, \mathbb{G}_m) & \xrightarrow{\mathrm{Inv}_F} & \mathbb{Q}/\mathbb{Z} \\ {\scriptstyle \mathrm{Cor}}\uparrow & & \downarrow{\scriptstyle =} \\ \mathrm{H}^2(K, \mathbb{G}_m) & \xrightarrow{\mathrm{Inv}_K} & \mathbb{Q}/\mathbb{Z} \end{array}$$

Deduce that the corestriction is an isomorphism $\mathrm{H}^2(K, \mathbb{G}_m) \to \mathrm{H}^2(F, \mathbb{G}_m)$, and also that for each $n \geq 1$, it induces an isomorphism $\mathrm{H}^2(K, \mu_n) \to \mathrm{H}^2(F, \mu_n)$.

Part IV

Class field theory

Class field theory is a loosely defined body of results pertaining to the abelian extensions of either number fields (that is, finite extensions of $\mathbb{Q}$), or local number fields. Some authors may write "by class field theory, the Brauer group of a local number field is isomorphic to $\mathbb{Q}/\mathbb{Z}$", and so by that understanding of the name, we are already knee-deep in class field theory. More typically, though, the term refers to the theorems which we state in the first section of the next chapter, in the local case, before turning to the proofs. The arguments rely on almost all of the material previously introduced in the book.

In the last chapter we shall study some of the properties of number fields, thus taking a first step into "algebraic number theory". We shall present, without proof, some statements of class field theory in "the global case" (as it is called, a terminology we shall try to explain).

Many readers will already know some of the classical facts about number fields. Indeed, the prerequisites for algebraic number theory are about the same as those for the present book, and you may very well have been studying this subject at the same time as you were reading the preceding chapters, or even before. In an attempt to tell you something that you do not already know, we have emphasized *completions* much more than is usual (especially as a starting point). We are able to go quite far in relatively few pages, and we include the highly nontrivial *Kronecker–Weber theorem* about abelian extensions of $\mathbb{Q}$. Chapter 13 will pave the way with a "local Kronecker–Weber theorem", in which $\mathbb{Q}_p$ replaces $\mathbb{Q}$.

13 Local class field theory

In this chapter, we state and prove the main results of class field theory for local number fields. We also encounter *Tate duality* along the way, which is a significant manifestation of the theory.

We conclude with the most classical – and also spectacular – application, which is the *local Kronecker–Weber theorem*, asserting that any abelian extension of $\mathbb{Q}_p$ is contained in a cyclotomic extension.

Statements

A most definitive result is the following.

Theorem 13.1 (Fundamental theorem of local class field theory) *Let F be a local number field, and let $\overline{F}$ be an algebraic closure of F. There exists a one-to-one, order-reversing correspondence between the finite abelian extensions of F contained in $\overline{F}$ and the subgroups of finite index in $F^\times$. The correspondence maps L/F to $\mathrm{N}_{L/F}(L^\times)$.*

When the base field F is understood, it is convenient to use the notation $N_L := \mathrm{N}_{L/F}(L^\times)$. Note that, when A is a subgroup of finite index of $F^\times$, the unique field L such that $N_L = A$ is called the *class field* of A.

Since the bijection of the theorem is order-reversing, we deduce immediately a couple of identities. Suppose L_1 and L_2 are two abelian extensions of F. As the composite L_1L_2 is the smallest abelian extension containing both L_1 and L_2, the group $N_{L_1L_2}$ must be the largest group contained in both N_{L_1} and N_{L_2}, namely $N_{L_1} \cap N_{L_2}$. That is,

$$N_{L_1L_2} = N_{L_1} \cap N_{L_2} . \tag{13.1}$$

For similar, formal reasons, we have

$$N_{L_1 \cap L_2} = N_{L_1} N_{L_2} . \tag{13.2}$$

These results are complemented by:

Theorem 13.2 (Reciprocity isomorphism) *Let L/F be any finite Galois extension of local number fields. Then there exists an isomorphism*

$$\mathrm{Gal}(L/F)^{ab} \cong F^\times / \mathrm{N}_{L/F}\left(L^\times\right).$$

In particular, when L/F is abelian, we have $\mathrm{Gal}(L/F) \cong F^\times/N_L$, and the index of N_L in $F^\times$ is $[L : F]$.

This generalizes what we knew of unramified extensions (Proposition 8.4).

To establish these results, the road goes somewhat backwards. In this chapter, we shall first prove Tate's theorem, a fact about group cohomology, from which the existence of the reciprocity isomorphism will be immediate. Using this, and some extra information obtained along the way, we prove equation (13.1). Next, a version of the "fundamental theorem" follows, with "subgroups of finite index" replaced by "subgroups of the form N_L".

Tate duality, mentioned above, can then be stated and proved, and it is instrumental in the sequel. The final step is to prove the "Existence theorem", which for any subgroup A of finite index in $F^\times$ shows the existence of an abelian extension L/F with $N_L = A$.

Cohomological triviality and equivalence

At the heart of class field theory is the theorem of Tate (sometimes called the "Tate and Nakayama lemma", but this should normally refer to a strengthened version). It is purely a fact about the cohomology of finite groups, and indeed, emerges during the study of a natural question: What can we say of modules such that $\widehat{\mathrm{H}}^r(S,M) = 0$ for all r and all subgroups S of G? We give some answers to this now, which of course could have been included in the previous part of this book. However, many readers who were already familiar with group cohomology will have skipped that part altogether, so we found it more fitting here. Besides, and this is something to wonder about, even if the results to be presented now have a strong feel of generality and power, the only applications have been to class field theory.

Here we follow [Bab72], sections 36 and 37.

cohomologically trivial

Definition 13.3 Let G be a finite group, and let M be a $\mathbb{Z}[G]$-module. We call M **cohomologically trivial** when $\widehat{\mathrm{H}}^r(S,M) = 0$ for every subgroup S of G and every $r \in \mathbb{Z}$. ■

Our first goal is to prove the theorem of Nakayama and Tate, which asserts that, if there exist two successive integers $r, r+1$ such that $\widehat{\mathrm{H}}^r(S,M) = \widehat{\mathrm{H}}^{r+1}(S,M) = 0$ for all S, then M is cohomologically trivial.

Lemma 13.4 *Let G be a finite group and M be a $\mathbb{Z}[G]$-module.*

1. *If $\widehat{\mathrm{H}}^0(S,M) = \mathrm{H}^1(S,M) = 0$ for all subgroups S of G, then $\mathrm{H}^2(S,M) = 0$ for all S.*
2. *If $\mathrm{H}^1(S,M) = \mathrm{H}^2(S,M) = 0$ for all subgroups S of G, then $\widehat{\mathrm{H}}^0(S,M) = 0$ for all S.*

Proof. First we observe that it is enough to prove the result when G is an ℓ-group, where ℓ is an arbitrary prime. (We save the letter p for p-adic fields.) Indeed, suppose we had covered this case, and let us turn to the general situation; so suppose that the assumption of (1) is in force for the arbitrary finite group G, let S be an arbitrary subgroup, and let us show that $\mathrm{H}^2(S,M) = 0$. We can apply the lemma to S_ℓ, an ℓ-Sylow of S; when we do, we come to the conclusion that $\mathrm{H}^2(S_\ell,M) = 0$. Now, $\mathrm{H}^2(S,M)$ injects into the sum of the various $\mathrm{H}^2(S_\ell,M)$, by Corollary 12.22 and the subsequent remark, so $\mathrm{H}^2(S,M) = 0$ as required. We treat (2) similarly.

So suppose G is an ℓ-group, and proceed by induction on its order. There is nothing to prove if $|G| = 1$. Assume now the result of the lemma for all ℓ-groups of order less than that of G. We start with (1), so we assume that $\widehat{\mathrm{H}}^0(S,M) = \mathrm{H}^1(S,M) = 0$ for all subgroups S of G. If S is such a subgroup with $|S| < |G|$, the induction hypothesis applies, showing that $\mathrm{H}^2(S,M) = 0$. We only need to prove that $\mathrm{H}^2(G,M) = 0$ also.

Choose a normal subgroup N of G of index ℓ (by the basic theory of ℓ-groups, this is possible). We use the exact sequence from Lemma 12.32:

$$\widehat{\mathrm{H}}^0(N,M) \longrightarrow \widehat{\mathrm{H}}^0(G,M) \longrightarrow \widehat{\mathrm{H}}^0\left(G/N,M^N\right) \longrightarrow 0. \tag{*}$$

Here $\widehat{\mathrm{H}}^0(G,M) = 0$ by assumption, so $\widehat{\mathrm{H}}^0(G/N,M^N) = 0$. However, since G/N is cyclic, we also have $\widehat{\mathrm{H}}^2(G/N,M^N) = 0$, by periodicity (Example 10.5). Now as $\mathrm{H}^1(N,M) = 0$, also by assumption, we have another exact sequence at our disposal from Corollary 12.31, to wit:

$$0 \longrightarrow \mathrm{H}^2\left(G/N,M^N\right) \longrightarrow \mathrm{H}^2(G,M) \longrightarrow \mathrm{H}^2(N,M). \tag{**}$$

By induction, we have $\mathrm{H}^2(N,M) = 0$, and we have just established that $\mathrm{H}^2(G/N,M^N) = 0$, so we have proved (1).

The argument for (2) is similar. Suppose $\mathrm{H}^1(S,M) = \mathrm{H}^2(S,M) = 0$ for all subgroups S of G, by induction it remains to prove that $\widehat{\mathrm{H}}^0(G,M) = 0$. By (**) we have $\mathrm{H}^2(G/N,M^N) = 0$, so $\widehat{\mathrm{H}}^0(G/N,M^N) = 0$ by periodicity. Since $\widehat{\mathrm{H}}^0(N,M) = 0$ by induction, (*) shows that $\widehat{\mathrm{H}}^0(G,M) = 0$ as requested. □

We need a little lemma about "dimension shifting" for all subgroups of G simultaneously.

Lemma 13.5 *Suppose G is a finite group and M is a $\mathbb{Z}[G]$-module. Let $r \in \mathbb{Z}$ be an integer. Then there exists a $\mathbb{Z}[G]$-module $M(-r)$ with $\widehat{\mathrm{H}}^n(S,M(-r)) = \widehat{\mathrm{H}}^{n+r}(S,M)$ for any $n \in \mathbb{Z}$ and any subgroup S of G.*

Proof. Let us simply scrutinize the argument that allows us to perform dimension shifting, namely the proof of Corollary 12.9. To shift in one direction, we use a free $\mathbb{Z}[G]$-module M'' which maps onto M; by Lemma 12.1, the module M'' is also free as a $\mathbb{Z}[S]$-module for any S, so the shift works for all S at the same time.

To shift in the other direction, we embed M into $M' = \mathrm{Ind}_{\{1\}}^G(M)$. We have already observed, during the proof of Corollary 12.31, the elementary fact that M', viewed as a $\mathbb{Z}[S]$-module, is a direct sum of $[G:S]$ copies of $\mathrm{Ind}_{\{1\}}^{S}(M)$, so $\mathrm{H}^n(S,M')=0$ for all $n\in\mathbb{Z}$ by Shapiro. Hence, in this case as well as in the previous one, the dimension shift will work for all S simultaneously. □

As promised, we now have:

Theorem 13.6 (Nakayama–Tate) *Suppose there exists an integer $r\in\mathbb{Z}$ such that, for any subgroup S of G, one has $\widehat{\mathrm{H}}^r(S,M)=\widehat{\mathrm{H}}^{r+1}(S,M)=0$. Then M is cohomologically trivial.*

Proof. By the last lemma, we find a module $M(-r)$ such that, for any subgroup S, we have $\widehat{\mathrm{H}}^n(S,M(-r)) = \widehat{\mathrm{H}}^{n+r}(S,M)$ for all integers n. As a result $\widehat{\mathrm{H}}^0(S,M(-r))=\widehat{\mathrm{H}}^1(S,M(-r))=0$ for all S, and by (1) of Lemma 13.4, this extends to $\widehat{\mathrm{H}}^2(S,M(-r))=0$. Thus we see that $\widehat{\mathrm{H}}^n(S,M)=0$ for $n=r,r+1,r+2$. Iterating, we have $\widehat{\mathrm{H}}^n(S,M)=0$ for all $n\geq r$.

Similarly, using (2) of the lemma to extend on the left rather than on the right, we have $\widehat{\mathrm{H}}^n(S,M)=0$ for all $n\leq r$. □

cohomological equivalence

Definition 13.7 Let $f\colon M\to N$ be a homomorphism of $\mathbb{Z}[G]$-modules. We call f a **cohomological equivalence** when the induced map $f_r\colon \widehat{\mathrm{H}}^r(S,M)\to\widehat{\mathrm{H}}^r(S,N)$ is an isomorphism, for each $r\in\mathbb{Z}$ and each subgroup S of G. ■

Lemma 13.8 *Let $f\colon M\to N$ as above. Then there exists a $\mathbb{Z}[G]$-module Q and maps δ_*, g_* such that for any subgroup S of G, the sequence*

$$\cdots \longrightarrow \mathrm{H}^{r-1}(S,Q) \xrightarrow{\delta_{r-1}} \widehat{\mathrm{H}}^r(S,M) \xrightarrow{f_r} \widehat{\mathrm{H}}^r(S,N)$$

$$\xrightarrow{g_r} \widehat{\mathrm{H}}^r(S,Q) \xrightarrow{\delta_r} \widehat{\mathrm{H}}^{r+1}(S,M) \xrightarrow{f_{r+1}} \cdots$$

is exact.

Proof. Let $\iota\colon M\longrightarrow I$ be an embedding of M into an induced $\mathbb{Z}[G]$-module, which is also an induced $\mathbb{Z}[S]$-module for each subgroup S of G. The map $(f,\iota)\colon M\longrightarrow N\oplus I$ is also injective; we call M' its image, and define $Q=(N\oplus I)/M'$. Consider the long exact sequence associated with

$$0 \longrightarrow M \xrightarrow{(f,\iota)} N\oplus I \xrightarrow{q} Q \longrightarrow 0.$$

The connecting homomorphism δ_* from that long exact sequence is the one we want, and we merely need to point out that $\widehat{\mathrm{H}}^n(S,N\oplus I)\cong\widehat{\mathrm{H}}^n(S,N)$, although some care is needed as we really want the map f_* itself to show up in our exact sequence.

So observe that

$$\pi_n\colon \widehat{\mathrm{H}}^n(S,N\oplus I) \xrightarrow{\cong} \widehat{\mathrm{H}}^n(S,N),$$

induced by the projection $\pi: N\oplus I\to N$, is an isomorphism; indeed, this follows from the long exact sequence for the split sequence

$$0 \longrightarrow I \longrightarrow N \oplus I \longrightarrow N \longrightarrow 0,$$

keeping in mind that $\widehat{\mathrm{H}}^n(S,I) = 0$ for all n since I is induced. As we have $\pi \circ (f,\iota) = f$, we have $(f,\iota)_* = \pi_*^{-1} \circ f_*$. If we define $g_* = q_* \circ \pi_*^{-1}$, we are done. □

Lemma 13.9 (Tate's criterion) *Let $f\colon M \to N$ be a homomorphism. Then f is a cohomological equivalence if and only if there exists an integer $r \in \mathbb{Z}$ such that, for every subgroup S of G, we have*

1. $f_{r-1}\colon \widehat{\mathrm{H}}^{r-1}(S,M) \to \widehat{\mathrm{H}}^{r-1}(S,N)$ *is surjective,*
2. $f_r\colon \widehat{\mathrm{H}}^{r}(S,M) \to \widehat{\mathrm{H}}^{r}(S,N)$ *is an isomorphism,*
3. $f_{r+1}\colon \widehat{\mathrm{H}}^{r+1}(S,M) \to \widehat{\mathrm{H}}^{r+1}(S,N)$ *injective.*

Proof. The conditions are obviously necessary, so let us assume that they hold and prove that f is a cohomological equivalence. Let Q be as in the previous lemma, and let us show that Q is cohomologically trivial, which of course is enough. By the Nakayama–Tate theorem, it suffices to show that $\widehat{\mathrm{H}}^{r-1}(S,Q) = \widehat{\mathrm{H}}^{r}(S,Q)$ for each subgroup S.

Indeed, since f_{r-1} is surjective, the map g_{r-1} is 0; by exactness, the group $\widehat{\mathrm{H}}^{r-1}(S,Q)$ is isomorphic to the kernel of f_r. The latter being injective, we see that $\widehat{\mathrm{H}}^{r-1}(S,Q) = 0$. The same argument works with r replacing $r-1$, so we are done. □

Proposition 13.10 *Let M be a $\mathbb{Z}[G]$-module. Assume that for any subgroup $S \subset G$, the group $\widehat{\mathrm{H}}^0(S,M)$ is cyclic of order $|S|$, and $\widehat{\mathrm{H}}^{-1}(S,M) = 0$. Then there exists a homomorphism $f\colon \mathbb{Z} \to M$ such that f is a cohomological equivalence.*

More precisely, for any $m \in M^G = \mathrm{H}^0(G,M)$ which maps to a generator of $M^G/\mathrm{N}_G(M) = \widehat{\mathrm{H}}^0(G,M)$, the homomorphism f defined by $f(1) = m$ is a cohomological equivalence.

Proof. Let m and f be as proposed. We wish to use Tate's criterion with $r = -1$, in order to show that $f\colon \mathbb{Z} \to M$ is a cohomological equivalence. Two out of three verifications are trivial. Surely f_* is surjective in degree -1, since its target is the zero group by assumption; and surely f_* is injective in degree 2, since its source is $\mathrm{H}^1(S,\mathbb{Z}) = \mathrm{Hom}(S,\mathbb{Z}) = 0$ since S is finite. It remains to check that f_* is an isomorphism in degree 0.

Let μ be the image of m in $\widehat{\mathrm{H}}^0(G,M)$. First we claim that $\mu_S := \mathrm{Res}_{G,S}(\mu)$ is a generator of the cyclic group $\widehat{\mathrm{H}}^0(S,M)$, for any subgroup S. Indeed, suppose the order of μ_S is d, a divisor of $|S|$. Then, from the relation

$$0 = \mathrm{Cor}_{G,S}(d\mu_S) = d\,\mathrm{Cor}_{G,S} \circ \mathrm{Res}_{G,S}(\mu) = \frac{d|G|}{|S|}\mu,$$

as in Proposition 12.20, we deduce that $d = |S|$, proving the claim.

Now, the group $\widehat{\mathrm{H}}^0(S,\mathbb{Z})$ is cyclic of order $|S|$ generated by the image of $1 \in \mathbb{Z}$ (this is true of any group S), and $\widehat{\mathrm{H}}^0(S,M)$ is cyclic of the same order, generated by μ_S as just established. If we keep in mind that μ_S is none other than the image of m in $\widehat{\mathrm{H}}^0(S,M)$, then it is clear that f_* in degree 0 maps 1 to μ_S, so it is an isomorphism. □

Theorem 13.11 (Tate) *Let G be a finite group, and let M be a $\mathbb{Z}[G]$-module. Assume that for any subgroup $S \subset G$, the group $\mathrm{H}^2(S,M)$ is cyclic of order $|S|$, and $\mathrm{H}^1(S,M) = 0$. Then for any $r \in \mathbb{Z}$ and any subgroup S of G, there exists an isomorphism $\widehat{\mathrm{H}}^{r-2}(S,\mathbb{Z}) \cong \widehat{\mathrm{H}}^r(S,M)$. In fact we can find a family of isomorphisms*

$$\theta_{G,S,M,r} : \widehat{\mathrm{H}}^{r-2}(S,\mathbb{Z}) \xrightarrow{\cong} \widehat{\mathrm{H}}^r(S,M),$$

with the following compatibility: Whenever $S' \subset S \subset G$, we have a commutative diagram for each $r \in \mathbb{Z}$:

$$\begin{array}{ccc} \widehat{\mathrm{H}}^{r-2}(S',\mathbb{Z}) & \xrightarrow{\theta_{G,S',M,r}} & \widehat{\mathrm{H}}^r(S',M) \\ \mathrm{Cor}\downarrow & & \downarrow\mathrm{Cor} \\ \widehat{\mathrm{H}}^{r-2}(S,\mathbb{Z}) & \xrightarrow{\theta_{G,S,M,r}} & \widehat{\mathrm{H}}^r(S,M). \end{array}$$

Proof. Of course the idea is to shift dimensions twice, and apply the preceding proposition. That is, we invoke Lemma 13.5 to find a module $M(-2)$ such that we have isomorphisms $\mathrm{H}^n(S,M) \cong \mathrm{H}^{n-2}(S,M(-2))$. The previous proposition gives the desired isomorphisms between the groups $\mathrm{H}^{r-2}(S,\mathbb{Z})$ and $\mathrm{H}^{r-2}(S,M(-2))$.

Since the resulting isomorphisms $\mathrm{H}^{r-2}(S,\mathbb{Z}) \cong \mathrm{H}^r(S,M)$ are obtained as the composition of a homomorphism induced by some $f : \mathbb{Z} \to M(-2)$, followed by connecting homomorphisms, it is clear that they commute with corestriction. □

Remark 13.12 In fact, we can obtain more precise statements, with some care. First, consider the isomorphisms of Proposition 13.10. One can check that, for a subgroup S, the isomorphism is in fact given by the cup-product with the class of $m \in M^S$ in $\widehat{\mathrm{H}}^0(S,M)$; this is the element written μ_S during the proof. Indeed, this is a simple matter of checking the definitions (in an easy setting since we are dealing with a class of degree 0).

Similarly, let us consider Theorem 13.11. Let $\xi \in \mathrm{H}^2(G,M)$ be a generator. Then the isomorphism $\theta_{G,S}$ constructed in the proof is the cup-product with ξ_S, the restriction of ξ (simply review the relation between cup-products and connecting homomorphisms). In this book we have not studied cup-products in Tate cohomology (that is, not in negative degrees), so the novice should ignore this remark on a first reading. The subsequent results in this chapter do not depend on it. ■

The reciprocity isomorphisms

Let us explain at once how to deduce Theorem 13.2, the existence of the "reciprocity isomorphism", from Tate's theorem. For this, we let L/F be a finite Galois extension of local number fields, put $G = \mathrm{Gal}(L/F)$, and $M = L^\times$. Any subgroup is of the form $S = \mathrm{Gal}(L/K)$. The fact that $\mathrm{H}^1(\mathrm{Gal}(L/K),L^\times) = 0$ is Hilbert's Theorem 90, and that $\mathrm{H}^2(\mathrm{Gal}(L/K),L^\times) \cong \mathrm{Br}(L/K)$ is cyclic of order $|\mathrm{Gal}(L/K)|$ is Corollary 8.11. So Tate's theorem applies. For $S = G$ and $r = 0$ we obtain an isomorphism between

$$\widehat{\mathrm{H}}^0\left(\mathrm{Gal}(L/F),L^\times\right) = F^\times/\mathrm{N}_{L/F}\left(L^\times\right)$$

and

$$\widehat{\mathrm{H}}^{-2}(\mathrm{Gal}(L/F),\mathbb{Z}) = \mathrm{H}_1(\mathrm{Gal}(L/F),\mathbb{Z}) \cong \mathrm{Gal}(L/F)^{ab}.$$

This completes the proof of Theorem 13.2.

It is convenient to introduce some notation for the various reciprocity isomorphisms. Let the local number field F be fixed (and an algebraic closure $\overline{F}$ chosen). For any finite extension L/F, with $L \subset \overline{F}$, we apply Theorem 13.11 as just explained, and obtain *some* family of isomorphisms $\mathrm{Gal}(L/K)^{ab} \cong K^\times/\mathrm{N}_{L/K}(L^\times)$, for all intermediate fields K, with the compatibility described in that theorem. We comment below (see Remark 13.15) on the fact that this is not canonical, and depends on choices; none of our development is affected by this, though.

In fact, as it turns out, it is best to have a special symbol for the map

$$K^\times \longrightarrow K^\times/\mathrm{N}_{L/K}\left(L^\times\right) \longrightarrow \mathrm{Gal}(L/K)^{ab},$$

where the first arrow is the natural quotient map, and the second is the isomorphism just chosen. We follow the tradition and write it $x \mapsto (x,L/K)$. It is called the *local norm residue map*, while the element $(x,L/K)$, sometimes also written $\left(\dfrac{L/K}{x}\right)$, is called the *local norm residue symbol*, or *Artin symbol*, of x.

Lemma 13.13 *Suppose $F \subset K \subset L$, with L/F finite and Galois. Then there is a commutative diagram*

$$\begin{array}{ccc} K^\times & \xrightarrow{x\mapsto(x,L/K)} & \mathrm{Gal}(L/K)^{ab} \\ {\scriptstyle \mathrm{N}_{K/F}}\downarrow & & \downarrow \\ F^\times & \xrightarrow{x\mapsto(x,L/F)} & \mathrm{Gal}(L/F)^{ab}, \end{array}$$

where the right vertical map is induced by the inclusion.

Proof. This is a direct translation of the naturality statement in Tate's theorem. □

Lemma 13.14 *Suppose $F \subset K \subset L$, with L/F and K/F both Galois, and L/F finite. Consider the diagram*

$$\begin{array}{ccc} F^\times & \xrightarrow{x\mapsto(x,L/F)} & \mathrm{Gal}(L/F)^{ab} \\ =\downarrow & & \downarrow \\ F^\times & \xrightarrow{x\mapsto(x,K/F)} & \mathrm{Gal}(K/F)^{ab}, \end{array}$$

where the right vertical map is induced by the restriction. Then the two homomorphisms $F^\times \to \mathrm{Gal}(K/F)^{ab}$ obtained from this diagram have the same kernel.

Proof. Both homomorphisms are surjective, and $\mathrm{Gal}(K/F)^{ab}$ is certainly finite, so by comparing indices we see that we only need to prove an inclusion between the kernels. Going down and then right, we see that the kernel of $x \mapsto (x, K/F)$ is N_K. By the previous lemma, the group N_K is taken by $x \mapsto (x, L/F)$ to that subgroup of $\mathrm{Gal}(L/F)^{ab}$ which is the image of $\mathrm{Gal}(L/K)^{ab} \to \mathrm{Gal}(L/F)^{ab}$; elements in that subgroup restrict to the identity of $\mathrm{Gal}(K/F)^{ab}$. Thus N_K is included in the kernel of the homomorphism obtained by going right first and then down. □

Remark 13.15 Following up on Remark 13.12, we can be a little more precise. After a translation in the language of field theory, what we have seen is that it is enough to choose a generator $\xi \in \mathrm{H}^2(\mathrm{Gal}(L/F), L^\times)$ to obtain, via the cup-products with appropriate restrictions of ξ, a family of isomorphisms $\mathrm{Gal}(L/K)^{ab} \cong K^\times / \mathrm{N}_{L/K}(L^\times)$ for all intermediate fields $F \subset K \subset L$. In this book we have defined an explicit isomorphism

$$\mathrm{H}^2\left(F, \overline{F}^\times\right) \longrightarrow \mathbb{Q}/\mathbb{Z},$$

and if $n = [L : F]$, then $\mathrm{H}^2(\mathrm{Gal}(L/F), L^\times)$ was naturally identified with the n-torsion subgroup of $\mathbb{Q}/\mathbb{Z}$, that is, $\frac{1}{n}\mathbb{Z}/\mathbb{Z}$. So we may pick ξ to be the element of $\mathrm{H}^2(\mathrm{Gal}(L/F), L^\times)$ mapping to the generator $\frac{1}{n}$.

This choice is not necessarily universal (see [Ser79], chapter X, section 5, remark after lemma 1), but at least it gives consistency. Indeed, it follows from Theorem 8.10 that the "canonical choice" of generator for L/F, as just defined, restricts to the "canonical choice" for K/F, for any intermediate field K. In the notation of Theorem 13.11, we have $\theta_{G,S} = \theta_{S,S}$, when we systematically make this choice.

As the reader will have guessed, when we have arranged things this way, the diagram in Lemma 13.14 actually commutes. However, proving this takes a surprising amount of work. The interested reader can consult [Ser79], where this is presented as a consequence of chapter XI, proposition 2; the proof of the proposition relies on the appendix to that chapter, and on material on group cohomology which we have not developed. By contrast, our proof of Lemma 13.14 was trivial.

In this book, we do present a (complete) construction of canonical reciprocity isomorphisms – see the discussion after Theorem 13.29. However, this is done after the main theorems of local class field theory have been established. ■

Norm subgroups

We continue to assume that F is a local number field. Any subgroup of $F^\times$ of the form N_L for some finite Galois extension L/F will be called a *norm subgroup*. Such a subgroup has finite index in $F^\times$, by Theorem 13.2.

Lemma 13.16 *Let L/F be finite and Galois, and let L^{ab}/F be the largest abelian extension contained in L. Then $N_L = N_{L^{ab}}$. As a result, norm subgroups can be defined using abelian extensions only.*

Proof. Apply Lemma 13.14 with $K = L^{ab}$, noting that, almost by definition, $\mathrm{Gal}(L^{ab}/F) = \mathrm{Gal}(L/F)^{ab}$. The two homomorphisms in the lemma have, as their kernels, the groups N_L and $N_{L^{ab}}$ respectively. □

Lemma 13.17 *Let L_1 and L_2 be finite abelian extensions of F, both contained in $\overline{F}$. Then*

$$N_{L_1L_2} = N_{L_1} \cap N_{L_2}.$$

Proof. The inclusion $N_{L_1L_2} \subset N_{L_1} \cap N_{L_2}$ is obvious, since L_1 and L_2 are subfields of L_1L_2, by the transitivity of norms.

So suppose $x \in N_{L_1} \cap N_{L_2}$. Apply Lemma 13.14 with $L = L_1L_2$ and $K = L_i$, for $i = 1$ or 2. Observing that L_1L_2/F is abelian, we conclude that $(x, L_1L_2/F) \in \mathrm{Gal}(L_1L_2/F)$ restricts to the trivial element of $\mathrm{Gal}(L_i/F)$. However, an element of $\mathrm{Gal}(L_1L_2/F)$ is determined by its action on L_1 and L_2, so $(x, L_1L_2/F) = 1$. It follows that $x \in N_{L_1L_2}$. □

Proposition 13.18 *The exists a one-to-one, order-reversing correspondence between the finite abelian extensions of F contained in $\overline{F}$ and the norm subgroups of $F^\times$. It takes L/F to N_L.*

Proof. Lemma 13.16 shows that $L \mapsto N_L$ is surjective, from abelian extensions to norm subgroups. Clearly $L_1 \subset L_2$ implies $N_{L_2} \subset N_{L_1}$. We claim that, conversely, $N_{L_2} \subset N_{L_1}$ implies $L_1 \subset L_2$. Granting this, we see that $N_{L_1} = N_{L_2}$ implies $L_1 = L_2$, so that the proposed map is injective, and the proposition is established.

To prove the claim, suppose $N_{L_2} \subset N_{L_1}$ and draw

$$N_{L_1L_2} = N_{L_1} \cap N_{L_2} = N_{L_2}$$

by the last lemma. As a result, $x \mapsto (x, L_1L_2/F)$ and $x \mapsto (x, L_2/F)$ have the same kernel. Using Lemma 13.14 with $L = L_1L_2$ and $K = L_2$, we conclude that the restriction map $\mathrm{Gal}(L_1L_2/F) \to \mathrm{Gal}(L_2/F)$ is injective; however, the kernel of this map is also $\mathrm{Gal}(L_1L_2/L_2)$, so $L_1L_2 = L_2$ and $L_1 \subset L_2$. □

Corollary 13.19 *Any subgroup of $F^\times$ containing a norm subgroup is a norm subgroup.*

Proof. The subgroups of $F^\times$ containing N_L are in bijection with the subgroups of the finite group $F^\times/N_L$. We can assume that L/F is abelian (by Lemma 13.16), and so $F^\times/N_L \cong \mathrm{Gal}(L/F)$, via the reciprocity isomorphism. The subgroups of $\mathrm{Gal}(L/F)$ are themselves in bijection with the (abelian) extensions K/F with $K \subset L$. It follows, for example by a cardinality argument, that the groups containing N_L are the various groups N_K for $F \subset K \subset L$. □

Tate duality

We consider very temporarily a general field F. In Definition 11.10 we defined the "symbol" $(\chi,f) \in \mathrm{H}^2(F,\mathbb{G}_m)$ when $\chi\colon \mathrm{Gal}(\overline{F}/F) \to \mathbb{Q}/\mathbb{Z}$ is character (that is, a continuous homomorphism), and $f \in F^\times$. A general fact is that the kernel of $f \mapsto (\chi,f)$ is N_L, in our current notation, where L/F is the extension corresponding to the kernel of χ via the Galois correspondence. Returning (already!) to the case when F is a local number field, we emphasize that:

Lemma 13.20 *If $(\chi,f) = 0$ for all $f \in F^\times$, then χ is the trivial (=zero) character.*

Proof. The assumption is that $N_L = F^\times$, where L is the field corresponding to the kernel of χ. By Proposition 13.18 we have $L = F$ (since $N_L = N_F$), so the kernel of χ is all of $\mathrm{Gal}(\overline{F}/F)$. □

"Tate duality", or at any rate the simple version which we present now, is a rephrasing of this in a striking particular case. Assume that F contains a primitive ℓth root of unity, for some prime ℓ (which may, or may not, be equal to p, the characteristic of the residue field $\mathbb{F}$). Recall that, when $a,b \in F^\times$, we have defined in Definition 11.14 the Hilbert symbol $(a,b) = (\chi_a,b)$ where $\chi_a\colon \mathrm{Gal}(\overline{F}/F) \to \mathbb{Z}/\ell\mathbb{Z} \cong \frac{1}{\ell}\mathbb{Z}/\mathbb{Z} \subset \mathbb{Q}/\mathbb{Z}$ is the character associated to a under the isomorphism $\mathrm{H}^1(F,\mathbb{Z}/\ell\mathbb{Z}) \cong F^\times/F^{\times\ell}$ (all of this being dependent on the choice of a specific primitive root of unity). Here (a,b) is also an element of $\frac{1}{\ell}\mathbb{Z}/\mathbb{Z}$, when we identify the cohomological Brauer group of our local field with $\mathbb{Q}/\mathbb{Z}$. To give all this a linear algebra feel, we identify $\mathbb{Z}/\ell\mathbb{Z}$ and $\frac{1}{\ell}\mathbb{Z}/\mathbb{Z}$ with $\mathbb{F}_\ell$ in the obvious way, and state:

Theorem 13.21 (Tate duality) *Let F be a local number field containing a primitive ℓth root of unity. The bilinear pairing*

$$\mathrm{H}^1(F,\mathbb{F}_\ell) \times \mathrm{H}^1(F,\mathbb{F}_\ell) \longrightarrow \mathbb{F}_\ell$$

given by $[a],[b] \mapsto (a,b)$ is nondegenerate.

Here we write $[a]$ for the image of $a \in F^\times$ under $F^\times \to F^\times/F^{\times\ell} \to \mathrm{H}^1(F,\mathbb{F}_\ell)$.

Proof. Bilinearity was established in Corollary 11.12. Saying that the pairing is nondegenerate means that given a nonzero class $[a] \in \mathrm{H}^1(F,\mathbb{F}_\ell)$, there exists $[b] \in \mathrm{H}^1(F,\mathbb{F}_\ell)$ such that $(a,b) \neq 0$, which is what the last lemma says, and the same with the roles and a and b exchanged, which is obvious by antisymmetry (see (2) of Lemma 11.18).

In other words, the map $[a] \mapsto (a,-)$ is an injection of $\mathrm{H}^1(F,\mathbb{F}_\ell)$ into its own dual (over $\mathbb{F}_\ell$). By Lemma 4.10, the dimension of $\mathrm{H}^1(F,\mathbb{F}_\ell)$ is finite, so we really have an isomorphism, hence the name "Tate duality". □

In practice, we shall use this in the form: if $(a,b) = 0$ for all $a \in F^\times$, then b has an ℓth root in F (since $[b] = 0$).

As an application, we shall study the group of "universal norms":

universal norms

Definition 13.22 The group of **universal norms** of the local number field F is the subgroup D_F of $F^\times$ which is the intersection of all the norm groups N_L, for all finite extensions L/F.

We sense trouble in this definition. If the "Existence theorem", mentioned in the introduction, is to be true at all, that is if any subgroup of finite index in $F^\times$ is of the form N_L, then surely D_F ought to be trivial: Indeed, as an exercise the reader will find a family of subgroups of finite index with trivial intersection. As a matter of fact, we proceed to show directly that $D_F = \{1\}$, and in turn, this will be instrumental in proving the Existence theorem.

Proposition 13.23 *For any finite extension E/F, we have $\mathrm{N}_{E/F}(D_E) = D_F$.*

As announced, we shall eventually see that both groups in this proposition are trivial, so it is perhaps surprising that the argument below should be so delicate.

Proof. ([Ser79], chapter XI, proposition 5.) Let $x \in D_E$ and $y = \mathrm{N}_{E/F}(x)$. For any finite extension L/F, consider the composite EL (all extensions in this proof can be viewed as contained in a fixed algebraic closure $\overline{F}$, so it makes sense to talk about the composite). Then $E \subset EL$, so x is a norm from EL, and so is y by the transitivity of norms; as $L \subset EL$, we see that y is a norm from L. So $y \in D_F$.

The other inclusion is harder. Let $a \in D_F$. For any extension L/E, we put

$$\mathrm{K}(L) = \mathrm{N}_{L/E}\left(L^\times\right) \cap \mathrm{N}_{E/F}^{-1}(a),$$

which is the set of elements in $E^\times$ whose norm is a, and which are norms from L. We want to establish precisely that the intersection of all the sets $\mathrm{K}(L)$, for all L/E, is nonempty. For this, the crucial point is to show that $\mathrm{K}(L)$ is *compact* for the usual topology on $E^\times$.

First we make a few points about the norm map $\mathrm{N}_{E/F}\colon E^\times \to F^\times$. Recall from Corollary 2.26 that the Galois action on $E^\times$ is by isometries, so in particular, continuous maps; it follows easily that $\mathrm{N}_{E/F}$ is itself continuous. Further, for $n \in \mathbb{Z}$ we consider the subset $F_n^\times \subset F^\times$ of elements of valuation n, and point out that these form an open cover of $F^\times$. If f_n is any element of valuation n, we have $F_n^\times = f_n F_0^\times = f_n \mathcal{O}_F^\times$, from which we also deduce that each $F_n^\times$ is compact (recall Proposition 4.1). Similarly, each subset $\mathrm{N}_{E/F}^{-1}(F_n^\times)$ is either empty or a translate of $\mathcal{O}_E^\times$, so it is always compact. Now, if C is any compact subspace of $F^\times$, then it is covered by finitely many $F_n^\times$, and we see that $\mathrm{N}_{E/F}^{-1}(C)$ is closed and contained in a compact space, so it is itself compact. Our conclusion is that $\mathrm{N}_{E/F}$ is *proper*, a term meaning precisely that it is continuous, and that the inverse image of a compact subspace is compact.

We see already that $\mathrm{N}_{E/F}^{-1}(\{a\})$ is compact. Moreover, an elementary result from topology asserts that a proper map $f\colon X \to Y$ between metric spaces must have a closed image. (If $(f(x_n))_{n\geq 0}$ is a sequence in the image of f converging to $y \in Y$, consider the compact subset $C = \{f(x_n) : n \geq 0\} \cup \{y\}$; the fact that $f^{-1}(C)$ is compact shows easily that y is in the image of f, so this image is closed.) Applying this to $f = \mathrm{N}_{L/E}$, which is proper by our analysis, we see that $\mathrm{N}_{L/E}(L^\times)$ is closed in $E^\times$. We have finally established that each $\mathrm{K}(L)$ is compact.

We can conclude. Suppose the intersections of all the sets $\mathrm{K}(L)$ were empty. Then the sets $\mathrm{K}(E) \cap \mathrm{K}(L)$, where L runs through the finite extensions of E, are closed subsets of the compact space $\mathrm{K}(E)$ having an empty intersection, so there must exist a finite collection of these with empty intersection. That is, we can find $L_1, \ldots, L_k$ such that

$$\mathrm{K}(E) \cap \mathrm{K}(L_1) \cap \cdots \cap \mathrm{K}(L_k) = \emptyset .$$

Now choose a single L_0 containing all the L_i. As an inclusion $L \subset L'$ implies $\mathrm{K}(L') \subset \mathrm{K}(L)$, it is enough to prove that $\mathrm{K}(L_0)$ is nonempty to reach a contradiction. Indeed, each set $\mathrm{K}(L)$ is nonempty, as a is a norm from any L. □

Proposition 13.24 *For any local number field F, the group D_F of universal norms is trivial.*

Proof. Let ℓ be a prime number, and let F_ℓ be obtained by adjoining the ℓth roots of unity to F. Let $b \in D_E$ be a universal norm, for some E containing F_ℓ. Then b is in the kernel of $e \mapsto (\chi, e)$ for any character χ (since this kernel is a norm subgroup). In particular, $(a,b) = 0$ for any $a \in E^\times$. By Tate duality, we conclude that $b = x^\ell$ for some $x \in E^\times$.

Now fix $f \in D_F$, a universal norm. By the previous proposition, we have $f = \mathrm{N}_{E/F}(b)$ for some $b \in D_E$ as above. We take norms on both sides of the equality $b = x^\ell$, and obtain $f = y^\ell$, where $y = \mathrm{N}_{E/F}(x)$. If we put, in the spirit of the previous proof,

$$\mathrm{K}(E) = \left\{ y \in F^\times : y^\ell = f \right\} \cap N_E ,$$

then we have just established that each $\mathrm{K}(E)$ is nonempty (for any E large enough, and so for any E at all). It is also a finite set. An argument similar to that in the previous proof shows that the intersection of all the sets $\mathrm{K}(E)$ is nonempty. In other words, we have $f = y^\ell$ for some $y \in D_F$.

This being true of all primes, we see that D_F is a divisible subgroup of $F^\times$. By Lemma 4.12, we have $D_F = \{1\}$. □

The Existence theorem

We are finally in a position to prove:

Theorem 13.25 (the Existence theorem) *Let A be a subgroup of finite index in $F^\times$. Then there exists an abelian extension L/F such that $A = N_L$.*

Together with Proposition 13.18, this completes the proof of Theorem 13.1.

Proof. ([Ser79], chapter XI, section 5, Theorem 2) Suppose first that A contains $\mathcal{O}_F^\times$. Since the latter is the kernel of the valuation map $v\colon F^\times \to \mathbb{Z}$, we see that A must be $v^{-1}(n\mathbb{Z})$ for some $n \geq 1$. By Proposition 8.4, we draw $A = N_{E_n}$ where E_n is the unramified extension of degree n. So the theorem holds in this case. We turn to the general case.

Consider any norm subgroup N. Then $A \cap N$ has finite index in $F^\times$, since A and N both do. The group $A_N = (A \cap N) \cdot \mathcal{O}_F^\times$, comprised of all elements of the form $a = a'a''$ with $a' \in A \cap N$ and $a'' \in \mathcal{O}_F^\times$, must also have finite index since it contains $A \cap N$. Moreover, by the case just treated, we see that A_N is a norm subgroup; as a result, so is $N \cap A_N$.

We claim that we can choose N so that $N \cap A_N \subset A$. This implies the theorem by Corollary 13.19. In fact, we claim more precisely that we can find N such that $N \cap \mathcal{O}_F^\times \subset A$; notice that in this case, since an element of $N \cap A_N$ can be written $a = a'a''$ as above, we have $a'' = a(a')^{-1} \in N \cap \mathcal{O}_F^\times$, and thus $a'' \in A$ and $a \in A$.

So we look for N such that $N \cap \mathcal{O}_F^\times \subset A$. We certainly have

$$\bigcap_N N \cap \mathcal{O}_F^\times = \{1\},$$

where the intersection is taken over *all* norm subgroups, since the left-hand side is contained in the group of universal norms D_F, which is trivial by Proposition 13.24. By Lemma 4.11, each norm subgroup is open-and-closed, and A is also open-and-closed. Let C be the complement of A in $F^\times$, which is closed, and observe that the various closed subsets $C \cap N \cap \mathcal{O}_F^\times$ of the compact space $\mathcal{O}_F^\times$ have empty intersection. It follows that there are finitely many norm subgroups, say $N_1, \ldots, N_s$, such that

$$C \cap N_1 \cap \cdots \cap N_s \cap \mathcal{O}_F^\times = \emptyset.$$

For the norm subgroup $N = N_1 \cap \cdots \cap N_s$, we have $N \cap \mathcal{O}_F^\times \subset A$. □

Example 13.26 It is natural to wonder what extension L/F satisfies $N_L = F^{\times n}$, where $n \geq 1$ is some integer (this is a subgroup of finite index by Lemma 4.10). We can at least answer this question when F contains the nth roots of unity, and this illustrates the interplay between Kummer theory and class field theory. Indeed, the reciprocity isomorphism guarantees that $\mathrm{Gal}(L/F) \cong F^\times/F^{\times n}$, so L/F is a Kummer extension, which must then be of the form $L = F[\sqrt[n]{A}]$ for some subgroup $A \subset F^\times/F^{\times n}$. By cardinality, since the order of A is $[L : F]$ by Kummer theory, but is also $[F^\times : F^{\times n}]$ by class field theory, we have simply $L = F[\sqrt[n]{F^\times}]$, the field obtained by adjoining all possible nth roots. ■

The local Kronecker–Weber theorem

As an application of local class field theory, we now prove:

Theorem 13.27 (Kronecker–Weber) *Let p be a prime number, and let $L/\mathbb{Q}_p$ be an abelian extension. Then L is contained in a cyclotomic extension of $\mathbb{Q}_p$.*

(Obviously this does not hold with $\mathbb{Q}_p$ replaced by an arbitrary local number field, lest we should conclude by Theorem 2.58 that any extension of such fields is contained in a cyclotomic extension, and in particular, is abelian.)

We can restrict attention to the extensions contained in a fixed algebraic closure $\overline{\mathbb{Q}}_p$. By the fundamental theorem of class field theory, we only need to prove a reverse inclusion between norm subgroups. As a result, we must before all determine the norm subgroups of cyclotomic extensions.

Part of this has already been done. Indeed, suppose the integer N is prime to p, and let $L = \mathbb{Q}_p[\mu_N]$. By Proposition 2.46, the extension $L/\mathbb{Q}_p$ is unramified of degree n, which is the smallest integer such that N divides $p^n - 1$. By Proposition 8.4, we have $N_L = v_p^{-1}(n\mathbb{Z}) = \langle p^n \rangle \times \mathbb{Z}_p^\times$. Recall that, conversely, any ramified extension is obtained by adjoining roots of unity of order prime to p.

It remains to treat the following case:

Lemma 13.28 *Let $L = \mathbb{Q}_p[\mu_{p^n}]$ for some $n \geq 1$. Then $N_L = \langle p \rangle \times U^{(n)}$.*

(Here of course $U^{(n)} = U^{(n)}_{\mathbb{Q}_p}$, and generally speaking $\mathbb{Q}_p$ plays the role of F throughout this section.)

Proof. By Proposition 2.48, we have $[L : \mathbb{Q}_p] = p^{n-1}(p-1)$, and this must be also the index of N_L in $\mathbb{Q}_p^\times$. However, the subgroup $\langle p \rangle \times U^{(n)}$ of $\mathbb{Q}_p^\times$ has the same index (it is a general fact that the order of $U^{(1)}/U^{(n)}$ is $|\mathbb{F}|^{n-1}$ and that $F^\times \cong \mathbb{Z} \times \mathbb{Z}/(|\mathbb{F}|-1)\mathbb{Z} \times U^{(1)}$, where $\mathbb{F}$ is the residue field, cf. Lemma 2.56 and Proposition 4.2). So we only need to prove the inclusion $\langle p \rangle \times U^{(n)} \subset N_L$. Since the same Proposition 2.48 gives us an element, namely $1 - \zeta \in L$, whose norm is p, it suffices to prove that $U^{(n)} \subset N_L$.

Suppose first that p is odd. Then the logarithm and exponential establish isomorphisms between $U^{(m)}$ and $\mathfrak{p}^m = (p^m)$ as soon as $m \geq 1$, by Proposition 4.4. Multiplication by $p-1$ is an isomorphism on $(p^m) \cong \mathbb{Z}_p$, while multiplication by p takes (p^m) to (p^{m+1}); in particular $p^{n-1}(p-1)(p) = (p^n)$. Translated into multiplicative notation using the exponential, we have $(U^{(1)})^{p^{n-1}(p-1)} = U^{(n)}$. So every element of $U^{(n)}$ is a norm from L, and in fact it is the norm of an element in $U^{(1)}$.

Now suppose $p = 2$. The lemma is obviously true if $n = 1$, so we assume $n \geq 2$. Here we only have isomorphisms $U^{(m)} \cong (2^m)$ for $m \geq 2$. Proceeding as above only guarantees that $(U^{(2)})^{2^{n-2}} = U^{(n)}$. Let $y \in U^{(n)}$ and $x \in U^{(1)}$ be such that $x^{2^{n-2}} = y$. If x is a square, then y is a 2^{n-1}-st power, so it is a norm from L and we are done. From Example 4.9, we know that the subgroup of squares in $U^{(2)}$ is just $U^{(3)}$, which has index 2. The number 5 is not 1 mod 8, so it is not in $U^{(3)}$, and as a result it is not a square in $U^{(2)}$ (or equivalently, in $\mathbb{Q}_2^\times$). So if x is not a square, then $x = 5z^2$ and $y = 5^{2^{n-2}} z^{2^{n-1}}$. The real question, then, is whether $5^{2^{n-2}}$ is a norm from $L = \mathbb{Q}_p(\mu_{2^n})$.

The number -1 is not a square in $\mathbb{Q}_2$, since it is not 1 mod 8 either. So let $i = \sqrt{-1}$, which is an element of L, and $[\mathbb{Q}_2[i] : \mathbb{Q}_2] = 2$. We have

$$\mathrm{N}_{L/\mathbb{Q}_2}(2+i) = \left(\mathrm{N}_{\mathbb{Q}_2[i]/\mathbb{Q}_2}(2+i)\right)^{2^{n-2}} = [(2+i)(2-i)]^{2^{n-2}} = 5^{2^{n-2}}.$$

This completes the proof. $\square$

We have all we need for the Kronecker–Weber Theorem. Let $L/\mathbb{Q}_p$ be an abelian extension. Innocent-looking Lemma 4.11 shows that N_L contains a subgroup of the form $\langle p^m\rangle \times U^{(n)}$ for some integers n and m. We write this as the intersection

$$\langle p^m\rangle \times U^{(n)} = \left(\langle p^m\rangle \times \mathbb{Z}_p^\times\right) \cap \left(\langle p\rangle \times U^{(n)}\right) = N_{L_1} \cap N_{L_2}$$

where L_1 is the unramified extension of degree m, that is $L_1 = \mathbb{Q}_p[\mu_N]$ for $N = p^m - 1$, and $L_2 = \mathbb{Q}_p[\mu_{p^n}]$. By the fundamental theorem of class field theory, we have $N_{L_1} \cap N_{L_2} = N_{L_1L_2} \subset N_L$, and so $L \subset L_1L_2 = \mathbb{Q}_p[\mu_{(p^m-1)p^n}]$. QED.

A concise reformulation

There are no problems at the end of this chapter, but instead, we invite the reader to fill in the details in this (optional) last section, where some proofs are only sketched. We are going to improve Tate duality (Theorem 13.21) by removing the requirements on roots of unity. Then we shall see that the implications of the new statement are much stronger than is immediately apparent.

Theorem 13.29 (Tate duality, strong version) *Let F be a local number field, and $n \geq 1$ be an integer. The pairing*

$$\mathrm{H}^1(F, \mathbb{Z}/n\mathbb{Z}) \times \mathrm{H}^1(F, \mu_n) \longrightarrow \mathrm{H}^2(F, \mu_n)$$

induced by the symbol (χ, b) from Chapter 11 is nondegenerate.

A word of explanation is in order. We may identify $\mathrm{H}^1(F, \mathbb{Z}/n\mathbb{Z})$ with $\mathrm{Hom}(G_F, \mathbb{Z}/n\mathbb{Z})$, where here and elsewhere $G_F = \mathrm{Gal}(\overline{F}/F)$, and an element of this cohomology group may also be viewed as taking its values in $\frac{1}{n}\mathbb{Z}/\mathbb{Z} \subset \mathbb{Q}/\mathbb{Z}$. Also, we identify $\mathrm{H}^1(F, \mu_n)$ with $F^\times/F^{\times n}$. We see easily that the symbol (χ, b) makes sense for $\chi \in \mathrm{H}^1(F, \mathbb{Z}/n\mathbb{Z})$ and $b \in \mathrm{H}^1(F, \mu_n)$, and takes its values in the n-torsion of $\mathrm{H}^2(F, \mathbb{G}_m)$, which is none other than $\mathrm{H}^2(F, \mu_n)$. So the pairing described in the theorem makes sense.

As a first exercise, the reader can show that we are in fact talking about the cup-product operation; indeed, a routine modification of the proof of Proposition 12.42 will suffice.

Finally, "nondegenerate" is understood as in Definition 1.19, of course.

Proof. Suppose we pick χ such that $(\chi, b) = 0$ for all b; then $\chi = 0$ by Lemma 13.20.

On the other hand, suppose b is such that $(\chi, b) = 0$ for all χ. Remembering that the kernel of $b \mapsto (\chi, b)$ is N_L, where L is the field corresponding to the subgroup $\ker(\chi)$ of G_F, we see that b belong to the intersection of all the groups N_L, for all extensions L/F with $\mathrm{Gal}(L/F)$ cyclic of order dividing n. If we let K denote the compositum of all these Ls (within $\overline{F}$), then the Fundamental Theorem guarantees that $b \in N_K$. However, the group $\mathrm{Gal}(K/F)$ is abelian of exponent dividing n, and K/F is the maximal extension with this property, so by the fundamental theorem again, the group N_K is minimal with the property that $F^\times/N_K$ has exponent n. We conclude that $N_K = F^{\times n}$. Thus $b = 0$ in $F^\times/F^{\times n}$. □

Thus we have deduced the strong form of Tate duality from the fundamental theorem, and from the existence of reciprocity isomorphisms, that is, from Theorems 13.1 and 13.2. In hindsight, we realize that both of these results, and essentially all the facts proved in this chapter, are derived using the existence of Artin symbols $(x, L/K)$ with the properties stated in Lemmas 13.13 and 13.14. So it may come as a surprise that, as we proceed to show, we can prove the existence of such symbols from the statement of strong Tate duality. Thus Theorem 13.29 is in principle equivalent[1] to everything else in the chapter.

Sheer curiosity is not our sole motivation for this: The symbols we are about to construct will be free of choices, and so are an improvement on the previous ones. Also, we will derive precise information about the *abelianization* of G_F.

We begin by defining this term, generalizing the discussion before Lemma 10.20 to profinite groups. If G is profinite, we define $[G, G]$ to be the *closure of* the group generated by all the commutators $ghg^{-1}h^{-1}$, and we put $G^{ab} = G/[G, G]$, calling it the abelianization of G. This is an abelian profinite group (or an inverse limit of finite abelian groups, which is the same thing). If A is an abelian profinite group, it is immediate that $\mathrm{Hom}(G, A) = \mathrm{Hom}(G^{ab}, A)$, where continuous homomorphisms are meant; if A has exponent n, we even have $\mathrm{Hom}(G, A) = \mathrm{Hom}(G^{ab}/(G^{ab})^n, A)$. So one of the two factors appearing in Tate duality is really

$$\mathrm{H}^1(F, \mathbb{Z}/n\mathbb{Z}) = \mathrm{Hom}\left(G_F^{ab}/\left(G_F^{ab}\right)^n, \mathbb{Z}/n\mathbb{Z}\right).$$

Pontryagin duality, as discussed in Chapter 1, can be applied to profinite groups. However, we are in a very particular case, so we only need a fraction of that machinery. Namely, let Γ be an abelian profinite group of exponent dividing n, and suppose that $\Gamma' = \mathrm{Hom}(\Gamma, \mathbb{Z}/n\mathbb{Z})$ is finite. This applies to $\Gamma = G_F^{ab}/(G_F^{ab})^n$ by Tate duality, since we know that $F^\times/F^{\times n}$ is finite. We note that Γ injects into $\Gamma'' = \mathrm{Hom}(\Gamma', \mathbb{Z}/n\mathbb{Z})$ using $g \mapsto \mathrm{ev}_g$ (the evaluation at g), and so we deduce that Γ is itself finite. By Lemma 1.18, the map just considered is in fact a canonical isomorphism $\Gamma \cong \Gamma''$.

Thus we see that Tate duality involves $F^\times/F^{\times n}$ and the dual Γ', with $\Gamma = G_F^{ab}/(G_F^{ab})^n$. Lemma 1.20 and the remarks just made together imply:

Corollary 13.30 *There is a canonical isomorphism*

$$F^\times/F^{\times n} \cong G_F^{ab}/\left(G_F^{ab}\right)^n.$$

If L/F is an abelian extension with $\mathrm{Gal}(L/F)$ *of exponent divising n, then the image of N_L in $F^\times/F^{\times n}$ corresponds, under this isomorphism, to the image of* $\mathrm{Gal}(\overline{F}/L)$ *in* $G_F^{ab}/(G_F^{ab})^n$.

[1] This was pointed out to me by Olivier Wittenberg.

The second statement is left as an exercise (first reduce to the case when $\mathrm{Gal}(L/F)$ is cyclic).

Taking inverse limits, we obtain:

Corollary 13.31 *There are canonical isomorphisms*

$$G_F^{ab} \cong \widehat{F}^\times \cong \widehat{\mathbb{Z}} \times \mathcal{O}_F.$$

Here the notation refers to profinite completions, as in Problem 3.7.

Proof. Since G_F^{ab} is profinite, we see easily that

$$G_F^{ab} = \lim_n G_F^{ab} / \left(G_F^{ab}\right)^n,$$

where the limit is taken over the poset of integer under the divisibility relation. On the other hand, we have $F^\times \cong \mathbb{Z} \times \mathcal{O}_F$ by Theorem 4.8, and $\mathcal{O}_F$ is profinite, so we draw

$$\lim_n F^\times / F^{\times n} \cong \left(\lim_n \mathbb{Z}/n\mathbb{Z}\right) \times \mathcal{O}_F.$$

Thus this corollary follows from the previous one. □

In particular, we have a canonical map $F^\times \longrightarrow G_F^{ab}$ whose image is dense. For a fixed abelian extension L/F, we consider the composition

$$F^\times \longrightarrow G_F^{ab} \longrightarrow \mathrm{Gal}(L/F).$$

We write it $x \mapsto [x, L/F]$.

Lemma 13.32 *The map $x \mapsto [x, L/F]$ is surjective, with kernel N_L. In particular, we recover the reciprocity isomorphism*

$$F^\times / N_L \cong \mathrm{Gal}(L/F).$$

What is more, the diagrams in Lemmas 13.13 and 13.14, with $[x, L/F]$ replacing $(x, L/F)$, both commute.

At this point, this can be left as an easy exercise. The commutativity in Lemma 13.14 is, this time, the trivial one. That in Lemma 13.13 follows from the "projection formula" (property (4) in Proposition 12.40).

As promised, we have thus constructed new "Artin symbols" $[x, L/F]$, which are, this time, completely canonical. The previous symbols $(x, L/F)$ depended on certain choices – and to be sure, it is in fact possible to make these choices in such a way that $(x, L/F) = [x, L/F]$, but proving this completely requires care, and is tedious. Knowing that the two types of symbols can be made to coincide is also not terribly useful. Instead, we may just take the new definition as the official one (dropping the notation $[x, L/F]$ which is not standard, in favor of $(x, L/F)$).

We have already mentioned that we can find a way back to the fundamental theorem, relying just on the last lemma, and so Tate duality, in the strong form given here, implies all of local class field theory. The work required to unravel the rest of the chapter from Theorem 13.29 is of course nontrivial, and we are not suggesting that this result is the only one to commit to memory. However, having such a concise encapsulation is great: Local class field theory says that certain cup-products are nondegenerate (purely a statement in Galois cohomology). Also, it explains why various generalizations of class field theory are stated as duality statements (and not, say, as the existence of a bijection as in the fundamental theorem).

14 An introduction to number fields

In this last chapter, we study some of the properties of number fields, and we shall insist on those results which can be naturally obtained by looking at their *completions*, which are p-adic fields. (In the literature, the "completions" to be defined below are sometimes called "non-Archimedean completions".)

This chapter is written in such a way that it can be understood without any prior knowledge of number fields, and we hope it can be an illuminating first introduction, although many readers will already know the basics.

We conclude the book with a proof of the Kronecker–Weber theorem, and some statements of class field theory for number fields.

Number fields are examples of "global fields". We shall give some explanation of the term, as well as finally tell you why p-adic fields are called "local" number fields.

Number fields and their completions

number field

Definition 14.1 A **number field** is a field F which is finite-dimensional over $\mathbb{Q}$. ■

Lemma 14.2 *Let F be a number field, let F_p be a p-adic field for some prime number p, and let $\iota \colon F \to F_p$ be a field homomorphism. The following conditions are equivalent.*

1. *$\iota(F)$ is dense in F_p, for its natural topology.*
2. *The homomorphism $\iota \otimes \mathrm{id} \colon F \otimes_{\mathbb{Q}} \mathbb{Q}_p \longrightarrow F_p$ is surjective.*

Proof. The image of $\iota \otimes \mathrm{id}$ is a finite-dimensional $\mathbb{Q}_p$-subspace of F_p, which is then complete for the natural topology by Theorem 3.28, and hence closed. If we assume (1), then (2) must certainly follow.

Now assume (2). Since $\mathbb{Q}$ is dense in $\mathbb{Q}_p$, we see that $F = F \otimes_{\mathbb{Q}} \mathbb{Q}$ is dense in $F \otimes_{\mathbb{Q}} \mathbb{Q}_p$, so that (1) follows. □

Definition 14.3 A homomorphism $\iota \colon F \to F_p$ satisfying the conditions of the lemma is called a **completion of F over p** (or at p, or above p).

completion of F over p

equivalent Two completions $\iota\colon F \to F_p$ and $\iota'\colon F \to F'_p$ are called **equivalent** when there is a $\mathbb{Q}_p$-linear isomorphism $\lambda\colon F_p \to F'_p$ such that $\iota' = \lambda \circ \iota$. ■

Sometimes we may speak of F_p as a completion of F, as a shorthand for $\iota\colon F \to F_p$, but this is normally an abuse of the term, and there are situation where this matters, as we shall see.

We proceed to describe the algebra $F \otimes \mathbb{Q}_p$, for some fixed prime p (all tensor products are now over $\mathbb{Q}$). Using the lemma, we will deduce all the possible completions of F. Our first step is to invoke the primitive element theorem, and to pick an element $\alpha \in F$ such that $F = \mathbb{Q}(\alpha)$. When working out specific examples, the choice of α will of course matter, but we will also make general observations which are independent of this choice.

Let $P \in \mathbb{Q}[X]$ be the minimal polynomial of α over $\mathbb{Q}$, so that $F \cong \mathbb{Q}[X]/(P)$. Write the factorization of P over $\mathbb{Q}_p$:

$$P = P_1 P_2 \cdots P_g\,,$$

with P_i an irreducible polynomial in $\mathbb{Q}_p[X]$ (the use of the letter g for the number of factors is standard). Since P is separable (characteristic 0!), the polynomials P_i have no root in common, and so are coprime. Now let us use the Chinese Remainder theorem, providing us with an isomorphism

$$F \otimes \mathbb{Q}_p \cong \frac{\mathbb{Q}_p[X]}{(P_1 \cdots P_g)} \longrightarrow \bigoplus_i \frac{\mathbb{Q}_p[X]}{(P_i)}.$$

Let $\overline{\mathbb{Q}}_p$ be a fixed algebraic closure of $\mathbb{Q}_p$, and pick $\alpha_i \in \overline{\mathbb{Q}}_p$ which is a root of P_i, so that $\mathbb{Q}_p[X]/(P_i) \cong \mathbb{Q}_p(\alpha_i)$. We now have an isomorphism

$$F \otimes \mathbb{Q}_p \cong \frac{\mathbb{Q}_p[X]}{(P_1 \cdots P_g)} \longrightarrow \bigoplus_i \mathbb{Q}_p(\alpha_i)$$

taking $Q \in \mathbb{Q}_p[X]$ to $(Q(\alpha_1), Q(\alpha_2), \ldots, Q(\alpha_g))$.

For $0 \le k \le r$, consider the homomorphism $\iota_k\colon F = \mathbb{Q}(\alpha) \to \mathbb{Q}_p(\alpha_k)$ defined by $\iota_k(\alpha) = \alpha_k$, which is well-defined because $P(\alpha_k) = 0$. It is obvious that the image of ι_k is dense in $\mathbb{Q}_p(\alpha_k)$, so we have a completion of F. Alternatively, ι_k is given by the composition

$$F \xrightarrow{x \mapsto x \otimes 1} F \otimes \mathbb{Q}_p \xrightarrow{\cong} \frac{\mathbb{Q}_p[X]}{(P_1 \cdots P_g)} \xrightarrow{Q \mapsto Q(\alpha_k)} \mathbb{Q}_p(\alpha_k)\,.$$

Thus $\iota_k \otimes \mathrm{id}\colon F \otimes \mathbb{Q}_p \longrightarrow \mathbb{Q}_p(\alpha_k)$ is just the composition of the last two arrows; if we think of $F \otimes \mathbb{Q}_p$ as $\bigoplus_i \mathbb{Q}_p(\alpha_i)$, then the map in question is simply the projection onto the kth factor, and it is visibly surjective.

Lemma 14.4 *Any completion of F at p is equivalent to some ι_k. When $k \neq \ell$, the completions ι_k and ι_ℓ are not equivalent.*

Proof. Let $\iota : F \to F_p$ be any completion. The surjective map $\iota \otimes \mathrm{id}$ defines a surjective homomorphism

$$h\colon \bigoplus_i \mathbb{Q}_p(\alpha_i) \longrightarrow F_p\,.$$

Let $A = \bigoplus_i \mathbb{Q}_p(\alpha_i)$, and let $e_i \in A$ denote the unit of $\mathbb{Q}_p(\alpha_i)$. Since $e_i^2 = e_i$, we have $h(e_i)^2 = h(e_i)$, so $h(e_i) = 0$ or 1 (as F_p is a field). Moreover, $\sum_i e_i = 1$ (the unit of A), so $\sum_i h(e_i) = 1$. We conclude that there is just one index, say k, such that $h(e_k) = 1$, while $h(e_i) = 0$ otherwise.

For $i \neq k$, the restriction of h to $\mathbb{Q}_p(\alpha_i)$ is thus 0 (as $h(x) = h(xe_i) = h(x)h(e_i) = 0$ for $x \in \mathbb{Q}_p(\alpha_i)$). On the other hand, the restriction of h to $\mathbb{Q}_p(\alpha_k)$ is a field homomorphism, so it is injective, and it is surjective because h is. It follows easily that ι is equivalent to ι_k.

Should ι_k and ι_ℓ be equivalent, we would have an isomorphism $\mathbb{Q}_p(\alpha_k) \to \mathbb{Q}_p(\alpha_\ell)$ which takes α_k to α_ℓ, and is $\mathbb{Q}_p$-linear. But the minimal polynomials of α_k and α_ℓ are P_k and P_ℓ respectively, and these are coprime unless $k = \ell$. □

Example 14.5 Take $F = \mathbb{Q}[\sqrt{2}]$. The minimal polynomial of $\alpha = \sqrt{2}$ is of course $P = X^2 - 2$, and given a prime number p, we must decide whether P splits over $\mathbb{Q}_p$ as a product of two linear factors or not. In other words, we must decide whether 2 is a square in $\mathbb{Q}_p$. We have worked this out in Example 4.9: The number 2 is not a square in $\mathbb{Q}_2$ since it is not 1 mod 8, and 2 is a square in $\mathbb{Q}_p$ with p odd if and only if it is one in $\mathbb{F}_p$. In turn, this can be checked quickly, as the condition amounts to $2^{\frac{p-1}{2}} = 1 \bmod p$ (besides, a classical result, which is part of the full statement of the "quadratic reciprocity law", asserts that this happens precisely when $p = \pm 1$ mod 8).

So for example, for $p = 2$ or 3, the polynomial $X^2 - 2$ is irreducible over $\mathbb{Q}_p$, there is only one completion above p (up to equivalence), and it is of the form $\iota : \mathbb{Q}[\sqrt{2}] \to \mathbb{Q}_p[\alpha_1]$ with $\iota(\sqrt{2}) = \alpha_1$, where $\alpha_1 \in \overline{\mathbb{Q}_p}$ is a square root of 2 (of course in practice one writes $\sqrt{2}$ for α_1).

For $p = 7$ on the other hand, we have $f = X^2 - 2 = (X - \alpha_1)(X + \alpha_1)$ over $\mathbb{Q}_p$, there are two (equivalence classes of) completions of $\mathbb{Q}[\sqrt{2}]$ over p, and they have the form $\iota_\pm : \mathbb{Q}[\sqrt{2}] \to \mathbb{Q}_p$, with $\iota_\pm(\sqrt{2}) = \pm\alpha_1$. Here again $\alpha_1 \in \mathbb{Q}_p$ could be denoted by $\sqrt{2}$ without much risk of confusion, but we give it a different name for supreme clarity. Note how the dimension over $\mathbb{Q}_p$ of a completion of F may be different from the dimension of F over $\mathbb{Q}$. ■

Example 14.6 Let $\alpha \in \mathbb{C}$ be a primitive nth root of unity for some $n \geq 1$, for example $\alpha = e^{\frac{2\pi i}{n}}$. A completion ι_k of $F = \mathbb{Q}(\alpha)$ takes values in $\mathbb{Q}_p(\alpha_k)$ where $\alpha_k \in \overline{\mathbb{Q}_p}$ is a root of $P = \min(\mathbb{Q}, \alpha) = \Phi_n$, the nth cyclotomic polynomial. In particular, α_k is a root of $X^n - 1$, and in fact simply using the fact that ι_k is a homomorphism of fields taking α to α_k, we see that α_k is a *primitive* nth root of unity.

To study the minimal polynomial of α_k over $\mathbb{Q}_p$, Propositions 2.46 and 2.48 are relevant. To take an extreme case, if $n = p^m$ is a power of p, then Proposition 2.48

shows that Φ_n remains irreducible over $\mathbb{Q}_p$. Thus there is, up to equivalence, just one completion $\iota\colon \mathbb{Q}(\alpha) \to \mathbb{Q}_p(\alpha_1)$, taking α to α_1, and $\dim_{\mathbb{Q}_p} \mathbb{Q}_p(\alpha_1) = \dim_{\mathbb{Q}} \mathbb{Q}(\alpha)$.

At the other end of the spectrum, if n divides $p-1$, then all the roots of $X^n - 1$ are in $\mathbb{Q}_p$. There are n equivalence classes of completions $\iota_k\colon \mathbb{Q}(\alpha) \to \mathbb{Q}_p$, taking α to the various nth roots of unity $\alpha_1, \ldots, \alpha_n$. ■

In these examples, the extension $F/\mathbb{Q}$ was Galois. We turn our attention to this situation in general, resuming all the notation from our discussion above. When $\mathbb{Q}(\alpha)/\mathbb{Q}$ is Galois, we have

$$P = (X - Q_1(\alpha)) \cdots (X - Q_n(\alpha))$$

for some polynomials $Q_i \in \mathbb{Q}[X]$. Applying $\iota_k\colon F \to \mathbb{Q}_p(\alpha_k)$, we obtain

$$P = (X - Q_1(\alpha_k)) \cdots (X - Q_n(\alpha_k)) .$$

So $\mathbb{Q}_p(\alpha_k)$ is a splitting field for $P \in \mathbb{Q}_p[X]$. In particular, all the extensions $\mathbb{Q}_p(\alpha_k)/\mathbb{Q}$ are isomorphic; in fact, as we have been careful to choose the numbers α_k in a common algebraic closure of $\mathbb{Q}_p$, we see that the subfield $F_p := \mathbb{Q}_p(\alpha_k)$ of $\overline{\mathbb{Q}}_p$ does not depend on k. Similarly, the subfield $F'_p = \mathbb{Q}(\alpha_k)$ of F_p is the same, whatever the value of k. (Still, there can be several, nonequivalent completions, as the examples show.) The polynomials P_i, appearing in the factorization of P, all have the same degree.

Let us write $G = \mathrm{Gal}(F/\mathbb{Q})$, and $G_p = \mathrm{Gal}(F_p/\mathbb{Q}_p)$. Any $\sigma \in G_p$ must preserve F'_p, since $F'_p/\mathbb{Q}$ is isomorphic to $F/\mathbb{Q}$, and hence is Galois. The completion $\iota_k\colon F \to F_p$, which restricts to an isomorphism $F \to F'_p$, thus gives a map

$$\iota_k^*\colon G_p \longrightarrow G ,$$

defined by $\iota_k^*(\sigma) = \iota_k^{-1} \circ \sigma \circ \iota_k$. It is clear that ι_k^* is injective, since $F_p = \mathbb{Q}_p(\alpha_k)$ and we can read $\sigma(\alpha_k)$ from $\iota_k^*(\sigma)$.

Lemma 14.7 *When ι_k is replaced by another completion ι_ℓ, which may or may not be equivalent, there is an element $\tau \in G$ such that*

$$\iota_\ell^*(\sigma) = \tau \iota_k^*(\sigma) \tau^{-1} ,$$

for all $\sigma \in G_p$. In particular, the groups $\iota_k^*(G_p)$ and $\iota_\ell^*(G_p)$ are conjugate.

When ι_ℓ is equivalent to ι_k, the element τ can be taken of the form $\tau = \iota_k(\rho)$ for some $\rho \in G_p$. In particular, $\iota_k^*(G_p) = \iota_\ell^*(G_p)$ in this case.

Proof. The first statement is obtained with $\tau = \iota_\ell^{-1} \circ \iota_k \in G$. For the second one, if $\iota_\ell = \lambda \circ \iota_k$, then λ can be seen as an element of G_p, and if we put $\rho = \lambda^{-1}$ we get the announced property. □

Example 14.8 During the course of the proof of Corollary 14.36, we shall establish the following fact. Let $F = \mathbb{Q}[e^{\frac{2i\pi}{n}}]$, and let p be a prime which does not divide n.

Then for any completion $\iota\colon F \to F_p$, the group $\iota^*(\mathrm{Gal}(F_p/\mathbb{Q}_q))$ is the subgroup generated by p in $(\mathbb{Z}/m\mathbb{Z})^\times \cong \mathrm{Gal}(F/\mathbb{Q})$. The reader can try to prove this now, as an exercise.

decomposition group

Definition 14.9 The subgroup $\iota^*(\mathrm{Gal}(F_p/\mathbb{Q}_p))$ of $\mathrm{Gal}(F/\mathbb{Q})$ is called the **decomposition group** of the completion $\iota\colon F \to F_p$. We only define it when $F/\mathbb{Q}$ is Galois.

The preceding lemma thus asserts that the various decomposition groups, corresponding to the various completions over a fixed prime p, are conjugate (and in fact, one sees easily that they form a complete conjugacy class of subgroups).

Next we consider ramification indices. Let $e_k = e(\mathbb{Q}_p(\alpha_k)/\mathbb{Q}_p)$ and $f_k = f(\mathbb{Q}_p(\alpha_k)/\mathbb{Q}_p)$, as in Definition 2.34. When $F/\mathbb{Q}$ is Galois, the numbers $e_1, \ldots, e_g$ are all equal, and their common value will be written e; likewise, f will be the common value of $f_1,\ldots,f_g$ when $F/\mathbb{Q}$ is Galois.

The next lemma is rather obvious at this stage, but its identities are classic, so it deserves to be emphasized.

Lemma 14.10 *Let $n = [F : \mathbb{Q}]$. Then*

$$n = \sum_{i=1}^{g} e_i f_i.$$

When $F/\mathbb{Q}$ is Galois, this reads

$$n = efg.$$

Proof. We have $n = \dim_{\mathbb{Q}_p} F \otimes \mathbb{Q}_p$, while

$$F \otimes \mathbb{Q}_p \cong \bigoplus_i \mathbb{Q}_p(\alpha_i).$$

Thus $n = \sum_i [\mathbb{Q}_p(\alpha_i) : \mathbb{Q}_p]$. Theorem 2.38 finishes the proof. □

unramified prime

Definition 14.11 We say that the prime number p is **unramified prime** in the number field F when $e_1 = e_2 = \cdots = e_g = 1$, or in other words, when the extensions $\mathbb{Q}_p(\alpha_i)/\mathbb{Q}_p$ are all unramified. Otherwise, we say that p **ramifies in F**.

Remark 14.12 Sometimes, extra terminology is introduced. When $g = 1$ and $f_1 = 1$, one may say that p is *totally ramified in F*. When $g = 1$ and $e_1 = 1$, the prime p is said to *remain inert in F*. When $g = n = [F : \mathbb{Q}]$, so that $e_1 = \cdots = e_g = f_1 = \cdots = f_g = 1$ by the lemma, the prime p is said to *split completely in F*.

Example 14.13 We return to the example of $F = \mathbb{Q}(\alpha)$ with $\alpha = e^{\frac{2i\pi}{n}}$. We claim that, when p is odd, then p ramifies in F if and only if p divides n. First, we are in the Galois case, and the image of any completion is $\mathbb{Q}_p(\mu_n)$, where $\mu_n = \mu_n(\overline{\mathbb{Q}_p})$ as usual. If n is prime to p, then Proposition 2.46 shows that $\mathbb{Q}_p(\mu_n)/\mathbb{Q}_p$ is unramified.

Now suppose conversely that $n = p^r m$ with m prime to p and $r > 0$. Then $\mathbb{Q}_p(\mu_{p^r}) \subset \mathbb{Q}_p(\mu_n)$, and $\mathbb{Q}_p(\mu_{p^r})/\mathbb{Q}_p$ is totally ramified by Proposition 2.48, so $\mathbb{Q}_p(\mu_n)/\mathbb{Q}_p$ is certainly not unramified.

For $p = 2$, this argument only breaks at the very end, when $r = 1$, for $\mathbb{Q}_2(\mu_2) = \mathbb{Q}_2$, so $\mathbb{Q}_2(\mu_2)/\mathbb{Q}_2$, while totally ramified, is also unramified! The amended statement is that 2 ramifies in F if and only if 4 divides n. ■

For the next definition, we refer to Lemma 2.45, where the inertia subgroup was introduced.

Definition 14.14 The subgroup $\iota^*(I(F_p/\mathbb{Q}_p))$ of $\mathrm{Gal}(F/\mathbb{Q})$ is called the **inertia subgroup** of the completion $\iota\colon F \to F_p$. ■

inertia subgroup

Since the inertia subgroup $I(F_p/\mathbb{Q}_p)$ is a normal subgroup of $\mathrm{Gal}(F_p/\mathbb{Q}_p)$, we see from Lemma 14.7 that the inertia subgroup of a completion only depends on its equivalence class; and when ι runs through the various completions of F above p, the corresponding inertia subgroups are conjugate. As an exercise in the notation, note that the order of an inertia subgroup is e.

In the important particular case when $\mathrm{Gal}(F/\mathbb{Q})$ is abelian, each prime p defines just one decomposition subgroup, usually written G_p, and just one inertia subgroup, usually written I_p.

Remark 14.15 In general, there is a lot to say about extensions K/F of number fields, and the way the completions of F extend to completions of K. We will not look into this, but at least we can make the following remark. Let $\iota\colon F \to F_p$ be a completion of F, with $F_p \subset \overline{\mathbb{Q}}_p$, our favorite algebraic closure of $\mathbb{Q}_p$. Then, if K is another number field with $F \subset K$, we can extend ι into a completion $\iota'\colon K \to K_p \subset \overline{\mathbb{Q}}_p$. Indeed, choose α such that $K = \mathbb{Q}(\alpha)$, so $K = F(\alpha)$, let $P = \min(F, \alpha)$, which divides $\min(\mathbb{Q}, \alpha)$, and pick a root $\alpha_1 \in \overline{\mathbb{Q}}_p$ of the polynomial $\iota(P)$. Since $K \cong F[X]/(P)$, there is a well-defined embedding of K into $\overline{\mathbb{Q}}_p$, extending ι and taking α to α_1. It is one of the completions of K constructed by the method above (since α_1 is a root of $\min(\mathbb{Q}, \alpha)$).

The next lemma is an example of use. ■

Lemma 14.16 *Let $F/\mathbb{Q}$ be Galois, and let $L \subset F$ be the fixed field of the subgroup of* $\mathrm{Gal}(F/\mathbb{Q})$ *generated by all the inertia subgroups, where we consider all primes p and all completions above them. Then no prime p ramifies in L.*

Proof. Recall that L is the intersection of the fixed fields of all the individual inertia subgroups. If $\iota\colon F \to F_p$ is a completion, the definitions imply that $\iota(L)$ is fixed by $I(F_p/\mathbb{Q}_p)$. By the preceding remark (with the extension F/L playing the role of K/F), any completion $L \to L_p$ takes its values in an unramified extension of $\mathbb{Q}_p$. By density, we see that $L_p/\mathbb{Q}_p$ is itself unramified (the largest unramified extension of $\mathbb{Q}_p$ contained in L_p is certainly closed). □

The work of the next section will serve to prove that no number field, other than $\mathbb{Q}$ itself, has the property stated in the lemma. So $L = \mathbb{Q}$, and $\mathrm{Gal}(F/\mathbb{Q})$ is generated by the various inertia subgroups.

The discriminant

We want to study the primes p which ramify in a given number field F. Ultimately, we shall prove that there is an integer Δ_F, called the *discriminant* of F, whose prime divisors are exactly the primes ramifying in F. Hence, there are finitely many of these. The discriminant will also be a good excuse to introduce several concepts of independent interest.

We shall then quote Minkowski's theorem, to the effect that when $F \neq \mathbb{Q}$, the discriminant satisfies $|\Delta_F| > 1$, and so there are some primes which actually do ramify.

In this section, we write $\overline{\mathbb{Q}}$ for the algebraic closure of $\mathbb{Q}$ in $\mathbb{C}$ (for definiteness).

Proposition 14.17 *Let $x \in \overline{\mathbb{Q}}$. The following conditions on x are equivalent.*

1. *The minimal polynomial* $\min(\mathbb{Q}, x)$ *has integer coefficients.*
2. *There exists some $f \in \mathbb{Z}[X]$ which is monic and such that $f(x) = 0$.*
3. *The ring $\mathbb{Z}[x] \subset \mathbb{C}$ generated by x is finitely generated as an abelian group.*

The set of elements x satisfying these conditions is a subring of $\overline{\mathbb{Q}}$, sometimes written $\overline{\mathbb{Z}}$, and called the ring of algebraic integers.

Proof. (1) $\Longrightarrow$ (2) is trivial. Assuming (2), there is some $n \geq 1$ such that

$$x^n = c_{n-1}x^{n-1} + \cdots + c_1 x + c_0$$

with $c_i \in \mathbb{Z}$, and it follows readily that $\mathbb{Z}[x]$ is generated by $1, x, \ldots, x^{n-1}$ as an abelian group. So (2) implies (3).

Assume (3), let $R = \mathbb{Z}[x]$, and let R_k be the subgroup generated by $1, x, \ldots, x^{k-1}$, whenever $k \geq 1$. We have $R_k \subset R_{k+1}$. However, since we assume that R is a finitely generated abelian group, it does not have any infinite (strictly) increasing sequence of subgroups, so there is some n with $R_{n+1} = R_n$. The relation $x^{n+1} \in R_n$ gives a polynomial relation as in (2).

To show that (2) $\Longrightarrow$ (1), we rely on our work with local number fields. Assume (2), let $F = \mathbb{Q}(x)$, and put $\alpha = x$ (merely to recover the notation from the previous section). For any completion $\iota \colon F \to F_p$ over some prime p, it is clear from the definitions that $\iota(x)$ is in $\mathcal{O}_{F_p}$; here we use Lemma 2.33, which also implies that $\min(\mathbb{Q}_p, \iota(x)) \in \mathbb{Z}_p[X]$. As in the previous section, write $P = \min(\mathbb{Q}, \alpha)$ and let $P = P_1 P_2 \cdots P_g$ be its factorization over $\mathbb{Q}_p$. Here $P_k = \min(\mathbb{Q}_p, \alpha_k)$, and $\iota_k(\alpha) = \alpha_k$, so $P_k \in \mathbb{Z}_p[X]$, for all k. Thus $P \in \mathbb{Z}_p[X]$, and the p-adic valuation of each coefficient of P is ≥ 0. Since p is arbitrary in this argument, we have $P \in \mathbb{Z}[X]$.

We have proved the equivalence, and we turn to the last statement. Let x, y be algebraic integers. Then $\mathbb{Z}[x,y]$ is finitely generated as an abelian group, clearly. Since $\mathbb{Z}[x+y]$ and $\mathbb{Z}[xy]$ are subgroups of $\mathbb{Z}[x,y]$, they are also finitely generated, so $x+y$ and xy are also algebraic integers. It follows that $\overline{\mathbb{Z}}$ is indeed a ring. □

ring of integers

Definition 14.18 The ring $\overline{\mathbb{Z}} \cap F$, where F is a number field, is called the **ring of integers** of F, and is denoted by $\mathcal{O}_F$. ∎

Example 14.19 In the exercises, we will explore the fact that, for $F = \mathbb{Q}(\sqrt{3})$, the ring of integers is $\mathbb{Z}[\sqrt{3}]$, while for $F = \mathbb{Q}(\sqrt{5})$, the ring of integers is $\mathbb{Z}[\frac{1}{2}(1+\sqrt{5})]$. ∎

Proposition 14.20 *Let $n = [F : \mathbb{Q}]$. The ring $\mathcal{O}_F$ is a free abelian group of rank n.*

Proof. We merely have to adapt the proof of Proposition 2.36, so we only give a sketch. First note that, when the minimal polynomial of $x \in F$ is $X^m + c_{m-1}X^{m-1} + \cdots + c_0$, then the minimal polynomial of dx is $X^m + dc_{m-1}X + d^2c_{m-2} + \cdots + d^m c_0$, for any $d \in \mathbb{Z}$. By choosing d appropriately, we can certainly arrange for dx to be an algebraic integer.

In passing, this proves that F is the field of fractions of $\mathcal{O}_F$, but also that there is a $\mathbb{Q}$-basis for F contained in $\mathcal{O}_F$; so the free abelian group M spanned by this basis is of rank n, with $M \subset \mathcal{O}_F$. It is enough to show that there is a single integer d with $dx \in M$ for all $x \in \mathcal{O}_F$. For this, we use the same trick, using traces, as in Proposition 2.36 (using Proposition 14.17 to show that the minimal polynomials of algebraic integers have integer coefficients). □

Now fix the number field F, a choice of prime number p, and consider the various completions $\iota_k \colon F \to \mathbb{Q}_p(\alpha_k)$ above p, in the notation above. Let $\mathcal{O}_k$ be the ring of integers in $\mathbb{Q}_p(\alpha_k)$ (as in Definition 2.15, say). We have already observed that $\iota_k(\mathcal{O}_F) \subset \mathcal{O}_k$. We shall now be concerned with reducing mod p, that is, with $\mathcal{O}_F/p\mathcal{O}_F = \mathcal{O}_F \otimes_{\mathbb{Z}} \mathbb{F}_p$.

Proposition 14.21 *The induced map*

$$\mathcal{O}_F \otimes_{\mathbb{Z}} \mathbb{F}_p \longrightarrow \bigoplus_i \mathcal{O}_i \otimes_{\mathbb{Z}} \mathbb{F}_p$$

is an isomorphism.

Proof. Let $n = [F : \mathbb{Q}]$. By the previous proposition, the ring $\mathcal{O}_F$ is isomorphic to $\mathbb{Z}^n$ as an abelian group, so $\mathcal{O}_F \otimes \mathbb{F}_p$ is an $\mathbb{F}_p$-vector space of dimension n. Using Proposition 2.36, and counting dimensions using that $F \otimes_{\mathbb{Q}} \mathbb{Q}_p \cong \bigoplus_i \mathbb{Q}_p(\alpha_i)$, we see that $\bigoplus_i \mathcal{O}_i \otimes_{\mathbb{Z}} \mathbb{F}_p$ is also an $\mathbb{F}_p$-vector space of dimension n. (Implicit here is the fact that $\mathcal{O}_i \otimes_{\mathbb{Z}} \mathbb{F}_p = \mathcal{O}_i \otimes_{\mathbb{Z}_p} \mathbb{F}_p = \mathcal{O}_i/p\mathcal{O}_i$.) So, it suffices to show that the map is injective.

Let $x \in \mathcal{O}_F$ be such that $\iota_k(x) \in p\mathcal{O}_k$, for some completion ι_k of F; in other words, $\frac{\iota_k(x)}{p}$ is integral over $\mathbb{Z}_p$. For the duration of this proof, let us say that a polynomial $X^m + c_{m-1}X^{m-1} + \cdots + c_0 \in \mathbb{Z}_p[X]$ is p-divisible when

$$X^m + \frac{c_{m-1}}{p}X^{m-1} + \frac{c_{m-2}}{p^2}X^{m-2} + \cdots + \frac{c_0}{p^m} \in \mathbb{Z}_p[X],$$

or in other words, when p^i divides c_{m-i}. An elementary computation, left to the reader, shows crucially that a product of two p-divisible polynomials is again p-divisible.

In our situation, we see that $\min(\mathbb{Q}_p, \iota_k(x))$ is p-divisible, since the condition amounts here to $\min(\mathbb{Q}_p, \frac{\iota_k(x)}{p}) \in \mathbb{Z}_p[X]$, and this is our assumption.

Now suppose this holds for all indices k, so for all completions of F, and hence for all completions of $\mathbb{Q}(x)$ by Remark 14.15. In this situation, we see that $P = \min(\mathbb{Q}, x)$ is a product of p-divisible polynomials, namely the various minimal polynomials of $\iota(x)$ where ι runs through the nonequivalent completions of $\mathbb{Q}(x)$, and so P is itself p-divisible. As already observed, this amounts to saying that $Q = \min(\mathbb{Q}, \frac{x}{p}) \in \mathbb{Z}_p[X]$; here the coefficients of Q are of the form $\frac{r}{p^s}$ with $r, s \in \mathbb{Z}$, so that we really have $Q \in \mathbb{Z}[X]$. This means that $\frac{x}{p}$ is an algebraic integer, or in other words, that $x = 0 \in \mathcal{O}_F \otimes \mathbb{F}_p$. □

Now, we have seen with Lemma 2.35 that $p\mathcal{O}_k = \mathfrak{p}_k^{e_k}$, where $\mathfrak{p}_k$ is the unique maximal ideal of $\mathcal{O}_k$. Hence, we have an alternative: either $e_k = 1$ and $\mathcal{O}_k/p\mathcal{O}_k$ is a field, or $e_k > 1$ and $\mathcal{O}_k/p\mathcal{O}_k$ has nonzero nilpotent elements (for example, a generator for $\mathfrak{p}_k$). By the proposition just established, we have:

Corollary 14.22 *The prime p ramifies in F if and only if $\mathcal{O}_F/p\mathcal{O}_F$ has nonzero nilpotent elements.* □

Recall that our goal is to show that p ramifies if and only if it divides a certain integer. Now we introduce numbers which can detect, among other things, the presence of nilpotent elements in an algebra.

Definition 14.23 Let A be an algebra over the field F_0, such that $[A : F_0]$ is finite. Let $\varepsilon_1, \ldots, \varepsilon_n$ be a basis for A as an F_0-vector space. We define

$$\Delta_A(\varepsilon_1, \ldots, \varepsilon_n) = \det\left[\mathrm{Tr}_{A/F_0}(\varepsilon_i\varepsilon_j)\right]_{ij} \in F.$$

discriminant of a basis

We call it the **discriminant of a basis** $\varepsilon_1, \ldots, \varepsilon_n$.

Thus, we use the general definition of traces, which was given in the opening lines of this book (see page 281).

Lemma 14.24 *If $\varepsilon_1', \ldots, \varepsilon_n'$ is another basis for A, then*

$$\Delta_A(\varepsilon_1', \ldots, \varepsilon_n') = x^2\,\Delta_A(\varepsilon_1, \ldots, \varepsilon_n),$$

for some $x \in F_0^\times$.

Proof. Let $M \in \mathrm{M}_n(F_0)$ be an invertible matrix with

$$\begin{pmatrix} \varepsilon'_1 \\ \vdots \\ \varepsilon'_n \end{pmatrix} = M \begin{pmatrix} \varepsilon_1 \\ \vdots \\ \varepsilon_n \end{pmatrix}.$$

Now consider the following identity of matrices with coefficients in A:

$$\begin{aligned} [\varepsilon'_i\varepsilon'_j]_{ij} &= \begin{pmatrix} \varepsilon'_1 \\ \vdots \\ \varepsilon'_n \end{pmatrix} (\varepsilon'_1, \ldots, \varepsilon'_n) \\ &= M \begin{pmatrix} \varepsilon_1 \\ \vdots \\ \varepsilon_n \end{pmatrix} (\varepsilon_1, \ldots, \varepsilon_n)\, {}^tM \\ &= M \left[\varepsilon_i\varepsilon_j\right]_{ij} {}^tM\,. \end{aligned}$$

We apply $\mathrm{Tr} = \mathrm{Tr}_{A/F_0}$ coefficient by coefficient, in order to obtain an identity of matrices with coefficients in F_0. Keeping in mind that Tr is F_0-linear, we simply get

$$\left[\mathrm{Tr}(\varepsilon'_i\varepsilon'_j)\right]_{ij} = M \left[\mathrm{Tr}(\varepsilon_i\varepsilon_j)\right]_{ij} {}^tM\,.$$

Now apply det on both sides, to deduce the statement of the lemma with $x = \det(M)$. □

As a result, the discriminant of a basis is either always zero (for all bases) or never zero. This is an intrinsic property of the algebra A.

Definition 14.25 We denote by $\widetilde{\Delta}_A$ the class of $\Delta_A(\varepsilon_1, \ldots, \varepsilon_n)$, where we have picked a basis $\varepsilon_1, \ldots, \varepsilon_n$ of A over F_0, in

$$F_0/F_0^{\times 2} = \{0\} \cup F_0^{\times}/F_0^{\times 2}\,.$$

By the last lemma, this does not depend on the choice of basis. We call it the **discriminant class** of A.

discriminant class

Example 14.26 Suppose A/F_0 is a separable extension of fields. Then $(x,y) \mapsto \mathrm{Tr}_{A/F_0}(xy)$ is a nondegenerate bilinear form (see page 281). Thus the matrix $[\mathrm{Tr}_{A/F_0}(\varepsilon_i\varepsilon_j)]_{ij}$ is invertible, and the discriminant class of A is nonzero.

Now suppose on the contrary that A possesses a nonzero nilpotent element ε_1. Complete ε_1 to a basis $\varepsilon_1, \ldots, \varepsilon_n$ of A. Since $\varepsilon_1\varepsilon_j$ is nilpotent for all j, we have $\mathrm{Tr}_{A/F_0}(\varepsilon_1\varepsilon_j) = 0$. The discriminant class of A is thus 0, since we take the determinant of a matrix whose entries on the first line are all zero.

This example shows that the discriminant class is the perfect tool for deciding whether the $\mathbb{F}_p$-algebra $\mathcal{O}_k/p\mathcal{O}_k$ is a field, or has nonzero nilpotent elements, which is our alternative (keep in mind that, when $\mathcal{O}_k/p\mathcal{O}_k$ is a field, it is certainly a separable extension of $\mathbb{F}_p$).

discriminant of a number field

Definition 14.27 Let F be a number field, and let $\varepsilon_1,\dots,\varepsilon_n$ be a $\mathbb{Z}$-basis for $\mathcal{O}_F$ (which is also a $\mathbb{Q}$-basis for F). The **discriminant of a number field** of F is by definition

$$\Delta_F := \Delta_F(\varepsilon_1,\dots,\varepsilon_n).$$

Lemma 14.28 *The discriminant is well-defined. It is an integer.*

Proof. First, Proposition 14.20 shows that there exist bases as proposed in the definition. Second, the argument given in the proof of Lemma 14.24 show that Δ_F is well-defined up to $\det(M)^2$, where $M \in \mathrm{GL}_n(\mathbb{Z})$; such an M satisfies $\det(M) = \pm 1$, so the definition of Δ_F makes sense.

Further, each element $\varepsilon_i\varepsilon_j$ is an algebraic integer, so that its trace down to $\mathbb{Q}$, which can be read from the minimal polynomial, is in $\mathbb{Z}$. We conclude that $\Delta_F \in \mathbb{Z}$. □

Example 14.29 Let us revisit the number fields of Example 14.19. For $F = \mathbb{Q}(\sqrt{3})$, a possible basis is $\varepsilon_1 = 1$, $\varepsilon_2 = \sqrt{3}$. We have $\mathrm{Tr}(1) = 2$, $\mathrm{Tr}(3) = 6$, $\mathrm{Tr}(\sqrt{3}) = \sqrt{3} - \sqrt{3} = 0$, where $\mathrm{Tr} = \mathrm{Tr}_{F/\mathbb{Q}}$. The discriminant is thus

$$\begin{pmatrix} 2 & 0 \\ 0 & 6 \end{pmatrix} = 12.$$

For $F = \mathbb{Q}(\sqrt{5})$, the reader should compute that the discriminant is 5.

Theorem 14.30 *The prime number p ramifies in the number field F if and only if p divides Δ_F.*

Proof. Proposition 14.21 describes $\mathcal{O}_F \otimes \mathbb{F}_p$ as the product of the algebras $\mathcal{O}_i \otimes \mathbb{F}_p$. Here is an elementary exercise for the reader: If $\widetilde{\Delta}$ is the discriminant class of $\mathcal{O}_F \otimes \mathbb{F}_p$, and $\widetilde{\Delta}_i$ is the discriminant class of $\mathcal{O}_i \otimes \mathbb{F}_p$, then one has

$$\widetilde{\Delta} = \widetilde{\Delta}_1 \cdots \widetilde{\Delta}_g.$$

(This is a general fact about product algebras.)

We know that p ramifies in F if and only if there is an index k such that $\widetilde{\Delta}_k = 0$, and so this happens if and only if $\widetilde{\Delta} = 0$.

Now, pick a basis $\varepsilon_1,\dots,\varepsilon_n$ for F, comprised of algebraic integers, so that $\Delta_F(\varepsilon_1,\dots,\varepsilon_n) = \Delta_F$. It is obvious that $\varepsilon_1,\dots,\varepsilon_n$ give a basis for $\mathcal{O}_F \otimes \mathbb{F}_p$, so that the reduction of Δ_F mod p, as an element of $\mathbb{F}_p$, is a representative for $\widetilde{\Delta}$. The theorem follows. □

The integer Δ_F has been the object of much study. Particularly remarkable is:

Theorem 14.31 (Minkowski) *Let F be a number field, and $n = [F : \mathbb{Q}]$. There exists an integer $s \leq \frac{n}{2}$ such that*

$$\sqrt{|\Delta_F|} \geq \left(\frac{\pi}{4}\right)^s \frac{n^n}{n!}.$$

We shall not provide a proof of Minkowski's theorem here. One reason is that the techniques involved, such as the study of the geometry of convex bodies in euclidean space, are completely different from those introduced so far, and would make for a long digression.

Luckily, the proof is a largely self-contained affair. Apart from reading a few basic definitions about the so-called "fractional ideals" of $\mathcal{O}_F$ (which are certainly worth knowing, and are surely familiar to many readers), you can open [Stea] directly at section 7.1 (corollary 7.1.9 in particular), or [Steb] at section 5 (corollary 5.10 in particular), or [Neu99] at chapter I, section 5 (exercise 3 in particular). Mostly any textbook on algebraic number theory will supply a proof, and it is almost always the same proof, too.

Corollary 14.32 *If F is a number field and $F \neq \mathbb{Q}$, then some prime number p ramifies in F.*

Proof. Let $n = [F : \mathbb{Q}]$. It is an elementary exercise to show that

$$\left(\frac{\pi}{4}\right)^s \frac{n^n}{n!} > 1$$

when $n > 1$ and $s \leq \frac{n}{2}$, whence the result by Minkowski's theorem and Theorem 14.30. □

As promised, we deduce:

Corollary 14.33 *Suppose $F/\mathbb{Q}$ is Galois. Then the group* $\mathrm{Gal}(F/\mathbb{Q})$ *is generated by the various inertia subgroups, considered for all primes p and all completions above them.*

Proof. By Lemma 14.16, the fixed field L corresponding to the subgroup generated by the inertia subgroups has no ramification. By the previous corollary, we have $L = \mathbb{Q}$. □

The (global) Kronecker–Weber theorem

There is a version of the Kronecker–Weber Theorem for number fields, showing that any abelian extension of $\mathbb{Q}$ is contained in a cyclotomic extension, and we turn to the proof. We start with a simple remark about abelian extensions.

Lemma 14.34 *Let F be a number field which is Galois over $\mathbb{Q}$, and suppose that* $\mathrm{Gal}(F/\mathbb{Q})$ *is abelian. For any prime number p, let I_p be the corresponding inertia subgroup. Then*

$$|\mathrm{Gal}(F/\mathbb{Q})| \leq \prod_p |I_p|,$$

where the product is over all prime numbers.

Proof. We know now that for all prime numbers, except those dividing Δ_F, the inertia subgroup is trivial, so the product is really a finite one. Now let $p_1, \ldots, p_s$ be the primes that do ramify in F. By Corollary 14.33, the group $\mathrm{Gal}(F/\mathbb{Q})$ is generated by the subgroups $I_{p_1}, \ldots, I_{p_s}$, so

$$\mathrm{Gal}(F/\mathbb{Q}) = I_{p_1} \cdots I_{p_s},$$

since we are dealing with abelian groups. The inequality follows. □

Theorem 14.35 (Kronecker–Weber) *Let $F/\mathbb{Q}$ be a finite abelian extension. Then F is contained in a cyclotomic extension of $\mathbb{Q}$.*

Proof. ([Neu99], chapter V, section 1, theorem 1.10.) Let p be a prime number, and let $\iota\colon F \to F_p$ be a completion (with values in a fixed algebraic closure $\overline{\mathbb{Q}}_p$ of $\mathbb{Q}_p$). Then $\mathrm{Gal}(F_p/\mathbb{Q}_p)$ is abelian (indeed, it is isomorphic to a subgroup of $\mathrm{Gal}(F/\mathbb{Q})$ via ι^*). By the local Kronecker–Weber theorem, that is Theorem 13.27, there is an integer n_p such that

$$F_p \subset \mathbb{Q}_p\big(\mu_{n_p}(\overline{\mathbb{Q}}_p)\big).$$

When p does not ramify in F, we can even arrange to have n_p prime to p (Corollary 2.49), and we assume that we have done so. Keep in mind that, since $F/\mathbb{Q}$ is Galois, the field F_p (and so the integer n_p) does not depend on the choice of completion, but merely on p.

Now define $e_p = v_p(n_p)$ (so $e_p = 0$ when p does not ramify in F). Define

$$n = \prod_p p^{e_p}.$$

We shall prove that

$$F \subset \mathbb{Q}\big(\mu_n(\overline{\mathbb{Q}})\big).$$

For this, we put $M = F(\mu_n)$, and we seek to prove that $M = \mathbb{Q}(\mu_n)$. Certainly we have $\mathbb{Q}(\mu_n) \subset M$, so we wish to establish

$$[M:\mathbb{Q}] \leq [\mathbb{Q}(\mu_n):\mathbb{Q}] = \varphi(n).$$

Noting that $M/\mathbb{Q}$ is abelian, this will be achieved using the previous lemma. So the proof reduces to a study of the ramification in M.

Let $\iota\colon M \to M_p \subset \overline{\mathbb{Q}}_p$ be a completion of M, which restricts to a completion $F \to F_p$ of F. Certainly $\iota(M) \subset F_p(\mu_n)$, and this image is dense, so $M_p = F_p(\mu_n)$. Since $F_p \subset \mathbb{Q}_p(\mu_{n_p})$, we have

$$M_p \subset \mathbb{Q}_p(\mu_{n_p})(\mu_n) = \mathbb{Q}_p(\mu_{p^{e_p}m})$$

for some integer m prime to p. By Corollary 2.49, the ramification index $e(M_p)$ divides $\varphi(p^{e_p})$. On the other hand, of course, we have $\mathbb{Q}_p(\mu_{p^{e_p}}) \subset M_p$, so $e(M_p) = \varphi(p^{e_p})$.

We can conclude. Letting I_p be the inertia subgroup at p for the abelian extension $M/\mathbb{Q}$, we have proved that $|I_p| = \varphi(p^{e_p})$. By Lemma 14.34, we have

$$[M:\mathbb{Q}] = \mathrm{Gal}(M/\mathbb{Q}) \leq \prod_p \varphi\left(p^{e_p}\right) = \varphi(n),$$

which was what we wanted. □

Ultimately, in this proof, the passage from p-adic fields to number fields was made possible because we know that a number field in which no prime ramifies must be $\mathbb{Q}$ (it is the essential ingredient in Lemma 14.34 – try to trace it back). This is a "local-to-global" phenomenon. The next section expands upon the terminology. Before we turn to this, however, let us give the application mentioned in the preface:

Corollary 14.36 *Let $P \in \mathbb{Z}[X]$ be monic, and let L be the splitting field of P over $\mathbb{Q}$. Assume that $L/\mathbb{Q}$ is abelian. Then there exists an integer m with the following property: For a prime number p, the question of deciding whether the reduction of P mod p splits into a product of linear factors has an answer which depends only on p mod m, with finitely many exceptions.*

Proof. We may as well, and we do, assume that P is irreducible. First we discard the finitely many prime numbers which ramify in L. Then we apply the Kronecker–Weber theorem, providing us with an integer m such that $L \subset K = \mathbb{Q}(\mu_m)$. As we have seen in the proof of the theorem, we can arrange to have m prime to p whenever p is unramified in L, so that m is prime to all the ps still under consideration. (Alternatively, pick any m and exclude all its prime divisors from the discussion.)

Next, we need to borrow something from the theory of polynomial discriminants, a classical tool in Galois theory which we have not reviewed yet. Let $\alpha_1, \alpha_2, \ldots$ be the roots of P in $\overline{\mathbb{Q}}$ (none is repeated as P is irreducible), and let

$$d = \prod_{i<j} \left(\alpha_i - \alpha_j\right)^2.$$

Then from Galois theory, we see that $d \in \mathbb{Q}$. Moreover, from the elementary theory of symmetric functions, we see that d is given by an integral polynomial expression in the coefficients of P, so in fact $d \in \mathbb{Z}$. Moreover, that polynomial expression depends only on the degree of P, and would be the same for all polynomials of that degree, over all fields. From this observation, we see that the reduction $\overline{P}$ of P modulo a prime p has repeated roots in $\overline{\mathbb{F}}_p$ precisely when p divides d.

Thus we may exclude the divisors of d from the rest of the proof. The benefit is that now, from Hensel's lemma, the polynomial P can be factored into a product of linear factors over $\mathbb{F}_p$ if and only if this can be achieved over $\mathbb{Z}_p$.

We let α denote a root of P, and we put $F = \mathbb{Q}[\alpha]$, so that we have the inclusions $\mathbb{Q} \subset F \subset L \subset K$. Let p be one of the remaining primes, and let us study the completions of F at p. We have $n = efg$ where n is the degree of P, by Lemma 14.10, so $n = fg$ since there is no ramification at p. What we want, as follows from the description of the completions of F, is a condition on p which is equivalent to $g = n$, or to $f = 1$.

All the completions at p have the same image F_p; likewise, all the completions of K have the same image K_p, and $\mathbb{Q}_p \subset F_p \subset K_p \subset \overline{\mathbb{Q}}_p$. We have $f = [F_p : \mathbb{Q}_p]$, so we shall pay attention to the action of $\mathrm{Gal}(K_p/\mathbb{Q}_p)$ on F_p, which is trivial if and only if $f = 1$.

Since p is prime to m, we may apply Proposition 2.46, which asserts that $K_p/\mathbb{Q}_p$ is unramified (noticing that $K_p = \mathbb{Q}_p(\mu_m)$), and that $[K_p : \mathbb{Q}_p]$ is the order of p in the group $(\mathbb{Z}/m\mathbb{Z})^\times$. Moreover, Corollary 2.41 describes $\mathrm{Gal}(K_p/\mathbb{Q}_p)$ as the cyclic group generated by the Frobenius element σ, itself characterized by the property $\sigma(x) \equiv x^p \bmod p$ for all $x \in \mathcal{O}_{K_p}$. Now, we can see $\mathrm{Gal}(K_p/\mathbb{Q}_p)$ as a subgroup of $\mathrm{Gal}(K/\mathbb{Q}) \cong (\mathbb{Z}/m\mathbb{Z})^\times$ (we pick one completion and treat it as an inclusion, a very mild abuse of notation). When we do, we see that $\sigma(x) = x^k$ for some integer k for all $x \in \mu_m(K)$, by the basic theory of cyclotomic extensions of $\mathbb{Q}$. It is easily deduced that $k = p \bmod m$ (take for x a primitive mth root of unity; the cyclic group $\mu_m(K)$ of order m generated by x injects into $\mathbb{K}^\times$, where $\mathbb{K}$ is the residue field of K_p). We conclude that $\mathrm{Gal}(K_p/\mathbb{Q}_p)$ is identified with the subgroup of $(\mathbb{Z}/m\mathbb{Z})^\times$ generated by p (consistently with the calculation of its order).

We merely have to point out that F_p is fixed by $\mathrm{Gal}(K_p/\mathbb{Q}_p)$ if and only if F is, by density. Our conclusion is that $f = 1$, that is P is a product of linear factors when reduced mod p, if and only if the subgroup of $(\mathbb{Z}/m\mathbb{Z})^\times$ generated by p is contained in $\mathrm{Gal}(K/F)$, under the standard identification $(\mathbb{Z}/m\mathbb{Z})^\times \cong \mathrm{Gal}(K/\mathbb{Q})$. This condition blatantly depends only on p mod m. □

In applications, one usually follows the steps of the above proof, rather than relying on the statement of the corollary itself. For example, in Problem 14.3, we ask you to deduce the the celebrated *quadratic reciprocity law* in this way.

We point out that this type of result, which asserts that something happens mod p if and only if p has some property mod m, explains why "reciprocity" is seen everywhere in class field theory (as in the "reciprocity isomorphism", for example).

The local and global terminology

The reader may well be wondering where the expressions "local" and "global" come from, since we have not yet given any intuition behind it. The short explanation is this: When something has been proved of all p-adic fields, for all primes p, one can sometimes deduce an analogous property of number fields. The Kronecker–Weber theorem, with the proof we have given, is a paradigm. Another one is the Hasse–Minkowski theorem mentioned in the introduction to Chapter 2.

There is actually a rather detailed analogy. Try to think of a number field as an *irreducible algebraic curve*, if you know what that is; if not, the metaphor works great (as long as we do not seek too much precision) with connected, compact topological surfaces, and we shall employ this language. The field $\mathbb{Q}$ will play the role of the sphere S^2 (or $\mathbb{P}^1$ if you prefer).

To be a little more precise, a number field $F/\mathbb{Q}$ is placed in analogy with a (connected, compact, topological) surface Σ endowed with a continuous map $f\colon \Sigma \to S^2$. Moreover, we assume that f is a *ramified cover*; that is, around any point $x \in \Sigma$, we assume that f looks like $z \mapsto z^e$ for z around the origin in the complex plane. (Note that the integer $e = e_x$ at x is then uniquely defined.)

In the analogy, the prime numbers correspond to points of S^2, and the completions of F to points of Σ. An unramified prime "is" a point $y \in S^2$ such that, for any $x \in \Sigma$ with $f(x) = y$, the map f is locally a homeomorphism around x (in other words, $e_x = 1$). The other points are called ramification points, and the integer e_x defined in the previous paragraph is the ramification index at x.

The claim is then that this analogy is quite valid. That is, if we observe facts in the realm of surfaces and ramified covers, which have analogous statements in the world of number fields, then more often than not, the analogous statements should be true. Consider the following examples. Given $f\colon \Sigma \to S^2$ as above, there are finitely many ramification points on S^2, since they clearly form a closed, hence compact, discrete subset of S^2; after a considerable effort, we have proved the same thing about ramification in number fields. Similarly, we can also see that f must have some ramification points, unless $\Sigma = S^2$ and f is the identity. Indeed, in the absence of ramification points, we conclude that f is a covering map of S^2, which is simply connected; but a simply connected topological space has no nontrivial, connected covering spaces. Analogously, the only number field F such that no prime ramifies in $\mathbb{Q}$ is $\mathbb{Q}$ itself, as we have shown.

It is a good idea to keep this analogy in mind when trying to reason about number fields. One can also decide to push the analogy much further, and study the "scheme" defined by the ring $\mathcal{O}_F$. In this approach, one redefines what "curves" and "points" are, very abstractly, and in the end the ring $\mathcal{O}_F$ really becomes a curve, whose points really are the completions of F. This theory is extremely powerful, but depending on what you want to do with number fields, it may be largely overkill.

Some statements from global class field theory

We have called class field theory "a loosely defined body of results", and this applies to global class field theory just as much as it does in the local case. What is more, the statements now come in two flavors: There is the "ideal-theoretic" version, involving the ideals of the ring $\mathcal{O}_F$, and the "cohomological" version, which is more similar to the local theory, as presented in this book. The two are in principle equivalent, and translating from one to the other is of course possible, but absolutely nontrivial. Ideal-theoretic class field theory is needed for most of the applications to algebraic number theory, and it is noteworthy that elementary proofs exist for most of it: see [Chi09]. The trade-off is that the statements are a little awkward, and the proofs not very enlightening, especially in comparison to the more clean-cut cohomological theory.

Here we shall be content with stating one significant theorem from global class field theory. We say nothing of the proof, but the reader can take our word that the arguments have a lot in common with those we have given in the local case.

Definition 14.37 Consider the ring

$$A = \mathbb{R} \times \prod_p \mathbb{Q}_p,$$

where p runs through all the prime numbers, and the operations are defined coordinate-wise. When $a \in A$ and p is a prime, we write a_p for the image of a in $\mathbb{Q}_p$ under the canonical projection $A \to \mathbb{Q}_p$.

We define $\mathbf{A}_\mathbb{Q}$ to be the subring comprised by those elements $a \in A$ having the property that $a_p \in \mathbb{Z}_p$ for all but finitely many primes p. It is easily seen that $\mathbf{A}_\mathbb{Q}$ is a ring, called the **ring of adèles** of $\mathbb{Q}$.

ring of adèles

For example, when $a \in \mathbb{Q}$, it defines an adèle $(a, a, a, \ldots)$, clearly (when p does not divide the denominator of a, we have $a \in \mathbb{Z}_p$). The map $\mathbb{Q} \to \mathbf{A}_\mathbb{Q}$ is an injective ring homomorphism, and we shall often see $\mathbb{Q}$ as a subring of $\mathbf{A}_\mathbb{Q}$. Moreover, we can see $\mathbf{A}_\mathbb{Q}$ as a vector space over $\mathbb{Q}$, so that the next definition makes sense.

Definition 14.38 Let F be a number field. We define

$$\mathbf{A}_F := F \otimes_\mathbb{Q} \mathbf{A}_\mathbb{Q},$$

and call it the ring of adèles of F.

Given our earlier discussion of $F \otimes_\mathbb{Q} \mathbb{Q}_p$, it will not surprise the reader to hear that $\mathbf{A}_F$ can be expressed as a subring of the direct product of all the possible completions of F (including completions which are isomorphic to $\mathbb{R}$ or $\mathbb{C}$), though we shall not prove it here. We merely want to record that F can be seen as a subring of $\mathbf{A}_F$, under $f \mapsto f \otimes 1$.

Definition 14.39 The group of invertible elements (units) in the ring $\mathbf{A}_F$ is written $\mathbf{I}_F$ and called the **group of idèles** of F. (So an idèle is an invertible adèle.) The notation J_F is also common.

group of idèles

The quotient $\mathbf{I}_F / F^\times$ is written $\mathbf{C}_F$ and called the **idèle class group** of F.

idèle class group

Now suppose L/F is a Galois extension of number fields. The norm map $\mathrm{N}_{L/F}$ can be extended to

$$\mathrm{N}_{L/F} \otimes \mathrm{id} \colon \mathbf{A}_L \longrightarrow \mathbf{A}_F,$$

although it is common to call the extension again $\mathrm{N}_{L/F}$. It takes $L^\times$ into $F^\times$, and thus there is a map

$$\mathrm{N}_{L/F}\colon \mathbf{C}_L \longrightarrow \mathbf{C}_F.$$

It is possible to put a topology on $\mathbf{A}_F$ (which is *not* the product topology), turning it into a topological ring; then one defines a topology on $\mathbf{I}_F$ (which is *not* the subspace topology, but rather that defined in Problem 3.8) and subsequently on $\mathbf{C}_F$, turning these into topological groups.

We finish this book with the following beautiful theorem.

Theorem 14.40 (Fundamental theorem of global class field theory) *Let F be a number field. There exists a one-to-one, order-reversing correspondence between the finite, abelian extensions of F contained in $\overline{\mathbb{Q}}$ and the open subgroups of $\mathbf{C}_F$ (which all have finite index). The correspondence maps L/F to $\mathrm{N}_{L/F}(\mathbf{C}_L)$.*

Moreover, there is a "reciprocity isomorphism"

$$\mathrm{Gal}(L/F) \cong \mathbf{C}_F/\mathrm{N}_{K/F}(\mathbf{C}_L).$$

At the heart of the proof is Tate's theorem 13.11, and $\mathbf{C}_F$ is really a replacement for $F^\times$. The story is rather a long one, and will not be told here.

Problems

14.1. Let d be an integer which is not divisible by any nontrivial square, and let $F = \mathbb{Q}[\sqrt{d}]$.

1. Show that the minimal polynomial of $x = a + b\sqrt{d} \in F$ is $X^2 - 2aX + (a^2 - db^2)$. Here $a, b \in \mathbb{Q}$.
2. Deduce that x as above is an algebraic integer if and only if one has either (i) $a, b \in \mathbb{Z}$, or (ii) a, b are both half-integers, that is $a = r/2$ and $b = s/2$ with r, s odd, and $r^2 - ds^2$ is divisible by 4.
3. Show that the ring of integers $\mathcal{O}_F$ is strictly larger than $\mathbb{Z}[\sqrt{d}] \Leftrightarrow \frac{1+\sqrt{d}}{2} \in \mathcal{O}_F \Leftrightarrow \mathcal{O}_F = \mathbb{Z}\left[\frac{1+\sqrt{d}}{2}\right] \Leftrightarrow d = 1 \bmod 4$.
4. Show that the discriminant of F is d when $d = 1 \bmod 4$, and is $2d$ otherwise.

14.2. Let p be an odd prime. On the cyclic group $(\mathbb{Z}/p\mathbb{Z})^\times$, the unique nontrivial homomorphism

$$(\mathbb{Z}/p\mathbb{Z})^\times \longrightarrow \{\pm 1\}$$

is called the Legendre symbol, and written $x \mapsto \left(\frac{x}{p}\right)$. In other words, $\left(\frac{x}{p}\right) = 1$ if and only if x is a square.

This is often extended to $\mathbb{Z}/p\mathbb{Z}$ by setting $\left(\frac{0}{p}\right) = 0$, and by precomposing with $\mathbb{Z} \to \mathbb{Z}/p\mathbb{Z}$, it is possible to view the Legendre symbol as defined on $\mathbb{Z}$.

Let $\omega = \exp\left(\frac{2i\pi}{p}\right) \in \mathbb{C}$, and consider the *Gauss sum*

$$\alpha = \sum_k \left(\frac{k}{p}\right) \omega^k,$$

where the sum is taken over all $k \in (\mathbb{Z}/p\mathbb{Z})^\times$. We shall prove that $\alpha = \sqrt{\pm p}$.

1. Show that one may write

$$\alpha^2 = a_0 + a_1\omega + \cdots + a_{p-1}\omega^{p-1}$$

   where the a_i are integers, they sum to 0, and $a_0 = \left(\frac{-1}{p}\right) \times (p-1)$.
   Hint: The sum of all the $\left(\frac{k}{p}\right)$ *is* 0.
2. Show that the elements of $\mathrm{Gal}(\mathbb{Q}[\omega]/\mathbb{Q})$ take α to $\pm\alpha$, and deduce that $\alpha \notin \mathbb{Q}$ while $\alpha^2 \in \mathbb{Q}$.
3. From the above, conclude that

$$a_0 - \alpha^2 = a_1 = a_2 = \cdots = a_{p-1},$$

   and then that

$$\alpha^2 = \frac{p}{p-1} a_0 = \left(\frac{-1}{p}\right) \times p.$$

   Hint: A basis for $\mathbb{Q}[\omega]$ *as* $\mathbb{Q}$*-vector space is* $1, \omega, \ldots, \omega^{p-2}$.
4. As an application, show the following particular case of the Kronecker–Weber theorem: Any field of the form $\mathbb{Q}[\sqrt{d}]$ is contained in a cyclotomic extension of $\mathbb{Q}$.

14.3. Let p and q be distinct, odd primes. The *quadratic reciprocity law* asserts that

$$\left(\frac{p}{q}\right)\left(\frac{q}{p}\right) = (-1)^{\frac{p-1}{2}\frac{q-1}{2}}.$$

As a preliminary, show that this is equivalent to Gauss's historical formulation: If $q \equiv 1 \bmod 4$, then the congruence $x^2 \equiv p \bmod q$ has a solution if and only if the congruence $x^2 \equiv q \bmod p$ has one; if $q \equiv -1 \bmod 4$, then the congruence $x^2 \equiv p \bmod q$ has a solution if and only if the congruence $x^2 \equiv -q \bmod p$ has one.

Then, prove the quadratic reciprocity law by following the proof of Corollary 14.36 in the particular case of the polynomial $P = X^2 - q$. Use that, by the previous problem, the field $\mathbb{Q}\left[\sqrt{\left(\frac{-1}{q}\right)q}\right]$ is contained in $\mathbb{Q}\left[\exp\left(\frac{2\pi i}{q}\right)\right]$, so the integer m can be taken to be either q or $2q$.

Of course, many elementary proofs of the quadratic reciprocity law exist. One of the nicest is that given by Rousseau in [Rou91] (it is also possible to derive it by playing further with Gauss sums, see [Neu99] or [Ser73]). However, the sheer fact that the method used here generalizes to become Corollary 14.36 shows its depth.

14.4. Let F be a number field.

1. Let $A \subset \mathcal{O}_F$ be a subring, which has finite index as an abelian subgroup of $\mathcal{O}_F$. Define the discriminant of A and relate it to the discriminant of F.
2. Apply this to $F = \mathbb{Q}[\alpha]$ and $A = \mathbb{Z}[\alpha]$, where $\alpha \in \overline{\mathbb{Q}}$. Relate the discriminant of A to the discriminant of the minimal polynomial of α, as defined in the proof of Corollary 14.36.

   In turn, the proof of that corollary can be shortened a little bit, using this problem. Can you see how?

14.5. Let $P \in \mathbb{Q}[X]$, and let K denote a splitting field for P. Let n be the degree of P, and let $G = \mathrm{Gal}(K/\mathbb{Q})$. Assume that P has no repeated roots in $\overline{\mathbb{Q}}$ (for example P might be irreducible). We see G as a subgroup of the symmetric group S_n using its action on the roots of P.

1. Suppose p is a prime number not dividing the discriminant of P (the discriminant of a polynomial was defined in the proof of Corollary 14.36, and we assume that you have worked through the previous problem). Write the factorization of P over $\mathbb{F}_p$ as $\overline{P} = P_1 P_2 \dots P_r$, where each P_i is irreducible, of degree n_i, say. Show that the group G contains a permutation whose cycle type is $(n_1, n_2, \dots, n_r)$.

   (For example, if $\overline{P} = P_1 P_2$ with P_1 of degree 3 and P_2 of degree 2, then G contains a permutation of the form $(a,b,c)(d,e)$.)

   Hint: It is relevant that a compositum of unramified extensions is still unramified, and hence, cyclic.
2. It is useful to use a computer for this question. Take $P = X^6 + 2X + 2$. Compute its discriminant, and check that the primes dividing it are $2, 89$, and 227. Use the above question with $p = 7$ to show that $\mathrm{Gal}(K/\mathbb{Q})$ contains a permutation of cycle type $(3,2,1)$, and then take $p = 11$ to show that the same group contains a permutation of cycle type $(5,1)$. Prove that this implies that $\mathrm{Gal}(K/\mathbb{Q}) = S_6$, the full symmetric group.

   This example is taken from Leonard Soicher's PhD thesis, "The computation of Galois groups" (Concordia University), unpublished but available online.

An amazing result by Frobenius asserts that, as you try more and more primes and count the occurrences of a given cycle type, the frequency converges to the actual proportion of elements in the Galois group having this cycle type. This is usually deduced from the Chebotarev density theorem (see [Neu99]), which in turn requires global class field theory, but more elementary arguments are possible: see [LS96].

Appendix: background material

Norms and traces

Here we keep things brief, and refer the reader to Morandi [Mor96, chapter II, section 8] for details.

Let F be a field, and let K be a commutative ring with $F \subset K$; we also assume that K is finite-dimensional as a vector space over F. When $a \in K$, we let $m_a : K \to K$ be the F-linear homomorphism given by multiplication by a, so $m_a(x) = ax$. We define

$$\mathrm{N}_{K/F}(a) = \det(m_a), \qquad \mathrm{Tr}_{K/F}(a) = \mathrm{Tr}(m_a),$$

where Tr is just the usual trace operator. We call these the norm and trace of a; they are elements of F. It is clear that $\mathrm{Tr}_{K/F} : K \to F$ is F-linear, and $\mathrm{N}_{K/F}$ is multiplicative (as $m_{ab} = m_a \circ m_b$).

Suppose for example that a is nilpotent, that is, $a^k = 0$ for some $k > 0$. Then m_a is also nilpotent as an operator, and thus $\mathrm{Tr}_{K/F}(a) = 0$. Moreover, for any $x \in K$, the element ax is also nilpotent, and in the end we see that $x \mapsto \mathrm{Tr}_{K/F}(ax)$ is identically 0. This is chiefly the fact that we wanted to highlight, when K is not a field (roughly speaking, in the last chapter of this book, it will be convenient to characterize fields among rings by the non-vanishing of certain traces). From now on, we assume that K is a field.

If we write $\chi_a \in F[X]$ for the characteristic polynomial of m_a, we have thus

$$\chi_a = X^n - \mathrm{Tr}_{K/F}(a)X^{n-1} + \cdots + (-1)^n \mathrm{N}_{K/F}(a),$$

where $n = \dim_F K$. Now let $P = \min(F, a)$, the minimal polynomial of a in the sense of Galois theory, which is also the minimal polynomial of m_a in the sense of linear algebra. Let $m = \deg(P)$. The Cayley–Hamilton theorem asserts that P divides χ_a, and it is classical that these two polynomials have the same roots (in an algebraic closure of K). However, the polynomial P is irreducible (since K is a field!), so we can conclude that $\chi_a = P^{\frac{n}{m}}$. If we write

$$P = X^m + c_{m-1}X^{m-1} + \cdots + c_0 = (X - \alpha_1) \cdots (X - \alpha_m),$$

we have thus

$$\mathrm{Tr}_{K/F}(a) = -\frac{n}{m}c_{m-1} = \frac{n}{m}(\alpha_1 + \cdots + \alpha_m),$$

and

$$\mathrm{N}_{K/F}(a) = (-1)^n c_0^{\frac{n}{m}} = (\alpha_1 \cdots \alpha_m)^{\frac{n}{m}}.$$

When $K = F$ is Galois, it is now a simple exercise to show that the norm and trace are indeed given by

$$\mathrm{N}_{K/F}(a) = \prod_{\sigma \in \mathrm{Gal}(K/F)} \sigma(a), \qquad \mathrm{Tr}_{K/F}(a) = \sum_{\sigma \in \mathrm{Gal}(K/F)} \sigma(a)$$

as announced on page 2. Better yet, assuming only that $K = F$ is separable, but not necessarily Galois, we have

$$\mathrm{N}_{K/F}(a) = \prod_{\sigma} \sigma(a), \qquad \mathrm{Tr}_{K/F}(a) = \sum_{\sigma} \sigma(a),$$

where σ runs now through the distinct F-homomorphisms $\sigma : K \to \overline{K}$. From this, one deduces easily

$$\mathrm{N}_{L/F} = \mathrm{N}_{K/F} \circ \mathrm{N}_{L/K}, \qquad \mathrm{Tr}_{L/F} = \mathrm{Tr}_{K/F} \circ \mathrm{Tr}_{L/K},$$

when $F \subset K \subset L$, with L/F separable.

An interesting application is this: When K/F is separable, there always exist $a_0 \in K$ with $\mathrm{Tr}_{K/F}(a_0) \neq 0$. To see this, embed K in a field L with L/F finite and Galois; if we can find $b \in L$ with $\mathrm{Tr}_{L/F}(b) \neq 0$, then $a_0 := \mathrm{Tr}_{L/K}(b)$ works for us. However, Dedekind's lemma on the independence of characters shows that $\sum_{\sigma \in \mathrm{Gal}(L/F)} \sigma$ is not identically 0, and so indeed, there exists b with $\sum_{\sigma} \sigma(b) \neq 0$. (In characteristic 0, we could have taken $a_0 = 1$, so $\mathrm{Tr}_{K/F}(a_0) = n \neq 0$.)

As a result, when $a \in K$ is nonzero, the linear form $x \mapsto \mathrm{Tr}_{K/F}(ax)$ is never identically 0 (with K/F separable): simply take $x = a_0 a^{-1}$. In other parlance, the bilinear form $(a,x) \mapsto \mathrm{Tr}_{K/F}(ax)$ is, in this case, nondegenerate.

Tensor products

We provide the bare minimum to understand tensor products. For more, see [Lan02, chapter XVI]. Tensor products are not used in Part I; they are used heavily in Part II, but only in the case of vector spaces; in Part III, a generalization is required, and it is presented when needed.

Throughout, we write R for a commutative ring with 1. We first present tensor products as a way to turn bilinear maps into linear ones.

Proposition A.1 *Let A and B be R-modules. There exists an R-module written $A \otimes_R B$, and called the tensor product of A and B over R, such that:*

1. *there exists a bilinear map $A \times B \longrightarrow A \otimes B$, which on elements is denoted by $(a,b) \mapsto a \otimes b$;*
2. *if $f\colon A \times B \to C$ is a bilinear map to an R-module C, then there exists a unique linear map $g\colon A \otimes_R B \longrightarrow C$ such that $f(a,b) = g(a \otimes b)$.*

What is more, the module $A \otimes_R B$ and its map from $A \times B$ are unique in a canonical way, described in the proof.

Proof. Let $A \boxtimes B$ denote the free R-module on a basis made of symbols $a \boxtimes b$, one for each pair $(a,b) \in A \times B$. Let M be the submodule generated by all elements of the form

$$(a_1 + a_2) \boxtimes b - a_1 \boxtimes b - a_2 \boxtimes b, \qquad a \boxtimes (b_1 + b_2) - a \boxtimes b_1 - a \boxtimes b_2,$$
$$(ra) \boxtimes b - r(a \boxtimes b), \qquad a \boxtimes (rb) - r(a \boxtimes b).$$

Put $A \otimes_R B = (A \boxtimes B)/M$, and write $a \otimes b$ for the class of $a \boxtimes b$ in M. Properties (1) and (2) are true by construction.

Now suppose T is another R-module equipped with a bilinear map $A \times B \to T$ which satisfies property (2). Applying (2) twice, we derive the existence of certain maps $A \otimes_R B \to T$ and $T \to A \otimes_R B$; by the uniqueness statement in (2), these must be the identity. □

Example A.2 Suppose $R = \mathbb{Z}$, let A be any abelian group, and let $B = \mathbb{Z}/n\mathbb{Z}$. We claim that $A \otimes_{\mathbb{Z}} \mathbb{Z}/n\mathbb{Z} \cong A/nA$. Indeed, the map $A \times \mathbb{Z}/n\mathbb{Z} \longrightarrow A/nA$ taking (a,k) to ka is visibly (well-defined and) bilinear. Thus, there is a map

$$A \otimes_R \mathbb{Z}/n\mathbb{Z} \longrightarrow A/nA$$

taking $a \otimes k$ to ka. In the other direction, we may directly define

$$A \longrightarrow A \otimes_R \mathbb{Z}/n\mathbb{Z}$$

by $a \mapsto a \otimes 1$. This vanishes on nA, since it takes na to $(na) \otimes 1 = n(a \otimes 1) = a \otimes n = a \otimes 0 = 0$. The induced map $A/nA \to A \otimes_R \mathbb{Z}/n\mathbb{Z}$ is an inverse for the previous one. ■

The reader must keep in mind that an element of $A \otimes_R B$ is, in general, not of the form $a \otimes b$: We can merely be certain that it is a sum of such "tensors" (or "elementary tensors"). Quite often, you will encounter a map g defined on a tensor product, which is specified only by the values $g(a \otimes b)$: What is meant is that $(a,b) \mapsto g(a \otimes b)$ is bilinear (a verification often left to the reader), so that g is well-defined.

Proposition A.3 *The tensor product operation enjoys the following properties.*

1. *$A \otimes_R R \cong A$.*
2. *$A \otimes_R (B \oplus C) \cong A \otimes_R B \oplus A \otimes_R C$.*

3. *$A \otimes_R (B \otimes_R C) \cong (A \otimes_R B) \otimes_R C$. In fact, if B is also an S-module for some other commutative ring S, and if the two module structures commute ($rsb = srb$), then $A \otimes_R B$ is naturally an S-module with $s \cdot (a \otimes b) = a \otimes sb$, and $A \otimes_R (B \otimes_S C) \cong (A \otimes_R B) \otimes_S C$.*

We leave the proof to the reader.

Corollary A.4 *If F is a field, and if A and B are F-vector spaces of dimensions n and m respectively, then $A \otimes_F B$ has dimension nm. If $e_1, \ldots, e_n$ is a basis for A, and if $\varepsilon_1, \ldots, \varepsilon_m$ is a basis for B, then a basis for $A \otimes_F B$ is provided by the $e_i \otimes \varepsilon_j$.*

Proof. More generally, if B is isomorphic to R^m for some $m \geq 1$, then properties (1) and (2) guarantee that $A \otimes_R B$ is isomorphic to A^m. As a result, in the situation of the corollary, $A \otimes_F B$ is isomorphic to A^m and so has dimension nm. Since the $e_i \otimes \varepsilon_j$ visibly generate $A \otimes_F B$ (as we see, for example, by returning to the construction as a quotient of $A \boxtimes B$), they must form a basis. □

Proposition A.5 *There is a canonical isomorphism of R-modules*

$$\mathrm{Hom}_R(A \otimes_R B, C) \cong \mathrm{Hom}_R(B, \mathrm{Hom}_R(A, C)).$$

Proof. Both modules are isomorphic to the module of bilinear maps $A \times B \to C$, so the result is not surprising. To prove it, with $g\colon A \otimes_R B \to C$ we associate $b \mapsto g(a \otimes b)$, and conversely, starting from $f\colon B \to \mathrm{Hom}_R(A, C)$, we build the homomorphism $a \otimes b \mapsto f(b)(a)$. The latter is well-defined, since $(a, b) \mapsto f(b)(a)$ is bilinear, so factors through the tensor product. □

Consider again the case of vector spaces over a field F, and pick $C = F$. Let us write V^* instead of $\mathrm{Hom}_F(V, F)$. We have thus proved that

$$(A \otimes_F B)^* \cong \mathrm{Hom}_R(B, A^*). \tag{*}$$

However, we can also prove that

$$(A \otimes_F B)^* \cong A^* \otimes_F B^*. \tag{**}$$

Indeed, we associate $\varphi \otimes \psi \in A^* \otimes_F B^*$ with the map $a \otimes b \to \varphi(a)\psi(b)$, a well-defined element of $(A \otimes B)^*$. Let us check that this defines an isomorphism. For concreteness, say that A and B are finite-dimensional, let the e_i form a basis for A and the ε_j form a basis for B, so that the $e_i \otimes \varepsilon_j$ form a basis for $A \otimes_F B$, as already observed. We write e_i^*, ε_j^*, and $(e_i \otimes \varepsilon_j)^*$ for the elements of the dual bases. Then $e_i^* \otimes \varepsilon_j^*$ is taken, by our proposed map, to $(e_i \otimes \varepsilon_j)^*$. This establishes (**)

Now apply (*) with A^* and B, and then apply (**). Appealing to the canonical isomorphism between A and its bidual A^{**}, we derive

$$A \otimes_F B^* \cong \mathrm{Hom}_F(B, A).$$

This is used quite frequently. Note that the isomorphism takes $a \otimes \psi$ to the homomorphism $b \mapsto \psi(b)a$.

In Chapter 9, we encounter tensor products over non-commutative rings – see the discussion around Proposition 9.46. Here we point out that $A \otimes_R B$ can only be formed when A is a right R-module and B is a left R-module; also, $A \otimes_R B$ is only an abelian group in general. However, when R is commutative the new construction does agree with the old one.

Notes and further reading

Part I

Kummer theory is presented here much as it is in [Mor96]. Equivariant Kummer theory is originally just a remark by Waterhouse in the introduction to his paper [Wat94], although it is generally agreed that this was a "folk" result, known to many experts long before (perhaps known to Galois himself in some form or other!).

Our presentation of p-adic fields is a mixture of [Neu99], [Ser79], and [Bla72]. The term "local number field" is a recent one (perhaps introduced only with the LMFDB as in Example 2.32).

To learn more about p-adic fields, the reader can turn to [Ser73], where quadratic forms over $\mathbb{Q}_p$ are studied, and a proof of the Hasse–Minkowski theorem is given (among other things). Another excellent read is [Kob77], which will give the reader a very good sense of what p-adic analysis is.

It must be mentioned that there exist more general "local fields" than local number fields. For example, many of the results we have presented hold true for fields equipped with a non-Archimedean valuation, complete for the induced topology, and such that the residue field, that is

$$\{x : v(x) \geq 0\}/\{x : v(x) > 0\},$$

is finite (or even merely "quasi-finite"). A typical case is $F = \mathbb{F}_p((X))$, the field of formal Laurent series in one variable over $\mathbb{F}_p$, for which $v(f)$ is the least exponent of X appearing in f; its ring of integers is $\mathbb{F}_p[[X]]$, the ring of power series (involving only nonnegative powers of X), and its residue field is $\mathbb{F}_p$.

To go further with profinite groups, we recommend [DdSMS99].

I am told that the proof of Theorem 4.13 given here is new (see Problem 4.4).

Part II

This part follows closely Blanchard's wonderful little book [Bla72]. Another inspiration is [Alp86], where applications of semisimple algebras, radicals, etc., to representation theory are developed (over an algebraically closed field, mostly).

A lot of information about central, simple algebra is given in [GS06], which culminates with a proof of the Merkurjev–Suslin theorem. The reader will also find in this book the definition of cyclic algebras, written $(a,b)_\omega$ where ω is a pth root of unity in the base field, for a prime p; these are generalizations of quaternion algebras (which one recovers when $p = 2$). In a sense, we do include these algebras in this book, but only in the guise of the Hilbert symbol. See Chapter 11 and Remark 11.15 in particular.

Part III

The theory of Ext and Tor is classical now, and many good textbooks explore the subject further, for example [Wei94]. We have avoided the language of categories and functors, which can be very useful in other contexts, but not so much here (perhaps only the proof of Proposition 9.55 could have been replaced by an abstract argument).

Our presentation is simpler, or at any rate shorter, than is usual, because we have not stated, let alone proved, many "naturality" results (for example, the long exact sequences in Theorem 9.43 are natural, in some sense). As indicated at the end of Chapter 9, we can get away with this because in our chief application, group (co)homology, we are able to make definite choices, implying the commutativity of all required diagrams.

The book [NSW08] has been very useful, and should be on your shelf if you plan to study group cohomology in detail. Unlike what the title of that book may indicate, the first few chapters give a very general and complete treatment of the cohomology of profinite groups. The exposition is in fact so encyclopedic that it is, unfortunately, difficult to read [NSW08] page after page.

An obvious omission in our treatment is that we do not provide any duality statement relating homology and cohomology. For example, when G is a finite group and A is a $\mathbb{Z}[G]$-module, one has

$$\mathrm{H}^n(G,A') \cong \mathrm{H}_n(G,A)',$$

where the notation refers to Pontryagin duals; for this, see proposition 1.9.1 in [NSW08]. In fact section 1.9 of that book, which contains this proposition, defines Tate cohomology for profinite groups (rather than merely finite groups), another subject we have avoided. (It is in fact interesting that Galois cohomology can mostly avoid negative Tate cohomology of profinite groups.)

The book [Ser02] contains many results which we have left out. For example, it studies the p-cohomological dimension of a profinite group G, that is, the smallest integer n such that $\mathrm{H}^q(G,A) = 0$ for all $q > n$ and all modules A of p-primary torsion. When G is finite (and nontrivial), this dimension is $+\infty$; it is very much a phenomenon of Galois cohomology to have interesting examples with finite p-cohomological dimension. Chapter II, section 5 of [Ser02] will show that the absolute Galois group of a local number field has p-cohomological dimension 2 for all p, as promised in Example 11.9.

Chapter 11 borrows much from [Ser79]. We have taken the opportunity, with Corollary 11.4, to present Connell's method for deriving statements from Hilbert 90, which is not so well-known: see [Con65].

The Arason exact sequence appeared in [Ara75]. See also [Ser02, section I.2, exercise 2].

For the proof of the Bloch–Kato conjecture, see [Voe11] (a difficult read, of course!). The overview [Wei08] is also available.

Part IV

The proofs given in Chapter 13 take their ingredients from [Ser79] (as mentioned, we also followed [Bab72]). Another approach to the Existence theorem is to develop the theory of Lubin–Tate formal group laws, and the corresponding field extensions. For this, see [Neu99] or the paper by Serre in [CF10].

As we have presented things, the logical precedence in that chapter is as follows: first, Tate's theorem 13.11, then the definition of Artin symbols, then Tate duality, then the Existence theorem to conclude the proof of the fundamental theorem. We have then included a stronger version of Tate duality with Theorem 13.29, and indicated that the latter could be used to re-prove everything, in fact in a more precise manner. It is natural to wonder whether we could have proved the statement of strong Tate duality first, from first principles. It is indeed essentially possible, but requires a rather long sequence of intermediate results in group cohomology, which are perhaps poorly motivated. If things went according to plan, after reading this book you have now an incentive to work your way through the "hard", but rewarding, proof.

To read about this, we recommend [Ser02, chapter II, section 5]. You must first learn about cohomological dimension in the first part of this book, including "dualizing modules". Spectral sequences are briefly mentioned (for this, see [NSW08, chapter II, section 1]). And of course, results from Part I, II, and III of the present book are freely used. However, local class field theory is not reconstructed from Tate duality – there does not seem to be any published source where this alternative exposition is taken seriously.

For the reader who wants to investigate global class field theory, our advice is to refer to [NSW08, chapter VIII, section 1]. Taking for granted the computation of $\widehat{\mathrm{H}}^0(\mathrm{Gal}(K/F), \mathbf{C}_K)$ and $\mathrm{H}^1(\mathrm{Gal}(K/F), \mathbf{C}_K)$, where K/F is a *cyclic* extension of number fields, many consequences are derived (they require some facts about number fields [and adèles] not developed in this book, but many tools have been presented, such as Shapiro's lemma or Krasner's lemma). A spectacular "intermediate" result is the *Hasse principle*, that is, the existence of an exact sequence

$$0 \longrightarrow \mathrm{Br}(F) \longrightarrow \bigoplus \mathrm{Br}(F_p) \longrightarrow \mathbb{Q}/\mathbb{Z} \longrightarrow 0,$$

where the middle sum is over all (equivalence classes of) completions over all p, and the map emanating from it is the sum of all the Hasse invariants. (In fact, the completions whose target is $\mathbb{R}$ or $\mathbb{C}$ have to be taken into account, too.) Then the authors go on to prove that $\mathbf{C}_K$ satisfies the hypothesis of our Theorem 13.11. Applying it gives Artin symbols.

There remains to prove the analog of the Existence theorem in this new setting, and to compute the cohomology groups above. References for these are given in [NSW08].

Of course, explicit formulae are nice when they are available, and this is another story altogether; there are some in [NSW08] and some in [Ser79], for example. Also, we have pointed out in the introduction that [Chi09] provides an elementary proof of the main theorems of global class field theory, which avoids cohomology entirely (the results themselves are also phrased differently, and it is nontrivial to check that the subject matter is the same!). Another interesting reference is [Cox89], a book which starts with problems in number theory and gradually expresses the need to have class field theory to solve them.

References

[Alp86] J. L. Alperin. *Local representation theory*. Cambridge University Press, Cambridge, 1986.

[Ara75] J. K. Arason. Cohomologische invarianten quadratischer Formen. *J. Algebra*, 36(3):448–491, 1975.

[Bab72] A. Babakhanian. *Cohomological methods in group theory*. M. Dekker, New York, 1972.

[Bla72] A. Blanchard. *Les corps non commutatifs*. Presses Universitaires de France, Vendôme, 1972.

[Bre97] G. E. Bredon. *Topology and geometry*. Springer-Verlag, New York, 1997.

[CF10] J. W. S. Cassels and A. Fröhlich, editors. *Algebraic number theory*. London Mathematical Society, London, 2010. Papers from the conference held at the University of Sussex, Brighton, September 1–17, 1965, including a list of errata.

[Chi09] N. Childress. *Class field theory*. Springer, New York, 2009.

[Con65] I. G. Connell. Elementary generalizations of Hilbert's Theorem 90. *Canad. Math. Bull.*, 8:749–757, 1965.

[Cox89] D. A. Cox. *Primes of the form* $x^2 + ny^2$. Wiley & Sons, New York, 1989.

[CR06] C. W. Curtis and I. Reiner. *Representation theory of finite groups and associative algebras*. AMS Chelsea Publishing, Providence, RI, 2006. Reprint of the 1962 original.

[DdSMS99] J. D. Dixon, M. P. F. du Sautoy, A. Mann, and D. Segal. *Analytic pro-p groups*. Cambridge University Press, Cambridge, second edition, 1999.

[Gow86] R. Gow. Construction of some wreath products as Galois groups of normal real extensions of the rationals. *J. Number Theory*, 24(3):360–372, 1986.

[GS06] P. Gille and T. Szamuely. *Central simple algebras and Galois cohomology*. Cambridge University Press, Cambridge, 2006.

[Kob77] N. Koblitz. *p-adic numbers, p-adic analysis, and zeta-functions*. Springer-Verlag, New York, 1977.

[Lan02] S. Lang. *Algebra*. Springer-Verlag, New York, 2002.

[LS96] H. W. Lenstra and P. Stevenhagen. Chebotarëv and his density theorem. *Math. Intelligencer*, 18(2):26–37, 1996.

[Mar02] D. Marker. *Model theory: an introduction*. Springer-Verlag, New York, 2002.

[Mor96] P. Morandi. *Field and Galois theory*. Springer-Verlag, New York, 1996.

[Mun75] J. R. Munkres. *Topology: a first course.* Prentice-Hall, Englewood Cliffs, N, 1975.

[Mun84] J. R. Munkres. *Elements of algebraic topology.* Addison-Wesley, Menlo Park, CA, 1984.

[Neu99] J. Neukirch. *Algebraic number theory.* Springer-Verlag, Berlin, 1999.

[NSW08] J. Neukirch, A. Schmidt, and K. Wingberg. *Cohomology of number fields.* Springer-Verlag, Berlin, 2008.

[Rou91] G. Rousseau. On the quadratic reciprocity law. *J. Austral. Math. Soc. Ser. A*, 51(3):423–425, 1991.

[Ser73] J.-P. Serre. *A course in arithmetic.* Springer-Verlag, New York-Heidelberg, 1973. Translated from the French.

[Ser79] J.-P. Serre. *Local fields.* Springer-Verlag, New York, 1979.

[Ser02] J.-P. Serre. *Galois cohomology.* Springer-Verlag, Berlin, English edition, 2002.

[Stea] W. Stein. *Algebraic number theory, a computational approach.* Available online. `https://wstein.org/books/ant/ant.pdf`.

[Steb] P. Stevenhagen. *Number rings.* Available online. `http://websites.math.leidenuniv.nl/algebra/ant.pdf`.

[Voe11] V. Voevodsky. On motivic cohomology with $\mathbf{Z}/l$-coefficients. *Ann. Math.*, 174(1):401–438, 2011.

[Wat94] W. C. Waterhouse. The normal closures of certain Kummer extensions. *Canad. Math. Bull.*, 37(1):133–139, 1994.

[Wei94] C. A. Weibel. *An introduction to homological algebra.* Cambridge University Press, Cambridge, 1994.

[Wei08] C. Weibel. 2007 Trieste lectures on the proof of the Bloch–Kato conjecture. In *Some recent developments in algebraic K-theory*, pp. 277–305. Abdus Salam International Centre for Theoretical Physics, Trieste, 2008.

Index